MOBILE BROADBAND COMMUNICATIONS FOR PUBLIC SAFETY

MOBILE BROADBAND COMMUNICATIONS FOR PUBLIC SAFETY: THE ROAD AHEAD THROUGH LTE TECHNOLOGY

Ramon Ferrús and Oriol Sallent
Universitat Politècnica de Catalunya (UPC), Spain

WILEY

This edition first published 2015

Registered Office
John Wiley & Sons, Ltd, The Atrium, Southern Gate, Chichester, West Sussex, PO19 8SQ, United Kingdom

For details of our global editorial offices, for customer services and for information about how to apply for permission to reuse the copyright material in this book please see our website at www.wiley.com.

Library of Congress Cataloging-in-Publication Data

Ferrus, Ramon 1971–
Mobile broadband communications for public safety : the road ahead through LTE technology / Ramon Ferrus and Oriol Sallent.
pages cm
Includes bibliographical references and index.
ISBN 978-1-118-83125-0 (hardback)
1. Long-Term Evolution (Telecommunications) 2. Emergency communication systems. 3. Telephone–Emergency reporting systems. 4. Public safety radio service. I. Sallent, Oriol, 1969– II. Title.
TK5103.48325.F47 2015
363.10028′4–dc23

2015017270

A catalogue record for this book is available from the British Library.

Cover Image: TheImageArea/iStockphoto

Set in 10/12pt Times by SPi Global, Pondicherry, India

Printed and bound in Singapore by Markono Print Media Pte Ltd

1 2015

Contents

Preface

Nowadays, public protection and disaster relief (PPDR) agencies mainly rely on the use of private/professional mobile radio (PMR) technologies (e.g. TETRA, TETRAPOL and Project 25) that were conceived in the 1990s. While PMR systems offer a rich set of voice-centric services, with a number of features matched to the special requirements of PPDR, including push-to-talk and call priority, the data transmission capabilities of these PMR technologies are rather limited and lag far behind the technological advancements made in the commercial wireless domain. In this context, long-term evolution (LTE) technology for mobile broadband PPDR is increasingly backed as the technology of choice for future PPDR communications. Technical work is currently being undertaken within the 3rd Generation Partnership Project (3GPP), the organization in charge of LTE standardization, to add a number of improved capabilities and features to the LTE standard that will further increase its suitability for PPDR and other professional users, by meeting their high demands for reliability and resilience. While the convergence to common technical standards for the PPDR and commercial domains offers significant opportunities for synergies and economies of scale, the delivery of PPDR broadband services demands new approaches in the way that network capacity is deployed and managed. The current paradigm for PPDR communications, based on 'dedicated technologies, dedicated networks and dedicated spectrum', is no longer believed to constitute the main approach for the provision of PPDR broadband data communications. On this basis, this book provides a comprehensive view of the introduction of LTE technology for PPDR communications. In particular, the following topics are covered in the book:

- The fundamentals of PPDR services, their operational framework and associated communications systems
- An overview of the main communications technologies and standards used nowadays by PPDR practitioners
- The operational scenarios and emerging multimedia, data-centric applications in growing demand by PPDR practitioners due to their great potential to improve their operational efficiency

- A discussion on the main techno-economic drivers that are believed to be pivotal for a cost-efficient delivery of mobile broadband PPDR communications, such as the use of common technical standards with the commercial domain, the consideration of infrastructure sharing and multi-network-based solutions as well as dynamic spectrum sharing
- The formulation of a comprehensive system view for the delivery of mobile broadband communications for PPDR, including dedicated LTE-based wide area networks, roaming and priority access to commercial networks' capacity, fast deployable equipment and satellite access as key components
- An analysis of the capabilities and features of the LTE standard that are relevant for an improved support of mission-critical communications, such as group communications enablers and direct mode operation
- A discussion on the different network implementation options to deliver mobile broadband PPDR communications services over dedicated or commercial LTE-based networks, including the applicability of the mobile virtual network operator (MVNO) model and other hybrid models
- A description of the network architecture design and implementation aspects that are central to the realization of the different delivery models, including the interconnection with legacy networks and with deployables (e.g. cells on wheels and system on wheels) and satellite access
- The estimation of spectrum needs for future broadband PPDR systems, a review of the allocated and candidate spectrum bands for PPDR communications and the consideration of dynamic spectrum sharing solutions intended to provide additional capacity to, for example, cope with a surge of PPDR traffic demand

The book is organized into six chapters:

Chapter 1 addresses the fundamentals of PPDR services, their operational framework and associated communications systems. First, the terminology and key definitions of PPDR, public safety (PS) and emergency communications are provided, identifying the scope of these terms and categorizing the different types of communications relationships found in emergencies. Next, the main functions and services delivered by PPDR organizations are introduced, providing a view on the so-called first responder agencies as well as on the role that other entities, such as utilities and telecom operators, could also play in an emergency response. On this basis, a description of the operational framework for PPDR operations is presented. Such a description covers a classification of PPDR operational scenarios, some generic organizational and procedural aspects in incident-response management and the communications' reference points and key characteristics of the communications services demanded by PPDR practitioners. Following this, a review of the main communications technologies and systems currently in use for PPDR is provided. The review outlines the type of requirements usually bound to PPDR communications systems, describes a common classification of the technologies used within the PPDR sector and provides an overview of the most widely used digital radio communications standards for PPDR communications as of today (TETRA, TETRAPOL, DMR, and Project 25). The review also encompasses the identification of some of the major limitations found in today's PPDR communications landscape through the analysis of an illustrative, hypothetical incident. Finally, the chapter concludes with a description of the regulatory and standardization framework for PPDR communications.

Chapter 2 describes the various types of data-centric, multimedia applications deemed critical for on-scene PPDR operations. Special attention is given to the 'Matrix of Applications' developed by the Law Enforcement Working Party/Radio Communication Expert Group (LEWP/RCEG) of the EU Council, which provides a characterization of technical and operational parameters of a list of PPDR applications agreed by a significant number of European PPDR organizations and recognized by CEPT administrations as being representative in terms of future PPDR applications. Next, the chapter presents various estimates of the throughput requirements for the mobile broadband data applications in demand, outlining typical peak data rates, mean session duration and number of transactions in the busy hour in normal conditions to sustain typical PPDR needs. Finally, the chapter concludes with a quantitative assessment of the overall data capacity needed in a number of representative PPDR operational scenarios within the categories of day-to-day operations, large emergency/public events and disaster scenarios.

Chapter 3 starts with a discussion on the idea that a paradigm change in the delivery of mobile broadband is needed with respect to the prevailing model used nowadays for the provision of voice-centric and narrowband data PPDR services, which is largely characterized by the use of dedicated technologies, dedicated networks and dedicated spectrum. Next, the key techno-economic considerations that are fuelling this paradigm change towards more cost-efficient PPDR communications delivery models are identified and discussed across the dimensions of technology, network and spectrum. Grounded on these techno-economic considerations, a comprehensive system view of the future mobile broadband PPDR communications systems is then described, identifying the key underlying principles and building blocks. Finally, the chapter concludes with a review of some relevant initiatives that are currently shaping the way forward towards the delivery of next-generation mobile broadband PPDR communications.

Chapter 4 provides a description of the new capabilities and features that are being added to the LTE standard. While the LTE standard is already a suitable technology to support a rich number of mobile broadband applications for the PPDR community, including video delivery, work is underway within the 3GPP to improve the standard and turn it into a full mission-critical communications technology. First, the chapter outlines the standardization roadmap established within 3GPP and other relevant standardization bodies in the area of PPDR communications and introduces some of the fundamentals on LTE technology and networks. Next, the enhancements being introduced to fulfil PPDR needs are described, including enhanced group communications enablers and mission-critical push-to-talk (MCPTT) functionality, device-to-device communications (referred to as proximity services in 3GPP specifications), isolated LTE network operation, support of higher transmit power terminals and prioritization and Quality of Service (QoS) control features to cope with capacity congestion. In addition, the enhancements being introduced to LTE with regard to radio access network (RAN) sharing are also described as another potential technology enabler that could facilitate the deployment of shared LTE network models for PPDR and other uses.

Chapter 5 describes the network implementation options to deliver mobile broadband PPDR communications services over dedicated and/or commercial LTE-based networks. First, a number of introductory remarks on the defining elements in current PPDR communications delivery models, the possibility enabled by LTE to provision separately the services from the underlying network and the characteristics expected from a 'public safety grade' LTE network design are discussed. On this basis, the different options that can be adopted for the implementation of LTE-based mobile broadband PPDR networks are categorized and described, emphasizing pros and cons of each option. In particular, the deployment of dedicated networks and the use of public networks as well as hybrid combinations are considered. Finally, the chapter delves into some network architecture design and implementation aspects that are central for the realization of the different delivery models. In particular, the reference model developed by ETSI for the overall system intended to provide critical communications services, the interconnection between commercial and dedicated networks, the interworking of broadband and narrowband legacy platforms, the interconnection of deployables and the use of satellite communications and the connectivity services and frameworks within the underlying IP-based backbones are addressed. Additionally, an overview of an MVNO-based solution, which is the approach currently under consideration as a viable short-term solution in some European countries, is presented.

Chapter 6 is focused on the diverse facets related to radio spectrum for PPDR communications. First, the main regulatory and legal instruments that currently govern the use and management of spectrum at global, regional and national levels are discussed, together with the models and evolution of spectrum management practices. Next, the existing provisions at international regulations with regard to harmonized frequency ranges for PPDR communications are presented, together with next key milestones expected in this area. On this basis, the chapter then delves into the characterization of spectrum needs for future broadband PPDR systems, describing the methodologies used for the computation of spectrum needs and gathering a number of estimates carried out from different organizations worldwide. Afterwards, the current spectrum availability for PPDR communications is presented focusing on existing assignments as well as on the candidate bands under consideration in some regions for the delivery of mobile broadband PPDR communications. Finally, the chapter addresses the issue of dynamic spectrum sharing for PPDR communications as a way to complement a dedicated assignment. A classification of the possible sharing models is given, identifying the key principles in each model and discussing on their suitability for PPDR use. On this basis, two possible spectrum sharing solutions are further described: one based on the applicability of the Licensed Shared Access (LSA) regime and the other exploiting secondary access to TV white spaces.

List of Abbreviations

2G	second generation
3ES	three emergency services
3GPP	3rd Generation Partnership Project
3GPP2	3rd Generation Partnership Project 2
AC	access class
ACB	Access Class Barring
ACCOLC	Access Overload Control
ACELP	Algebraic Code Excited Linear Prediction
ACLR	adjacent channel leakage ratio
ACMA	Australian Communications and Media Authority
AES	Advanced Encryption Standard
AF	application function
AGA	air–ground–air
AH	Authentication Header
AI	air interface
AIE	air interface encryption
AMBR	Aggregate Maximum Bit Rate
AMR-WB	AMR Wideband
ANF	Additional Network Feature
ANPR	automatic number plate recognition
APCO	Association of Public-Safety Communications Officials
API	Application Programming Interface
APL	automatic personnel location
APN	Access Point Name
APN-AMBR	Access Point Name Aggregate Maximum Bit Rate
AppComm	Application Community
APT	Asia-Pacific Telecommunity
ARIB	Association of Radio Industries and Business

ARNS aeronautical radio navigation service
ARP Allocation and Retention Priority
ARQ Automatic Repeat reQuest
AS access stratum
ASA Authorised Shared Access
ASMG Arab Spectrum Management Group
ASP application service provider
ATIS Alliance for Telecommunications Industry Solutions
ATM Asynchronous Transfer Mode
ATU African Telecommunications Union
AuC authentication centre
AV authentication vector
AVL automatic vehicle location
BB broadband
BBDR broadband disaster relief
BM-SC Broadcast Multicast Service Centre
BoM bill of materials
BSO beneficial sharing opportunity
BS base station
BSSM base station spectrum manager
BTOP Broadband Technology Opportunities Program
BWT broadband wireless trunking
CA carrier aggregation
CAD computer-aided dispatching
CAI Common Air Interface
CAP Compliance Assessment Program
CAPEX capital expenditures
CATR China Academy of Telecommunication Research
CBRS Citizens Broadband Radio Service
CBS Citizens Broadband Service
CCA critical communications application
CCBG Critical Communications Broadband Group
CCC command and control centre
CCS critical communications system
CCSA China Communications Standards Association
CDIS coexistence discovery and information server
CDR charging data record
CE coexistence enabler
CE consumer electronics
cell ID cell identity
CFSI Conventional Fixed Station Interface
CGC complementary ground component
CISC Communications Interoperability Strategy for Canada
CITEL Inter-American Telecommunications Commission
CM coexistence manager
CO–CO contractor owned and contractor operated

COP	common operating picture
COTM	communications on the move
COTS	commercial off-the-shelf
COW	cell on wheel
CR	cognitive radio
CRS	cognitive radio systems
CRS	control room systems
CS	circuit switched
CSFB	Circuit-Switched Fallback
CSSI	Console Subsystem Interface
CUS	collective use of spectrum
D2D	device to device
DAS	distributed antenna systems
dB	decibel
DeNB	donor eNB
DFT	discrete Fourier transform
DGNA	dynamic group number assignment
DHS	Department of Homeland Security
DL	downlink
DM	device management
DMO	direct mode operation
DNS	Domain Name Service
DP	delivery partner
DR	disaster relief
DSA	Dynamic Spectrum Arbitrage
DSATPA	Dynamic Spectrum Arbitrage Tiered Priority Access
DTT	digital terrestrial television
DWDM	dense wavelength-division multiplexing
DySPAN	Dynamic Spectrum Access Networks
E2EE	end-to-end encryption
EAB	Extended Access Barring
EC	European Commission
ECA	European Common Allocation
ECC	Electronic Communications Committee
ECC	emergency control centre
ECCS	Emergency Communication Cell over Satellite
ECG	electrocardiogram
ECN&S	electronic communications networks and services
ECO EFIS	European Communications Office Frequency Information System
ECS	electronic communications services
EEA	European Economic Area
EHPLMN	Equivalent HPLMN
EIRP	equivalent isotropic radiated power
EMA	externally mounted antennas
eMBMS	evolved MBMS
eMLPP	Enhanced Multi-Level Precedence and Pre-emption

EMS	emergency medical services
eNB	evolved Node B
ENISA	European Union Agency for Network and Information Security
ENUM	E.164 NUmber Mapping
EPC	Evolved Packet Core
EPL	Ethernet private lines
EPS AKA	EPS Authentication and Key Agreement
EPS	Evolved Packet System
ESMCP	emergency services mobile communications programme
ESN	Emergency Services Network
ESO	European Standards Organization
ESP	Encapsulating Security Payload
ETS	emergency telecommunications services
ETSI TC TCCE	ETSI Technical Committee on TETRA and Critical Communications Evolution
ETSI	European Telecommunications Standards Institute
EU	European Union
E-UTRAN	Evolved UMTS Radio Access Network
FBI	Federal Bureau of Investigation
FCC	Federal Communications Commission
FDMA	frequency division multiple access
FirstNet	First Responder Network Authority
FM PT 53	Frequency Management Project Team 53
FM PT49	Frequency Management Project Team 49
FNO	fixed network operator
FS_IOPS	Feasibility Study on Isolated E-UTRAN Operation for Public Safety
GB	gigabytes
GBR	Guaranteed Bit Rate
GCS AS	GCS Application Server
GCS CA	GCS Client Application
GCS	group communications services
GCSE	Group Communications System Enablers
GETS	Government Emergency Telecommunications Service
GIS	Geographic Information System
GLDB	geo-location database
GMDSS	Global Maritime Distress and Safety System
GO	government owned
GO–CO	government owned and contractor operated
GO–GO	government owned and government operated
GPRS	General Packet Radio Service
GSC	Global Standards Collaboration
GSMA	Global System for Mobile Association
GSM-R	GSM-Railway
GUTI	Globally Unique Temporary Identifier
GW	gateway
GWCN	Gateway Core Network

HD	high definition
HetNet	heterogeneous network
HO	Home Office
H-PCRF	Home PCRF
HR	high resilience
HSS	Home Subscriber Server
HTS	high-throughput satellite
HTTPS	HTTP Secure
HVAC	heating, ventilation and air conditioning
IC	Industry Canada
ICS	incident command structure
ICT	information and communications technology
IDA	Info-Communications Development Authority
IDRA	Integrated Dispatch Radio
IEEE	Institute of Electrical and Electronics Engineers
IETF	Internet Engineering Task Force
IKEv1	Internet Key Exchange 1
IKEv2	Internet Key Exchange 2
IKI	Inter-Key Management Facility Interface
IM CN	IP Multimedia Core Network
IMS	IP Multimedia Subsystem
IMSI	International Mobile Subscriber Identity
IP ISI	IP-based Inter-System Interface
IP VPN	IP virtual private network
IPX	IP Packet Exchange
ISACC	ICT Standards Advisory Council of Canada
ISI	Inter-System Interface
ISP	Internet service provider
ISSI	Inter-RF Subsystem Interface
ITU	International Telecommunication Union
ITU-R	ITU Radiocommunication
JHA	Justice and Home Affairs
KCC	Korea Communications Commission
LAA	Licenced-Assisted Aggregation
LAN	local area network
LA-RICS	Los Angeles Regional Interoperable Communications System
LC	LSA controllers
LD/SC	liquidated damages/service credits
LEWP	Law Enforcement Working Party
LI	lawful interception
LIPA	Local IP Access
LMR	land mobile radio
LPG	liquid petroleum gas
LR	LSA repository
LSA	Licenced Shared Access
LTE	Long-Term Evolution

M2M	machine to machine
MAC	medium access control
MBMS	Multimedia Broadcast Multicast Service
MBMS-GW	MBMS Gateway
MBR	Maximum Bit Rate
MCC	mobile country code
MCPTT NMO	MCPTT Network Mode Operation
MCPTT	mission-critical push-to-talk
MDM	mobile device management
MEF	Metro Ethernet Forum
MFCN	mobile/fixed communications networks
MIC	Ministry of Internal Affairs and Communications
MIFR	Master International Frequency Register
MIMO	multiple-input/multiple-output
MME	Mobility Management Entity
MMI	man–machine interface
MNO	mobile network operator
MOA	memorandum of agreement
MOCN	Multi-Operator Core Network
MPLS	Multiprotocol Label Switching
MPS	Multimedia Priority Service
MPT	Ministry of Post and Telecommunication
MS	mobile service
MSC	mobile switching centre
MSS	mobile satellite service
MT	mobile termination
MTPAS	Mobile Telecommunication Privileged Access Scheme
MVNA	mobile virtual network aggregator
MVNE	mobile virtual network enabler
MVNO	mobile virtual network operator
NAS	non-access-stratum
NB	narrowband
NE	network entity
NEMA	National Emergency Management Agency
NeNB	nomadic eNB
NFV	network functions virtualization
NGN	next-generation network
NGO	non-governmental organization
NIST	National Institute of Standards and Technologies
NMS	network management system
NoI	Notice of Inquiry
NPSBN	National Public Safety Broadband Network
NPSTC	National Public Safety Telecommunications Council
NRA	national regulatory authority
NS/EP	national security and emergency preparedness
NSM	Network spectrum manager

NTFA	National Table of Frequency Allocations
NTIA	National Telecommunications and Information Administration
NTP	Network Time Protocol
OAM	operation, administration and maintenance
OCHA	Office for the Coordination of Humanitarian Affairs
OCS	Online Charging System
OFCS	Offline Charging System
OFDM	orthogonal frequency-division multiplexing
OFDMA	orthogonal frequency-division multiple access
OMA DM	Open Mobile Alliance Device Management
OMA	Open Mobile Alliance
OOBE	out-of-band emission
OPEX	operational expenditures
OSI	Open Systems Interconnection
OTA	over the air
OTAR	over-the-air rekeying
OTN	Optical Transport Network
OTT	over the top
P25 PTToLTE	P25 PTT over LTE
P25	Project 25
PAS	Publicly Available Specifications
PAWS	Protocol to Access White Space
PCC	Policy and Charging Control
PCEF	Policy and Charging Enforcement Function
PCPS	Push-to-Communicate for Public Safety
PCRF	Policy and Charging Rules Function
PD	packet data
PDB	packet delay budget
PDN	packet data network
PEI	Peripheral Equipment Interface
PELR	packet error loss rate
PEP	performance-enhancing proxy
P-GW	PDN Gateway
PIM	personal information manager
PKI	public key infrastructure
PLMN	Public land mobile network
PMN	public mobile network
PMR	professional/private mobile radio
PMSE	programme making and special event
PoC	Push-to-Talk over Cellular
PP	public protection
PPDR	public protection and disaster relief
PRD	Permanent Reference Document
ProSe	proximity-based services
PS	public safety
PSAC	Public Safety Advisory Committee

PSAP	public safety answering point
PSA	public safety agency
PSC	Public Safety Communications
PSCR	Public Safety Communications Research
PSD	power spectral density
PSG	public safety grade
PSN	public safety network
PSS	Public Safety and Security
PSTN	public switched telephone network
PTIG	Project 25 Technology Interest Group
PTT	push to talk
PWS	Public Warning System
QCI	QoS Class Identifier
QoE	quality of experience
QoS	quality of service
QPSK	quadrature phase-shift keying
RAN	radio access network
RAS	radio astronomy service
RAT	radio access technology
RB	Resource Block
RBS	radio base stations
RCC	Regional Commonwealth in the Field of Communications
RCEG	Radio Communications Expert Group
RCS	Rich Communications Suite
REM	Radio Environment Map
RF	radio frequency
RFI	request for information
RFID	radio frequency identity
RFP	request for proposals
RN	relay node
ROHC	Robust Header Compression
RR	Radio Regulations
RRC	Radio Resource Control
RRS	Reconfigurable Radio System
RSC	Radio Spectrum Committee
RSE	RAN Sharing Enhancements
RSPG	Radio Spectrum Policy Group
RSPP	Radio Spectrum Policy Programmes
RTP	Real-time Transport Protocol
SAGE	Security Algorithms Group of Experts
SA	security association
SAS	spectrum access system
SC	service code
SC	spectrum coordinator
SC-FDMA	single-carrier frequency-division multiple access
SCI	Subscriber Client Interface

SCPC	single channel per carrier
SDH	Synchronous Digital Hierarchy
SDK	software development kit
SDL	supplementary downlink
SDN	software-defined networking
SDO	standards development organization
SDP	service delivery platform
SDR	software-defined radio
SDS	short data service
SEG	Security Gateway
S-GW	Serving Gateway
SIB	System Information Block
SIM	Subscriber Identity Module
SIMTC	System Improvements to Machine-Type Communication
SIP	Session Initiation Protocol
SLA	service-level agreement
SLIGP	State and Local Implementation Grant Program
SMLA	Spectrum Manager Lease Agreement
SMS	Short Message Service
SN ID	serving network identity
SN	serving network
SONET	Synchronous Optical Networking
SOS	Spectrum Occupancy Sensing
SOW	system on wheel
SPR	Service Profile Repository
SPR	Subscriber Profile Repository
SPS	semi-persistent scheduling
SRDoc	system reference documents
SRVCC	Single Radio Voice Call Continuity
SSAC	Service Specific Access Control
SSAR	shared spectrum access right
STA	special temporary authority
SwMI	Switching and Management Infrastructure
TBCP	Talk Burst Control Protocol
TC	technical committee
TCCA	TETRA and Critical Communications Association
TCCE	TETRA and Critical Communications Evolution
TCO	total cost of ownership
TDM	time-division multiplexing
TDMA	time division multiple access
TE	terminal equipment
TEA	TETRA Encryption Algorithm
TEDS	TETRA Enhanced Data Service
TETRA ISI	TETRA Inter-System Interface
TETRA	Terrestrial Trunked Radio
TFA	Table of Frequency Allocations

TFT	Traffic Flow Template
TIA	Telecommunications Industry Association
TMGI	Temporary Mobile Group Identity
TMO	trunked mode operation
TRA	telecommunications regulatory authority
TTA	Telecommunications Technology Association
TTC	Telecommunication Technology Committee
TTI	Transmission Time Interval
TVBD	TV band devices
TVWS	TV white spaces
UAV	unmanned aerial vehicle
UAV	Unmanned aeronautical vehicle
UE-AMBR	UE Aggregate Maximum Bit Rate
UESM	UE spectrum manager
UL	uplink
UN	United Nations
UPS	uninterruptible power supply
USIM	Universal Subscriber Identity Module
UTC	Utilities Telecom Council
VC	virtual circuits
VIP	very important people
VoIP	Voice over IP
VoLTE	Voice over LTE
V-PCRF	Visited PCRF
VPN	virtual private network
VSAT	very small aperture terminal
WAN	wide area network
WB	wideband
WGET	Working Group on Emergency Telecommunications
WI	Work Item
WPS	Wireless Priority Service
WRAN	wireless regional area network
WRC	World Radiocommunication Conferences
WRC-03	World Radio Conference 2003
WS	white spaces
WSD	white space devices
WTDC	World Telecommunication Development Conferences
WTSA	World Telecommunication Standardization Assembly
XCAP	XML Configuration Access Protocol
XDMS	XML Document Management Servers
XML	Extensible Markup Language

1

Public Protection and Disaster Relief Communications

1.1 Background and Terminology

The public protection and disaster relief (PPDR) sector brings essential value to society by creating a stable and secure environment to maintain law and order and to protect the life and values of citizens. PPDR services such as law enforcement, firefighting, emergency medical services (EMS) and disaster recovery services are pillars of our society organization. The protection ensured by PPDR services covers people, property, the environment and other relevant values for the society. It addresses a large number of threats both natural and man-made. The PPDR sector is for most nations intimately connected to the public sector of society, either directly as part of the governmental structure or as a function which is outsourced under strict rules and intensively monitored by government's contracting ministry or department. Regulatory, organizational, operational and technical elements underpinning an effective PPDR preparedness can vary substantially from country to country, even between regions or municipalities in countries where local preparedness might be under the auspices of regional or local public authorities.

One important task of PPDR services is to deal with emergency and surveillance situations on land, sea and air. The most important part of this work is done in the field, so all the tools must match the needs accordingly. Radiocommunications are extremely important to PPDR organizations to the extent that PPDR communications are highly dependent upon it. At times, radiocommunication is the only form of communications available.

There are terminology differences between administrations and regions in the scope and specific meaning of PPDR and related radiocommunication services. PPDR is defined in ITU Radiocommunication (ITU-R) Resolution 646 in World Radio Conference 2003 (WRC-03)

Mobile Broadband Communications for Public Safety: The Road Ahead Through LTE Technology, First Edition.
Ramon Ferrús and Oriol Sallent.

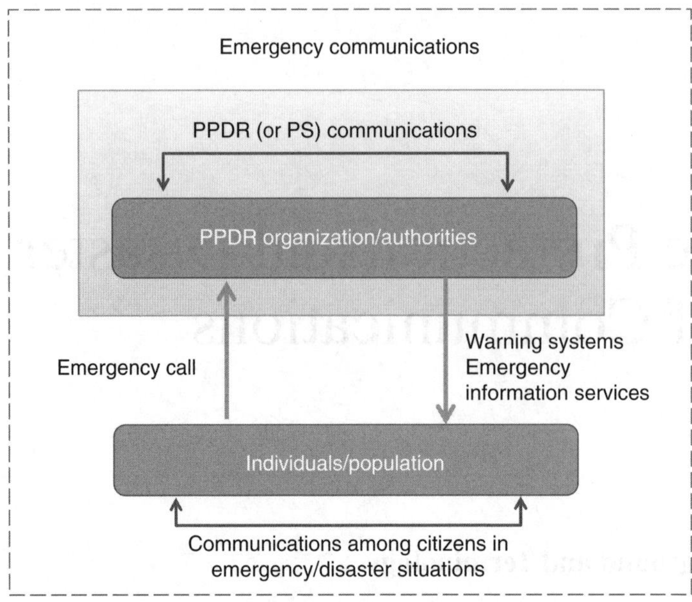

Figure 1.1 Scope of PPDR and emergency communications.

through a combination of the terms 'public protection (PP) radiocommunication' and 'disaster relief (DR) radiocommunication' [1]:

- **PP radiocommunication**. Radiocommunications used by responsible agencies and organizations dealing with maintenance of law and order, protection of life and property and emergency situations
- **DR radiocommunication**. Radiocommunications used by agencies and organizations dealing with a serious disruption of the functioning of society, posing a significant, widespread threat to human life, health, property or the environment, whether caused by accident, nature or human activity

A term also commonly used to refer to PPDR communications is public safety (PS) communications. These terms are often used interchangeably [2]. Another term related to PPDR communications is emergency communications. Broadly defined, emergency communications involves not only communications within and between PPDR agencies and public authorities involved in the management of an emergency case but also communications involving citizens. As illustrated in Figure 1.1, the generally agreed categories to be considered in the provision of emergency communications are [3]:

- **Communication between authorities/organizations**. Refers to communications within and among authorities/organizations. This is the category that fits with the scope of PPDR communications.
- **Communication from authorities/organizations to citizens**. Refers to communications from authorities/organizations with individuals, groups or the general public. Warning and information systems to alert the population are part of this category.

- **Communication of citizens with authorities/organizations**. Emergency call services (e.g. calls to emergency numbers such as 112 or 911 through public telephone networks) are part of this category.
- **Communication among citizens**. In case of a disaster, individuals may have a strong demand to communicate among themselves in order to ascertain/learn the state of relatives, property, etc., as well as coordinate actions of mutual interest. Particularly, new social media communications technologies can potentially enable citizens to more quickly share information, assist response and recovery in emergencies and mobilize for action in political crises.

In this context, it is also common to refer to PPDR organizations as emergency services or emergency response providers. In particular, an emergency service can be defined as an agency or service that provides immediate and rapid assistance in situations where there is a direct risk to life or limb, individual or public health or safety, private or public property, or the environment but not necessarily limited to these situations [4].

The focus of this book is on communications within and between PPDR organizations and authorities. In this regard, the terms PPDR, PS and emergency communications are used interchangeably within the book to refer to this type of communications.

1.2 PPDR Functions and Organizations

PPDR organizations or agencies are the ones responsible for the prevention and protection from events that could endanger the safety of the general public. The main functions and services provided by PPDR organizations are [5, 6]:

- **Law enforcement**. Law enforcement is the function to prevent, investigate, apprehend or detain any individual, which is suspected or convicted of offences against the criminal law. Law enforcement is a function usually performed by police organizations.
- **EMS**. The function of medical services is to provide critical invasive and supportive care of sick and injured citizens and the ability to transfer the people in a safe and controlled environment. Components of the EMS system include the following: medical first responders (people and agencies that provide non-transporting first aid care before an ambulance arrives on scene), ambulance services (basic and advanced life support), specialty transport services (helicopter, boat, snowmobile, etc.), hospitals (emergency, intensive, cardiac, neonatal care units, etc.) and specialty centres (trauma, burn, cardiac, drug units, etc.). The function of EMS includes also the function of 'disaster medicine', which is the provision of triage, primary aid, transportation and secondary care in major incidents. Doctors, paramedics, medical technicians, nurses or volunteers can supply these services.
- **Firefighting**. This is the function of putting out hazardous fires that threaten civilian populations and property. Hazardous fires can appear in urban areas (houses or buildings) or rural areas (forest fires). Professional and volunteer fire protection agencies supply this service.
- **Protection of the environment**. This is the function to protect the natural environment of a nation or a regional area, including its ecosystems composed by animals and plants. This function is limited to the everyday operation of protecting the environment like monitoring of the water, air and land. Forest guards, firefighters, volunteer organizations or public organizations are usually responsible for this activity.

- **Search and rescue**. This function has the objective to locate, access, stabilize and transport lost or missing persons to a place of safety. Search and rescue is one of the activities performed by different PS organizations such as firefighters or EMS.
- **Border security**. Control of the border of a nation or a regional area from intruders or other threats, which could endanger the safety and economic well-being of citizens. Covers areas such as verification of illegal immigration, verification of the introduction of illegal substances and verification of introduction of goods in offence of customs laws. Border security is usually performed by police organizations or specialized border security guard. Coastal guard is a special case of border security.
- **Emergency management**. Emergency management, also referred to as civil protection, is the organization and management of resources and responsibilities for dealing with all aspects of major emergencies/disasters, in particular prevention, preparedness, response and rehabilitation. Emergency management provides central command and control of PPDR agencies during emergencies. Emergency management involves plans, structures and arrangements established to engage the normal endeavours of government, voluntary and private agencies in a comprehensive and coordinated way to respond to the whole spectrum of major emergency needs. Emergency management includes also the recovery of the essential flows related to food, health, transportation, building material, electrical energy supply, telecommunications and daily stuff, situation awareness and communication.

The distribution of the above functions and services among PPDR organizations is not homogeneous across countries and regions. In Europe, similar organizations may not perform exactly the same functions in different countries due to the non-homogeneous historical development of PPDR services in each nation. Also, the organization and standard operating procedures can differ significantly among PPDR organizations that could span from volunteer organizations, which have received limited training, to specialist paramilitary organizations (e.g. explosives, hazardous materials specialists). Common types of PPDR organizations in Europe are described in the following list, identifying which is the main function or functions provided by each:

- **Police**. The main objective of the police is law enforcement creating a safer environment for its citizen. Functions: law enforcement.
- **Fire services**. With variations from region to region and country to country, the primary areas of responsibility of the fire services include structure firefighting and fire safety, wild land firefighting, life-saving through search and rescue, rendering humanitarian services, management of hazardous materials and protecting the environment, salvage and damage control, safety management within an inner cordon and mass decontamination. Functions: law enforcement, protection of the environment and search and rescue.
- **Border guard (land)**. Border guard comprises national security agencies which perform border control at national or regional borders. Their duties are usually criminal interdiction, control of illegal immigration and illegal trafficking. Functions: law enforcement and border security.
- **Coast guard**. Coast guard services may include, but not be limited to, search and rescue (at sea and other waterways), protection of coastal waters, criminal interdiction, illegal immigration and disaster and humanitarian assistance in areas of operation. Coast guard functions may vary with administrations, but core functions and requirements are generally

common globally. Functions: law enforcement, protection of the environment, search and rescue and border security.

- **Forest guards**. Type of police specialized in the protection of the forest environment. It supports other agencies in firefighting and law enforcement in rural and mountain environment. Functions: law enforcement, protection of the environment and search and rescue.
- **Hospitals and medical first responders**. These are the central components for the provision of EMS. They usually count on mobile units such as ambulances and other motorized vehicles such as aircraft helicopters and other vehicles. Functions: EMS and search and rescue.
- **Road transport police**. Transport police is a specialized police agency responsible for the law enforcement and protection of transportation ways like railroad, highways and others. Functions: law enforcement.
- **Railway transport police**. Railway transport police is a specialized police agency responsible for the law enforcement and protection of railways. In some cases, it is a private organization dependent on the railway service provider. Functions: law enforcement.
- **Custom guard**. An arm of a state's law enforcement body, responsible for monitoring people and goods entering a country. Given the removal of internal borders in the European Union (EU), customs authorities are particularly focused on crime prevention. Functions: law enforcement.
- **Airport security**. Airport enforcement authority is responsible for protecting airports, passengers and aircrafts from crime. Functions: law enforcement.
- **Port security**. Port enforcement authority is responsible for protecting port and maritime harbour facilities. Functions: law enforcement.
- **Volunteers organizations for civil protection**. Volunteer organizations are civilian with training on a number of areas related to PS and environment protection. They voluntarily enter into an agreement to protect environment and citizens without a commercial or monetary profit. Functions: protection of the environment and search and rescue.

In addition to the above-mentioned types of PPDR organizations, public authorities at different levels (local, regional, national) can also be directly involved in PPDR operations, leading or supporting emergency management functions. Public authorities are responsible for the establishment of a set of preparedness and contingency plans to handle emergency situations. Public authorities can be at the core of the response to most serious emergencies to put in place the emergency plans as well as provide advice and assistance to businesses and voluntary organizations about business continuity management.

Moreover, emergency response may also involve other public or private organizations such as departments of transportation, public works, utility companies (water, gas, electricity) and telecom operators. In the case of telecom operators, the emergency management plans may include a listing of emergency telecommunications facilities that need to be prepared for use in the event of a major emergency/disaster. The telecom operators have to support these plans where special operational modes may be predefined in a policy-based network management scheme and invoked in emergency situations (e.g. invocation or priority access schemes, rerouting calls to specific answering points).

Military forces can also support PPDR operations during major national emergencies where military authorities provide manpower and equipment to supplement PS resources. These incidents are frequently in response to natural forces (e.g. flooding, earthquakes). Military units can also give pre-planned support in major events (e.g. Olympic Games) as well as specialist response to man-made emergencies (e.g. terrorist attacks) where specialist military skills or equipment are necessary and may form an integral part of the emergency response.

Last but not least, some individuals can also belong to entities and organizations that have a role to play in emergency situations [7]. In particular, professionals and/or volunteers in non-governmental organizations (NGOs) and civic organizations may have a supporting role in handling emergencies. Their efficient involvement will highly depend upon their liaisons with the authorities organizing and steering the overall rescue plan. Providing them with tools to report their field observations or get the optimal information on the status of the crisis, they are involved in, can be crucial.

Also, the owners of the site, vessel, etc., where the emergency occurs, have certain obligations to fulfil. Site staff (or personnel) are supposedly fit to manage the site/plant and may participate in the rescue and clearance, as well as being affected individuals. Importantly, assistance in logistic coordination and utility provisioning may be also provided by providers of gas, electricity, electronic communications services and water supply. The utility owner, usually outside the emergency area, may represent control and control its action from a control centre. Utility staff may be directly working within the emergency area (or nearby) with the manual operations needed. Finally, the role of media (journalists, radio/TV news reporters) is also crucial in spreading information from the emergency scene and from the authorities to other affected individuals. Broadcasting can also be used for recruiting and coordinating new people to volunteer.

In this context, the term ‘first responder’ is commonly used to refer to law enforcement, emergency medical, firefighting and rescue services. In turn, the term ‘emergency responders’ is typically used with a wider scope than first responder, including in this case other entities such as electric, water and gas utilities; transportation; transit; search and rescue; hospitals; the Red Cross; and many others, which can be involved in diverse incident responses.

1.3 Operational Framework and Communications Needs for PPDR

PPDR organizations are required to manage emergencies and major incidents on a daily basis. These incidents may vary widely in terms of scale. The definitions of ‘major incident’, ‘emergency’ and similar terms are general in terminology and encompass significant degrees of latitude in their interpretation. Incidents may take on a greater degree of urgency or seriousness because of particular circumstances. For example, a public disorder incident in a town involving 500 people will be more serious in its potential when there are 5 officers to deal with it than where there are 50. Incidents may involve the interaction of multiple PPDR services (police, firefighters, ambulances, specialist units, etc.). In addition, since incidents do not respect administrative, regional and national or language boundaries, operational scenarios may include a variety of cross-border operational activities. According to Ref. [8], a ‘major incident’ is any emergency that requires the implementation of special arrangements by one or more of the emergency services and will generally include the involvement, either directly or indirectly, of large numbers of people. For example:

- The rescue and transportation of large numbers of casualties
- The large-scale combined resources of the emergency services

- The mobilization and organization of the emergency and support services such as local or regional authorities, to deal with the threats of homelessness, serious injury or death involving a large number of people
- The handling of a large number of enquiries generated from the citizen and the mass media, usually directed at the police

It is a strongly held view that requirements for incidents have a considerable degree of commonality. There will be issues of scalability, spatial and temporal considerations, as well as certain incident-specific demands such as cross-border governance procedures, operations to detect and capture offenders in terrorist or criminal incidents and so on.

Within the emergency services, it is both possible and indeed commonplace to develop contingency plans for known risks and where a significant number of values can be defined: counterterrorism plans for an attack on a VIP's residence, evacuation plans for a hospital and a major fire at a large retail centre, for example. However, there are many major incidents which cannot be so clearly defined or prepared for: the cause, location, scale, impact and medium and long-term implications are indeterminate. For this reason, emergency services and other authorities must necessarily build a considerable degree of flexibility into their thinking and operational practices to attempt to build a set of responses to cover every conceivable eventuality and to avoid that these could rapidly become bureaucratic in the extreme, unwieldy and completely unmanageable.

There is a vast literature describing operational scenarios involving PPDR agencies and personnel with the purpose of establishing guidance and best practices as well as deriving organizational, functional and technological (including communications) requirements and standards (e.g. [1, 5, 6, 9–12]). Based on these references, the following subsections provide a comprehensive vision of operational aspects concerning PPDR communications that includes a categorization of PPDR operational scenarios, a description of a generic operational framework, the identification of main components and communications' reference points and the identification of current and expected communications services that are central to PPDR operations.

1.3.1 Operational Scenarios

From the perspective of the use of radiocommunications means in PPDR operations, three distinct radio operating environments are usually defined that impose different requirements on the use of PPDR applications and their importance:

- **Day-to-day operations**. Day-to-day operations encompass the routine operations that PPDR agencies conduct within their jurisdiction. Typically, these operations are within national borders.
- **Large emergency and/or public events**. Large emergencies and/or public events are those that PP and potentially DR agencies respond to in a particular area of their jurisdiction. However, they are still required to perform their routine operations elsewhere within their jurisdiction. The size and nature of the event may require additional PPDR resources from adjacent jurisdictions, cross-border agencies or international organizations. In most cases, there are either plans in place, or there is some time to plan and coordinate the requirements.

A large fire encompassing three to four blocks in a large city or a large forest fire are examples of a large emergency under this scenario. Likewise, a large public event (national or international) could include the G8 Summit, the Olympics, etc.
- **Disasters**. Disasters can be those caused by either natural or human activity. For example, natural disasters include an earthquake, a major tropical storm, a major ice storm, floods, etc. Examples of disasters caused by human activity include large-scale criminal incidences or situations of armed conflict. Given the large numbers of people that may be impacted by a disaster, the considerable potential for property damage and the risk to social cohesion in the aftermath of a disaster, effectiveness of cross-border PPDR operation or international mutual aid could be largely beneficial.

The above operational scenarios are found in one or a number of the following domains, which also have an impact on the definition of requirements for the equipment including communications systems [6]:

- **Urban environment**. Identifies an area in a city or a densely urbanized area. It has usually high density of people and buildings. Emergency crisis and other types of PS scenarios in an urban environment are often characterized by a limited area of operation (hundreds of meters to few km), presence of man-made obstacles and need for a high reaction speed. Urban environment may have many facilities, but traffic congestion may limit the mobility of PPDR responders.
- **Rural environment**. Identifies an area, which is not densely urbanized like countryside, mountains, hills or forest areas. There may be natural obstacles like mountains and hills. An emergency crisis in a rural area may be quite large for the geographical extension (tens of square kilometer). A rural environment does not have usually an extensive communications infrastructure.
- **Blue and/or green borders**. Identifies the border between land and sea or a major lake (blue border) or between two and more different political regions in the land (green border). We can make a distinction between a border included in a single political or governmental region (i.e. national context) and a border across different political or governmental regions (i.e. cross-national context). Because different PPDR organizations are likely to operate in the second case, interoperability requirements may be more relevant.
- **Port or airport**. A port or airport has similar features to the urban environment as it is usually limited in size (few square kilometer). In comparison to a generic urban environment, there is a larger presence of critical facilities (e.g. traffic control centre) which should be protected or whose services should be maintained. Critical facilities like deposit of dangerous materials with inflammable or chemical substances may also be present.

The following dimensions are useful to capture the characteristics that are relevant to PPDR communications used in the different operational scenarios:

- **Geographical extension**. This dimension describes the size of the area involved in the emergency crisis.
- **Environment complexity**. This dimension represents the complexity of the emergency crisis in terms of number of entities involved, difficulty of the environment and so on.
- **Crisis severity**. This dimension represents the risk for the security of the citizens, infrastructures and environment.

Most day-to-day operations are conducted in low-/medium-coverage extensions and show a low environment complexity (i.e. personnel from a single agency involved) and the crisis severity is low. In turn, a natural disaster such as a flooding or an earthquake is likely to affect a large regional area and requires a complex emergency response (i.e. involving a number of PPDR agencies, volunteers and militaries), and infrastructures (e.g. transportation, communications) can be severely degraded or destroyed during a natural disaster.

1.3.2 Framework for PPDR Operations

PPDR operations usually involve [9]:

- **Intervention teams on the field**, composed of first responder officers carrying out their professional core missions and tasks (e.g. law enforcement, firefighting, medical assistance, search and rescue).
- **Intervention team leaders on the field** (integrated in intervention teams or in mobile command rooms). The leader must have the intelligence of the mission and the way to perform it. As to radiocommunications employed in the operation, team leaders master the radio scheme of their mission, for example, who has to speak with whom and on which talk groups.
- **Dispatchers or operators in the control rooms**, supporting the intervention teams on the field and their leaders in the execution of their professional core missions and tasks as well as by managing their radiocommunications (e.g. patching of talk groups). Control rooms, also referred to as emergency control centres (ECCs), are operational centres typically deployed per PPDR service/discipline (e.g. police, EMS, firefighting have separate control rooms). Control rooms house a number of operational systems including radio dispatcher terminals, computer-aided dispatching (CAD) systems for coordination and control, information systems (e.g. Geographical Information Systems (GIS)) and integrated communication control systems with connectivity to other control rooms and networks.
- **Back-office support teams** (e.g. network operator, manufacturers), which are not directly involved in the operations with first responders, but are responsible for the technical–operational conception, design and implementation of the radiocommunication systems that first responders use (e.g. data bases pre-provisioning, adjustment of technical–operational parameters, technical maintenance).

An emergency can be handled by a single PPDR agency, or it may require the participation of a number of them. Agencies involved in the emergency can be in charge of the same or different services (e.g. an emergency situation requiring only police forces vs. an emergency where police, fire and EMS are involved). Moreover, the involved agencies may be acting in their own jurisdiction or be displaced from other jurisdictions to assist in an incident (e.g. local, regional, national). The resulting combinations can lead to the following hierarchy levels of communications [5]:

- **Intra-agency communications**, thus involving a single PPDR service/single jurisdiction
- **Inter-agency communications**, involving (i) single PPDR service/multiple jurisdictions, (ii) multiple PPDR services/single jurisdiction and (iii) multiple PPDR services/multiple jurisdictions

These hierarchical levels define increasingly complex communications interactions and administration as the hierarchy moves from the single PPDR service/single jurisdiction situation to the multiple services/multiple jurisdictions events. Interoperability is key for inter-agency coordination. Interoperability impacts on the organizational and procedural aspects as well as on technical means (e.g. communications systems) used by the involved agencies. Technology provides tools to improve the effectiveness and efficiency when handling the tasks and procedures but can never replace the responsibility of the authorities and PPDR agencies and the correct application of their agreed procedures in the event of an incident. Inter-agency interoperability can be classified in the following three modes of operation [5]:

1. **Day to day**. This includes routine operations with neighbouring agencies to provide support or backup. This form of interoperability makes up the most of an individual first responder's multi-agency activities.
2. **Task force**. The task force mode defines a cooperative effort between specific agencies with extensive pre-planning and practice of the operation. This includes cooperative efforts among mixed yet specific agencies and services such as operations that are planned or scheduled and are proactive and operations that have a common goal, common leader and common communications.
3. **Mutual aid**. The mutual aid mode describes major events with large numbers of agencies involved, including agencies from remote locations. Their communications are not usually well planned or rehearsed. The communications must allow the individual agencies to carry out their missions at the event, but follow the command and control structure appropriate to coordinate the many agencies involved with the event. This could be needed in a major event that causes a large number of agencies to respond from multiple jurisdictions. Considerable coordination is required.

Considering that the majority (as much as 90% according to Ref. [13]) of the communications usage falls under the day-to-day operations mode, the communications systems must support the day-to-day operations with all the same performance features that may be required to support the other modes of operation. Unless the systems provide the first responders with seamless functionality regardless of the mode of operation, the first responders will not use their systems efficiently or effectively, especially when they need to operate in the task force and mutual aid modes.

PPDR operations follow a strict command and control hierarchy. In the context of an important event (disaster-like), incident command structures (ICS) are established. An ICS, regardless of national variations, can be typically divided into three discrete levels of management [8]: strategic overview, tactical management and operational implementation. The scope of each level is outlined in Table 1.1.

In various countries and indeed organizations within national boundaries, these structural levels have been designated with names to enable their easy identification: Gold/Silver/Bronze or Level 1/Level 2/Level 3 is the terminology commonly used. 'Gold', 'Silver' and 'Bronze' are titles of functions adopted by each of the emergency services and are role, not rank, related (i.e. titles do not convey seniority of service or rank, but depict the function carried out by that particular person). In high-profile incidents, particularly those such as terrorist incidents where there is a significant national/political involvement, it is not unusual for some to attempt to define a 'super-strategic' or 'platinum' level above the three levels described previously.

Table 1.1 Common levels of management in a command and control hierarchy.

Command and control levels	Functions
Strategic overview ('What should we do?')	Determines the strategic issues relevant to the incident Defines the critical issues defines/records the strategy for the incident Ultimately provides liaison between the emergency services and key authorities (local, regional and national) and other relevant bodies, dependent on the nature of the incident Delegates' management of the incident to the tactical level
Tactical management ('How do we achieve it?')	Defines and directs the tactical parameters for managing the incident Plans and coordinates tasks and agencies Obtains resources Remains detached from the incident itself – must not get involved at operational level
Operational implementation ('I'm doing it!')	Controls and deploys resources either by geographic delineation or functional role for their respective service

It is possible that, early on in the incident, members of one PPDR service will spontaneously carry out tasks normally under the responsibility of another. As soon as sufficient staff arrives, each service is expected to establish unequivocal command and control of functions for which it is normally responsible. Therefore, each of the first responder disciplines may have its own branch commanders at a large incident. As the incident progresses, it is essential that these branch commanders are able to coordinate, communicate and share information among them.

Command and control of functions are likely to be discrete in the early stages of an incident. As the incident progresses, at some stage they may move (certainly at strategic level) into one central location. At the tactical level, there may be a combination of discrete, technologically combined or physically co-located control centres, depending on a number of factors including technological systems in use, geographical location and the actual nature of the incident. The high command levels are usually operating outside the affected area.

The formation of both a Gold and Silver coordinating group can be of great value at all major incidents. Therefore, at some point during the early part of the operation, one or more tactical-level mobile control centres may be established at designated locations. Each of these should have direct voice/data communications links back to their respective permanent control centres. The staff with these mobile control centres would take control of the personnel dealing with the incident and inform the strategic-level command structure on the progress being made to address the strategy for the incident.

Some major incidents may be so quickly resolved that there is no requirement to convene a Gold coordinating group. Where a Gold coordinating group is convened, it initially consists primarily of the police, fire brigades and EMS services [8]. Additional Gold level representation from other agencies will be dependent upon the requirements of the incident (e.g. nature scale and dynamics).

The command and control structure should allow PPDR personnel to work seamlessly on situations that may begin small, but can evolve into large incidents requiring many resources

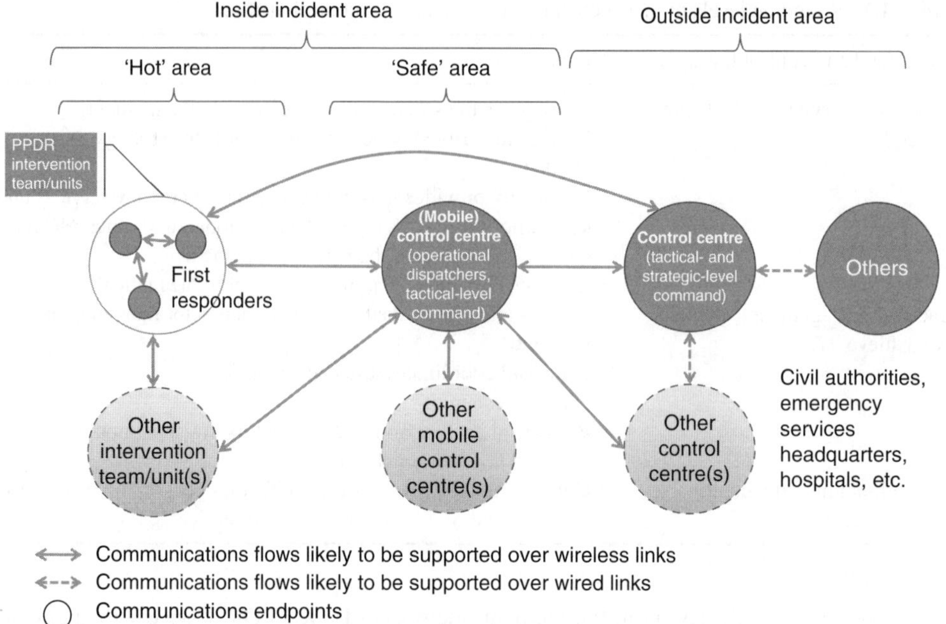

Figure 1.2 Illustrative view of potential communications flows in a common incident command structure.

and assistance from numerous jurisdictions. As an incident grows in magnitude, the incident commander has to know what resources and capabilities are becoming available for use and, if necessary, request the temporary assistance of personnel and equipment of other agencies. PPDR services are required to develop working arrangements according to circumstances.

Based on the earlier considerations, it becomes evident that multiple voice and data communications flows maybe required among intervention teams/units on the field, potentially belonging to different PPDR agencies, tactical-level mobile control centres established at designated locations within the incident area and tactical- and strategic-level control centres in remote locations outside the incident area. Figure 1.2 illustrates some of these potential communications flows in a typical ICS, distinguishing those likely to be supported over radio links from those provided over wired links and systems.

1.3.3 Communications' Reference Points in PPDR Operations

A comprehensive generic model that captures the major components and reference points between PPDR responders, authorities and other entities that may be involved in routine or emergency situations has been developed within the ETSI Special Committee on Emergency Communications (ETSI SC EMTEL) [3]. The model is illustrated in Figure 1.3.

A central component of the model is the ECC. ECCs are operational centres typically deployed per PPDR service/discipline (e.g. police, EMS, firefighting) that house a number of operational systems (e.g. dispatcher consoles, CAD and GIS tools) and are interconnected to other ECCs and a variety of networks (e.g. public telephone networks, Internet).

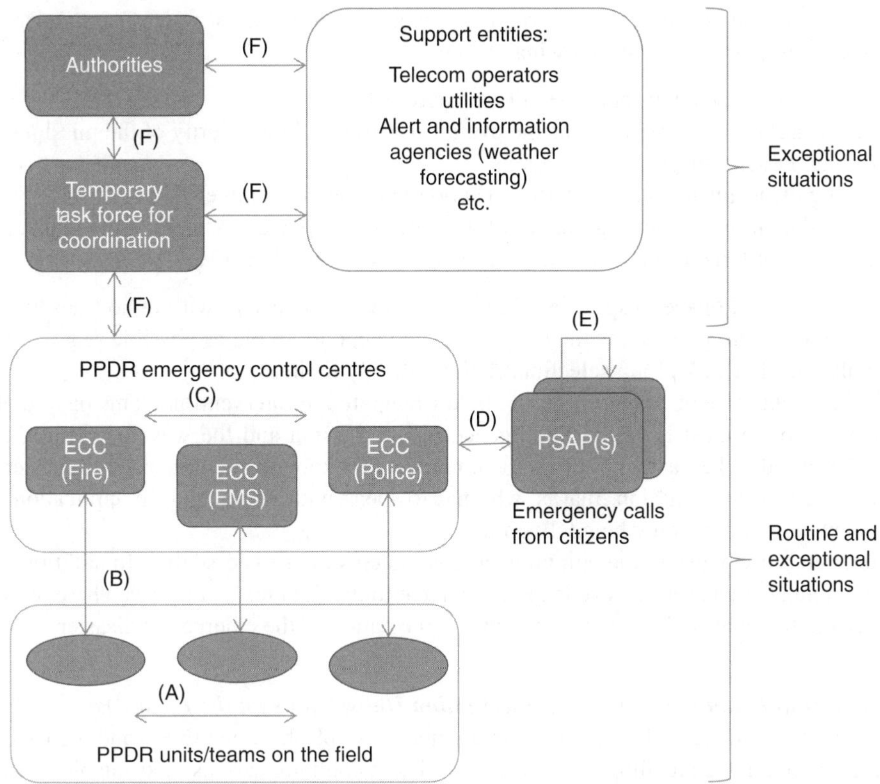

Figure 1.3 Components and reference points in PPDR communications.

Another central element in emergency handling is the so-called public safety answering point (PSAP). The PSAP is a physical location where emergency calls from citizens are received and, if necessary, forwarded to the competent ECC(s).

In exceptional situations (e.g. large emergencies and disasters), special task force or temporary headquarters can be established for emergency coordination so that communications are needed among these temporary facilities and ECCs and intervention teams on the field. Also, the involvement of public authorities (local, regional, national) may be required as well as the coordination with other non-PPDR entities that may also have a central role in the emergency response (e.g. telecom operators, utilities, information agencies such as weather forecasting, media, etc.).

The kind of actions that require communications and the main communications needs across the reference points depicted in Figure 1.3 (i.e. points tagged from (A) to (F)) are described in the following:

Communications among PPDR Unit/Teams on the Field: (A)

The intervention teams (also referred to as mobile teams) need facilities for communication among team members as well as for communications with other mobile teams. The need is for communication across the services involved, as well as within each service. Communications at this level mainly aim at the following:

- Management of the teams and operational coordination
- Reassessing on a continuous basis the overall situation and the priority of the missions
- Enabling the reporting within the teams
- Enabling the teams to call for additional support and other resources
- Exchanging information for guidance of the staff on the spot, assessment of the injuries and preparation of fixed rescue facilities before arrival of injured people

Officers on the field have to spend a minimum of time and capacity with radio transmission. Therefore, the radio procedures they apply must be kept as simple as possible (e.g. terminal manipulations, like talk group selection, shall be strictly limited).

The intervention team leader(s) on the field (integrated in intervention teams or in mobile command rooms) must have the intelligence of the mission and the way to perform it. In particular, besides the basic officer radiocommunication knowledge, they need to master the radio scheme of their mission, that is, who has to speak with whom and, in application of a functional radio model, on which talk groups.

Interoperability between the communications systems in use is essential. In addition, fall-back communications service needs to be available to the field teams in cases where network service is either unavailable or disturbed due to the nature of the emergency/disaster.

Communications among ECCs and Intervention Units/Teams on the Field: (B)
The access to permanent bidirectional communications links between ECCs and their mobile teams is crucial in the handling of emergencies. These communications links enable the same kind of functions discussed previously for communications among teams but now involve PPDR personnel on control rooms (e.g. operational dispatchers, tactical/strategic-level commanders) as well as the use of supporting tools and systems (e.g. GIS-based applications) that can greatly improve emergency response.

Dispatcher(s) or operators in the control rooms give support to the intervention teams on the field and their leaders in the execution of their professional core missions and tasks. Depending on the governance model, dispatchers will only assist and coordinate the teams on behalf of an end chief or will also have the lead on them. Dispatchers also give support by managing the radiocommunications used by intervention teams. Therefore, dispatchers must have a solid understanding of the functional radio model and the related operational radio procedures to face an evolving or an unexpected situation.

Through the ECC, communications can be established between all involved parties (mobile team members, control room staff, receiving and assisting units/institutions). Among the main features needed for this type of communications are:

- Seamless radio coverage throughout the affected area, including guaranteed availability of coverage also under exceptional conditions as well as means to maintain communication during network outage.
- Enough traffic capacity at the incident. The need for radio capacity is increasing during major incidents and accidents. Efforts have to be made to ensure as far as possible that sufficient communications facilities are available.
- Sufficient voice quality not to impair the understanding of the message.

- Specialized features at the disposal of dispatchers to regulate, in real time, the radiocommunications such as combining (patching) of groups, remotely programming extra groups on terminals and remotely selecting groups on terminals. In addition, access to the network shall be controlled by using functionalities such as assigning priority to potential users, thereby restricting some parties from access to the network under certain circumstances.

Communications among ECCs: (C)
ECCs need to have the facilities to collaborate with other ECCs, either within the same service or across services (e.g. between fire and EMS). Interconnection of ECCs may rely on the use of fibre-optic, microwave, and copper landline systems to provide backbone links for voice and data applications. Examples of cases where this is needed are:

- Callers are transferred to the wrong ECC so that the call needs to be forwarded to the correct ECC together with additional information (e.g. location data).
- Cases involving more than one ECC (e.g. fires with risks for human lives that typically involve fire, health and police, CBRN incidents (or suspected incidents), terrorism).
- The communications facilities exist to integrate the resources from two or more ECCs, in case of a larger emergency situation.

Communications requirements among ECCs must:

- Establish communications links to support a number of services, including speech and data.
- Conform to the relevant procedures established by the ECCs or their organizations.
- Support conference calls including external resources that may need to be set up and kept over a substantial amount of time. In contingencies, calls to external resources may be required.

Communications among ECCs and PSAPs: (D)
PSAP and ECC are two different functionalities that may or may not be integrated. The PSAP will, after reception of an emergency call from an individual/citizen, communicate without delay with the competent ECC and transmit the location and nature of the emergency of the calling party along with any other relevant information that may be available associated with the call.

For this purpose, reliable and pre-planned communications links among all of the ECCs in the competence zone of the emergency situation must enable to transmit voice and transfer all the data received at the PSAP (especially location data) or collected by the operator of the PSAP.

Communications among PSAPs: (E)
PSAPs normally work independently and their interrelation is not subject to special needs.

In cases where calls arrive at another PSAP than the one responsible for the area where the call is originated (e.g. mobile phones in the bordering area between different PSAPs), there may be a need to transfer the call together with additional information (e.g. location data). The need will depend upon the operation rules that have been established for these types of situation, for example:

- The call is handled by the receiving PSAP (e.g. the immediate help is a key point in the case, the case of PSAP backup, or load sharing).
- The call is immediately transferred to the normal PSAP, which handles all the case. In such a scenario, the location data must remain accessible to the normal PSAP, as for any received call.

- Depending on local procedures, the receiving PSAP may transfer the call directly to the relevant ECC, possibly together with information to the correct PSAP that the call has been transferred.

It is the responsibility of the PSAPs or their organization to predefine these rules of procedures.

Communications among Special Task Force or Temporary Headquarters and Permanent Entities in Special Conditions: (F)
For their efficient work in handling emergencies, special task force or temporary headquarters and ECCs are depending on access to permanent bidirectional links with the mobile teams and temporary headquarters. This access needs to be available for the duration of the emergency/disaster. The basic need is for configurable communications, to fulfil all contingency plans for, under the possible stages of escalation for a simple emergency situation, through a crisis to a regional or national disaster.

A much more extended description of the functional requirements for communications for all of the earlier reference points in PPDR communications can be found in Refs [3, 9].

1.3.4 Communications Services Needed for PPDR Operations

PPDR communications have to be effective, fast, reliable, secure and interoperable where possible, in order not to put at risk the success of PPDR operations. The efficiency of the emergency operation is dependent upon the ability of the communications network to deliver a real-time exchange of information between several authorized emergency personal. As discussed in the previous subsection, this can occur at various levels in the emergency situation, for example, between agents on the field, between officers in the ECCs and the agents on the field and, in major incidents, between any established temporary coordination headquarter and the involved ECCs.

To emphasize the criticality of PPDR communications needs, two types of operational situations addressed by PPDR organizations are typically defined [10]:

1. **Mission-critical situations**. The expression ‘mission critical’ is used for situations where human life, rescue operations and law enforcement are at stake and PPDR organizations cannot afford the risk of having transmission failures in their voice and data communications or for police in particular to be ‘eavesdropped’.
2. **Non-mission-critical situations**. This refers to situations where communication needs are non-critical: human life and properties are not at stake, administrative tasks for which the time and security elements are not critical, etc.

Therefore, it can be stated that mission-critical communications refer to any information transfer that becomes crucial to the successful resolution of a PPDR operation.[1] Table 1.2

[1] The distinction between ‘mission critical/non-mission critical’ is sometimes considered an oversimplification of what it could really be a hierarchy or continuum of criticality extending from minor incidents to international catastrophes [14]. Moreover, this terminology is not specific to the PPDR sector but also used in other areas such as the critical infrastructure sectors (e.g. energy, transportation, etc.). A more transversal definition of ‘mission critical’ is proposed in Ref. [14]: *‘A mission is “critical” when its failure would jeopardise one or more human lives or put at*

Table 1.2 High-level classification of required communications services.

Service	Interactivity level	Description
Voice	Interactive	Interactive voice communications between public safety practitioners and their supervisors, dispatchers, members of the task force, etc. require immediate and high-quality response and must meet much higher-performance demands than those required by commercial users of wireless communications. Commands, instructions, advice and information are exchanged that often result in life-and-death situations for public safety practitioners, as well as for the public.
	Non-interactive	Non-interactive voice communications occur when a dispatcher or supervisor alerts members of a group about emergency situations or acts to share information, without an immediate response being required or designed in the communications. In many cases, the non-interactive voice communications have the same mission-critical needs as the interactive service
Data	Interactive	Interactive data communications mean that there is query made and a response provided. Such communications can provide practitioners with maps, floor plans, video scenes, etc. A practitioner may not need to initiate the query and response; it can include automated queries or responses. A common form of interactive data communications is instant messaging. Commanders, supervisors, medical staff, etc., can make intelligent decisions more efficiently with data from field personnel. Similarly, personnel entering a burning building armed with information about the building, such as contents, locations of stairwells, hallways, etc., can also perform their duties more efficiently
	Non-interactive	Non-interactive data communications are one-way streams of data, such as the monitoring of firefighter biometrics and location, etc., which greatly increase the safety of the practitioners. This form of communications also makes command and control easier because a commander is aware of the condition and location of the on-scene personnel

provides a common classification of the communications services that public safety agencies (PSAs) require to handle mission-critical operations [5]. The classification distinguishes between voice and data services and, for each of them, the level of interactivity required. Voice is so far the most important communications mechanism for mission-critical operations, though it is worth noting that over time the definition of mission critical will remain ever changing and the demands of tomorrow's first responders may change. In fact, data communications are becoming increasingly important to PPDR practitioners to provide the information needed to carry out their missions so that some data applications will likely become mission critical in the future. Additional details on voice and data services are provided in the following subsection.

risk some other asset whose impairment or loss would significantly harm society or the economy. In such cases even small degradations of communication supporting the mission could have possibly dire consequences'. For instance, in the utility and transport sectors, preventing socioeconomic damage above some agreed levels could be defined as a 'critical mission' (in addition to preventing injury or loss of life), while damage below agreed levels might at most be 'business critical' (affecting specific individuals or firms but not enough to harm society or the wider economy).

1.3.4.1 Voice Services

The term mission-critical voice has been used within the PS community for decades, but there has been no one single complete definition of what, exactly, mission-critical voice is. In an effort to provide a basis for a common understanding of the meaning of and the multiple requirements of mission-critical voice, the following key elements for the definition of mission-critical voice have been recently identified [15]:

- **Direct or talk around**. This mode of communications provides PS with the ability to communicate unit to unit when out of range of a wireless network or when working in a confined area where direct unit-to-unit communications is required. In this way, responders can talk even when network infrastructure is unavailable.
- **Push to talk (PTT)**. This is the standard form of PS voice communications today: the speaker pushes a button on the radio and transmits the voice message to other units. When they are done speaking, they release the PTT switch and return to the listen mode of operation.
- **Group call or talk group**. This method of voice communications provides communications from one-to-many members of a group and is of vital importance to the PS community.
- **Full duplex voice systems**. This form of voice communications mimics that in use today on cellular or commercial wireless networks where the networks are interconnected to the public switched telephone network (PSTN).
- **Talker identification**. This provides the ability for a user to identify who is speaking at any given time and could be equated to the caller identification feature available on most commercial cellular systems today.
- **Emergency alerting**. This indicates that a user has encountered a life-threatening condition and requires access to the system immediately and is, therefore, given the highest level or priority.
- **Audio quality**. This is a vital ingredient for mission-critical voice. Audio quality must ensure voice to be intelligible in the difficult noise environments that first responders might encounter. The listener must be able to understand without repetition, identify the speaker, detect stress in a speaker's voice and hear background sounds as well without interfering with the prime voice communications.

1.3.4.2 Data Services

While voice communications will remain a critical component of PPDR operations, new data and video services are expected to play a key role increasingly. For instance, PPDR agencies today already use applications such as video for surveillance of crime scenes and of highways as well as to monitor and conduct damage assessment of wild land fire scenes from airborne platforms that can provide real-time video back to ECCs. In addition, there is a growing need for full motion video for many other uses such as situational awareness from intervention teams or the use of robotic devices in human life-threatening conditions.

Indeed, data services can be used to provide a large number of applications, which can have widely differing requirements in terms of capacity, timeliness and robustness of the underlying data communications service. As an illustrative example, Table 1.3 provides a list of

Table 1.3 Data applications and associated requirements.

Service	Throughput	Timeliness	Robustness
E-mail	Medium	Low	Low
Imaging (e.g. picture transfer)	High	Low	Variable
Digital mapping/geographical information services	High	Variable	Variable
Location services	Low	High	High
Video (real time)	High	High	Low
Video (slow scan)	Medium	Low	Low
Data base access (remote)	Variable	Variable	High
Data base replication	High	Low	High
Personnel monitoring	Low	High	High

potential data-centric applications for PPDR use as identified in Ref. [3], which have been characterized in terms of their impact on network throughput (i.e. data bit rate), timeliness (i.e. responsiveness) and robustness (i.e. reliability).

Some applications may be used with dedicated communications assets tuned to the particular needs of that application, although interfaces may be necessary to exchange data from such dedicated systems with other applications. Where appropriate, such applications should be based on appropriate standards to facilitate information exchange among them. Where data applications share the use of a data transmission capability, the provision of sufficient capacity and effective management must be provided to ensure application data is communicated appropriately.

1.4 Communications Systems for PPDR

This section describes the main communications technologies and systems currently in use for the delivery of the communications services needed for PPDR operations discussed in the previous section. In this regard, a review of the type of requirements usually bound to PPDR communications systems is provided in the first subsection. Afterwards, a common classification of the technologies based on achievable bit rates (narrowband (NB), wideband (WB) and broadband (BB)) is presented, pointing out the main standards that fit in each category. The classification is then followed by an overview of the most widely used digital radiocommunication standards for PPDR communications as of today (Terrestrial Trunked Radio (TETRA), TETRAPOL, DMR, P25). Finally, the section is concluded with the identification of some of the major limitations found in today's PPDR communications landscape through the analysis of an illustrative, hypothetical incident.

1.4.1 General PPDR Requirements on Communications Systems

Due to its unique operational requirements, PPDR has multiple complex communications technology needs. These requirements involve communications solutions that in some cases are unique to PPDR. General requirements for PPDR communication systems have been widely discussed [1, 3, 5]. A comprehensive view of the type of requirements that should be accounted for are given in the following:

- **Service capabilities and performance**. Central service capabilities required for PPDR are PTT operations, broadcasting/group communications and talk around (i.e. direct mode, terminal to terminal) for voice communications. There may be additional requirements in terms of data services (e.g. status messaging, short messaging, automatic vehicle location (AVL) and tracking) and supplementary services such as ambience listening, dynamic group number assignment (DGNA), call authorization by dispatchers, late entry and many others. Fast responsiveness and low latency requirements are typically required for these services (e.g. fast call set-up below 300 ms and end-to-end voice delay in the range of 200 ms). The equipment in use is typically required to support most of these service capabilities, while the user is in motion. The equipment may also require high audio output (to overcome high-noise environment); unique accessories, such as special microphones; operation while wearing gloves; operation in hostile environments (heat, cold, dust, rain, water, shock, vibration, explosive environments, etc.); and long battery life.
- **Strict control of the communications means**. This includes centralized dispatch for coordination and control of communications channels within the PPDR system together with the administration of terminal, subscriber and group call settings (e.g. group monitoring, dynamic regrouping, etc.). This also encompasses the support of prioritization mechanisms to make sure that important calls are always treated first in case of system congestion. The supported priorities may be required to reflect a hierarchy (among people or departments) but also an operational situation requiring special treatment of the call, whatever the hierarchy of the person. Prioritization may include pre-emptive emergency call capabilities, overriding if necessary ongoing low priority communications.
- **Security-related requirements**. Efficient and reliable PPDR communications within a PPDR organization and between various PPDR organizations, which are capable of secure operation, may be required. Security requirements may cover the need for subscriber/network authentication and support of encryption and integrity mechanisms over the radio interface and, in some cases, even require the use of end-to-end encryption (E2EE) between the endpoint terminals. Notwithstanding, there may be occasions where administrations or organizations, which need secure communications, bring specific equipment to meet their own security requirements. Furthermore, it should be noted that many administrations have regulations limiting the use of secure communications for visiting PPDR users.
- **Coverage**. The PPDR system is usually required to provide complete coverage for 'normal' traffic within the relevant jurisdiction and/or operation (national, provincial/state or at the local level). This coverage is required 24 h/day, 365 days/year. Usually, the systems supporting PPDR organizations are designed for peak loads and wide fluctuations in use. Additional resources, enhancing system capacity, may be added during an incident by techniques such as reconfiguration of networks with intensive use of direct mode and vehicular repeaters, which may be required for coverage of localized areas. Systems supporting PPDR are also usually required to provide reliable indoor and outdoor coverage, coverage in remote areas and coverage in underground or inaccessible areas (e.g. tunnels, building basements). Coverage requirements can be specified as, for example, 99.5% (outdoor mobile) and 65% or better (indoor mobile). Appropriate redundancy to continue operations when the equipment/infrastructure fails is extremely beneficial. PPDR systems are not generally installed

inside numerous buildings. PPDR entities do not have a continuous revenue stream to support installation and maintenance of an intensive variable density infrastructure. Urban PPDR systems are designed for highly reliable coverage of personal stations outdoors with limited access indoors by direct propagation through the building walls. Subsystems may be installed in specific building or structures (e.g. tunnels) if penetration through the walls is insufficient. PPDR systems tend to use larger radius cells and higher-power mobile and personal stations than commercial service providers.

- **Capacity**. Very low call blocking levels are typically required in PPDR systems (e.g. well below 1% under worst-case dimensioning assumptions). The system capacity must be sufficient to manage the anticipated traffic and yet be flexible enough in its functional design to also support communication during 'surge' conditions which exceed the anticipated traffic, for example, by using additional transportable switch and base stations. Sufficient data bandwidth may be also required to support a wide variety of data applications. Noting the extreme circumstances that may be in force during an emergency, it may be desirable for networks to degrade gracefully when user requirements exceed the normal levels of service.
- **High levels of service availability**. Service availability relates to the amount of time (usually per year) that a service is up and running. It is commonly measured in 'nines' of availability (e.g. 99.9 or 99.99% availability at all times is a typical requirement for PPDR applications). The achievement of high levels of service availability may require several layers of redundancy as well as resilient and robust equipment (e.g. hardware, software, operational and maintenance aspects).
- **System reconfiguration**. A rapid dynamic reconfiguration of the system serving PPDR may be required. This includes robust operation, administration and maintenance (OAM) systems offering status and dynamic reconfiguration. For instance, system capability to reprogramme field units over the air could be extremely beneficial.
- **Interconnection**. While PPDR systems are mainly intended to provide private, in-system communications, appropriate levels of interconnection to public telecommunications networks may also be required. The decision regarding the level of interconnection (i.e. all mobile terminals vs. a percentage of terminals) may be based on the particular PPDR operational requirements. Furthermore, the specific access to the public telecommunications network (i.e. directly from mobile or through the PPDR dispatch) may also be based on the particular PPDR operational requirements.
- **Interoperability**. Communications interoperability might be required at different levels of a PPDR operation, from the most basic level, that is, a firefighter of one organization communicating with a firefighter of another, up to the highest levels of command and control. Usually, coordination of tactical communications between the on-scene or incident commanders of multiple PPDR agencies is required. Remarkably, the achievement of the desired interoperability is not only a technical matter but spans from organizational and operational aspects to regulatory and legal frameworks. Mainly from a technical perspective, various options are available to facilitate communications interoperability between multiple agencies. These include, but are not limited to, the use of common frequencies and equipment, communication via dispatch centres/patches, interconnection of the PPDR networks via wire line interfaces or utilizing technologies such as radio gateways (e.g. 'back-to-back' gateways or relays) or more advanced software-defined radio (SDR) equipment.

- **Spectrum usage and management**. Depending on national frequency allocations, PPDR users must share with other terrestrial mobile users for some applications (e.g. point-to-point radio links). The detailed arrangements regarding sharing of the spectrum vary from country to country. Furthermore, there may be several different types of systems supporting PPDR operating in the same geographical area. Therefore, interference to systems supporting PPDR from non-PPDR users should be minimized as much as possible. Depending on the national regulations, the systems supporting PPDR may be required to use specific channel spacing between mobile and base station transmit frequencies. Each administration has the discretion to determine suitable spectrum for PPDR.
- **Regulatory compliance**. The systems supporting PPDR should comply with the relevant national regulations. In border areas (near the boundary between countries), suitable coordination of frequencies should be arranged, as appropriate. The capability of the systems supporting PPDR to support extended coverage into the neighbouring countries should also comply with regulatory agreements between the neighbours. For DR communications, administrations are encouraged to adhere to the principles of the Tampere Convention. Flexibility should be afforded to PPDR users to employ various types of systems (e.g. HF, satellite, terrestrial, amateur, Global Maritime Distress and Safety System (GMDSS)) at the scene of the incident in times of large emergencies and disasters.
- (Last but not least) **Cost-related requirements**. Cost-effective solutions and applications are extremely important to PPDR users. The deployment of dedicated PPDR networks is usually very demanding for PPDR organizations from an economic point of view. A national or regional network is usually an investment for 10–15 years or more. This can be facilitated by open standards, a competitive marketplace and economies of scale. Furthermore, cost-effective solutions that are widely used can reduce the deployment and upgradeability costs of permanent network infrastructure. Administrations should consider the cost implications of interoperable equipment since this requirement should not be so expensive as to preclude implementation within an operational context.

It is noted that individual administrations or PPDR organizations may have their own requirements for PPDR that go beyond those described herein and that each standard would need to be evaluated on a case-by-case basis against those requirements.

1.4.2 Technologies in Use for PPDR Communications

A common classification of technologies used for PPDR communications is based on the average data rates required by the supported PPDR applications [1]:

- **Narrowband (NB)**. It refers to technologies mainly intended to deliver voice-centric communications and low-speed data applications. Typical data rates are up to a few tenths of kilobits per second, operating on radio-frequency channels of up to 25 kHz. Current state of the art in the PPDR sector are digital radio trunking technologies such as TETRA, TETRAPOL and Project 25 (P25), which are commonly used to deploy wide area coverage networks. These technologies are typically referred to as professional/private mobile radio (PMR) technologies (the term land mobile radio (LMR) is also common in North America), though it's worth noting that PMR technologies are not only used in the

PPDR sector but also adopted in many other markets such as transportation, utilities, industry, private security and even military. A further insight into the main NB PMR standards used for PPDR communications is addressed in Section 1.4.3

- **Wideband (WB)**. It refers to technologies that can deliver application data rates of several hundreds of kilobits per second (e.g. in the range of 384–500 kb/s). Systems for WB applications to support PPDR have been or are underdevelopment in various standards organizations. The most significant example in the context of PPDR is the evolution of TETRA, known as TETRA Enhanced Data Service (TEDS), that adds support for more efficient modulations and the use of radio-frequency channels of up to 150 kHz wide. Nevertheless, WB technologies have not been yet widely deployed as their NB counterparts. The point is that WB technology is currently regarded as not sufficient to meet future PPDR demands so that the natural upgrade from NB to WB technology is likely to be bypassed. A migration path from NB to BB and their coexistence is as of today the most likely migration scenario.
- **Broadband (BB)**. It refers to technologies that enable an entirely new level of functionality with additional capacity to support higher-speed data communications than WB, likely including high-resolution video transmission. Initially, the use of BB technologies was mainly intended to support PPDR operations in localized areas (e.g. 1 km^2 or less), providing indicative data rates in the range of several megabits per second. Localized operational scenarios open up numerous new possibilities for PPDR applications, including tailored area networks, hot spot deployment and ad hoc networks. In this regard, specialized communications gear such as tactical networking equipment [16] intended to fulfil PPDR responders' needs is already available, though its adoption by PPDR practitioners is very limited (its use is basically confined to the military domain). These communications systems are typically based on Wi-Fi-like radio interfaces and can operate in open bands, such as the 2.4- and 5.8-GHz Wi-Fi bands, and/or on restricted bands, such as the 4.4-GHz (allocated to military use in some countries) and 4.9-GHz bands (allocated to PPDR use in some countries). However, in addition to localized service, the demand for BB applications is now importantly shifting towards wide area coverage. In this context, the current mainstream commercial Long-Term Evolution (LTE) technology ecosystem is consolidating as the 'de facto' standard for the delivery of mobile BB PPDR applications.

Table 1.4 gives an overview of the various PPDR applications alongside the particular feature provided and specific PPDR examples of use, as developed in Report ITU-R M.2033 [1]. The applications are grouped under the NB, WB and BB headings to indicate which technologies are most likely to be required to supply the particular application and their features. Application types listed in the table can be used in any of the operational environments described in Section 1.3.2. The detailed choice of PPDR applications and features to be provided in any given area by PPDR is a national or operator-specific matter.

1.4.3 Current NB PMR Standards Used in PPDR

The requirements of the PPDR community for mission-critical voice and data services are currently satisfied by a range of voice-centric NB technologies such as TETRA, TETRAPOL, DMR and P25. These are all NB digital trunking systems able to offer a wide range of

Table 1.4 PPDR applications and examples.

Application	Feature	PPDR example
1. Narrowband		
Voice	Person to person	Selective calling and addressing
	One to many	Dispatch and group communication
	Talk-around/direct mode operation	Groups of portable to portable (mobile–mobile) in close proximity without infrastructure
	Push to talk	Push to talk
	Instantaneous access to voice path	Push to talk and selective priority access
	Security	Voice
Facsimile	Person to person	Status, short message
	One to many (broadcasting)	Initial dispatch alert (e.g. address, incident status)
Messages	Person to person	Status, short message, short e-mail
	One to many (broadcasting)	Initial dispatch alert (e.g. address, incident status)
Security	Priority/instantaneous access	Man down alarm button
Telemetry	Location status	GPS latitude and longitude information
	Sensory data	Vehicle telemetry/status Electrocardiograph (EKG) in field
Database interaction (minimal record size)	Forms-based records query	Accessing vehicle license records
	Forms-based incident report	Filing field report
2. Wideband		
Messages	E-mail possibly with attachments	Routine e-mail message
Data talk-around/ direct mode operation	Direct unit-to-unit communication without additional infrastructure	Direct handset-to-handset, on-scene localized communications
Database interaction (medium record size)	Forms and records query	Accessing medical records Lists of identified person/missing person Geographical Information Systems (GIS)
Text file transfer	Data transfer	Filing report from scene of incident Records management system information on offenders Downloading legislative information
Image transfer	Download/upload of compressed still images	Biometrics (fingerprints) ID picture Building layout maps
Telemetry	Location status and sensory data	Vehicle status
Security	Priority access	Critical care
Video	Download/upload compressed video	Video clips Patient monitoring (may require dedicated link) Video feed of in-progress incident
Interactive	Location determination	Two-way system Interactive location data

Table 1.4 *(continued)*

Application	Feature	PPDR example
3. Broadband		
Database access	Intranet/Internet access	Accessing architectural plans of buildings, location of hazardous materials
	Web browsing	Browsing directory of PPDR organization for phone number
Robotics control	Remote control of robotic devices	Bomb retrieval robots, imaging/video robots
Video	Video streaming, live video feed	Video communications from wireless clip-on cameras used by in building fire rescue
		Image or video to assist remote medical support
		Surveillance of incident scene by fixed or remote controlled robotic devices
		Assessment of fire/flood scenes from airborne platforms
Imagery	High-resolution imagery	Downloading Earth exploration-satellite images

Reproduced from Ref. [1].

voice-centric services and features but limited data capability. In addition to these digital systems, analogue systems, both conventional[2] and trunked, still remain operational in some places (e.g. VHF FM radios, MPT1327 systems) [17]. Nevertheless, as digital standards become more mature and more manufacturers release low-cost digital products, there is a steady migration to digital technologies, further facilitated by the fact that some of the digital standards can operate in dual mode to maintain analogue and digital compatibility and ensure an easy migration.

Within digital systems, both frequency division multiple access (FDMA) and time division multiple access (TDMA) technologies are used. Most current FDMA and TDMA products can offer the equivalent of one voice channel per 6.25 kHz of RF channel bandwidth. FDMA systems can be typically programmed to use either 6.25- or 12.5-kHz channels, while TDMA typically offers only a set of two or four voice channels (slots) carried inside a 12.5- or 25-kHz RF channel. FDMA and TDMA each have certain advantages for specific uses. For instance, an FDMA system using a single 6.25-kHz RF channel has an extra 3 decibels (dB) of sensitivity

[2] Conventional systems, also sometimes referred to as 'non-trunked' systems, possess no centralized management of subscriber operation or capability. Conventional systems allow users to operate on fixed RF channels without the need for a control channel. All aspects of system operation are under control of the system users. Operating modes within non-trunked systems include both direct (i.e. radio-to-radio) and repeated (i.e. through an RF repeater) operation. Users simply select the appropriate channel in their radios and communicate immediately with no repeater set-up time. Conventional systems may enough to meet the needs of agencies for cost-effective, low-density communications systems. On the other hand, trunked systems provide for management of virtually all aspects of radio system operation, including channel access and call routing. Most aspects of system operation are under automatic control, relieving system users of the need to directly control the operation of system elements. Unlike conventional operation in which a radio channel is dedicated to a particular user group for communications, trunking provides users access to a shared collection of radio channels. Trunked systems may be particularly attractive to agencies in communities that want to join together to form shared large-scale systems.

and less noise than other offerings in 12.5 or 25 kHz. Therefore, if there is a need to operate a voice channel over a long distance and in a noisier-than-normal RF environment, a single 6.25-kHz FDMA is likely to outperform the other radios in range and noise tolerance. On the other hand, if the requirements include mostly data communications, a TDMA radio like TETRA can offer the highest data bandwidth by aggregating the four voice channels into a single 25-kHz data channel.

PMR technologies are used in the PPDR domain to deploy from small-scale systems with only one or a few sites serving the needs of an individual agency to nationwide networks shared by multiple PPDR organizations. Indeed, in Europe, national multi-agency networks with countrywide coverage have been (some are still being) deployed in most countries based on TETRA and TETRAPOL technologies. In turn, a more complex environment is found in the United States where there is a plethora of independent systems deployed at different jurisdictional levels (municipalities, counties) and significant efforts are being undertaken in some states towards improving interoperable communications with the deployment of statewide P25 systems. Some further details on the TETRA, TETRAPOL, DMR and P25 technologies and network deployments are given in the following subsections, including a comparative view chart of their key features in Table 1.5. A more extended description of the technical and operational characteristics of several digital PMR technologies introduced throughout the world in the PPDR and other sectors can be found in the International Telecommunication Union (ITU) Report ITU-R M.2014 [18].

Table 1.5 Comparison chart of the main digital PMR technologies used for PPDR.

Feature	TETRA	TETRAPOL	DMR	P25
Technology	Four-slot TDMA	FDMA	Tier I: FDMA Tier II and III: Two-slot TDMA	Phase 1: FDMA Phase 2: TDMA
Frequencies	VHF, UHF, 800 MHz	VHF, UHF, 800 MHz	Frequency bands 66–960 MHz	VHF, UHF, 700, 800, 900 MHz
Channel bandwidth	25 kHz	12.5 kHz	12.5 kHz	Phase 1: 12.5 kHz Phase 2: 25 kHz
Data rate	28 kb/s(up to ~500 kb/s over 150 kHz channels with TEDS)	<8 kb/s	<8 kb/s	9.6 kb/s
Modulation	$\pi/4$ DQPSK	GMSK	Four-level FSK	C4FM
Vocoder	ACELP	RP-CELP	AMBE+2	AMBE+2
Encryption	TEA algorithms for AIE+specific algorithms for E2EE	E2EE. Algorithms may comply with TETRAPOL specifications or specific algorithms can be used	AES	AES/DES
More information	TETRA+Critical Communications Association (TCCA)(http://www.tandcca.com/)	TETRAPOL Forum(http://www.tetrapol.com/)	DMR Association(http://dmrassociation.org/the-dmr-standard/)	P25 Technology Interest Group (PTIG)(http://project25.org/)

1.4.3.1 TETRA

TETRA is a mature and established TDMA digital voice trunked and data radio technology used around the world. TETRA is an open standard developed by the European Telecommunications Standards Institute (ETSI). The TETRA standard was finalized in 1995 and introduced in the market in 1997. TETRA equipment is available in the VHF, UHF and 800-MHz bands. TETRA standard confers interoperability between radio equipment from multiple vendors. The TETRA and Critical Communications Association (TCCA) was established in December 1994 (known then as the TETRA MoU Association) to create a forum to act on behalf of all parties interested in TETRA technology, representing users, manufacturers, application providers, integrators, operators, test houses and telecom agencies. A TETRA Interoperability Certification process is managed by this organization to enable an open multivendor market for TETRA equipment and systems.

The TETRA standard supports many voice-centric services and facilities. Examples are group call, pre-emptive priority call (also called 'emergency call'),call retention, priority call, busy queuing, direct mode operation (DMO), DGNA, ambience listening, call authorized by dispatcher, late entry and many others. TETRA uses the Algebraic Code Excited Linear Prediction (ACELP) vocoder. Up to four voice channels can be supported over a single RF 25-kHz channel (four-slot TDMA).

The TETRA standard also provides some data transmission capabilities. Status messaging and short data service (SDS) are simple services that allow for the exchange of short strings of bytes. In addition to sending human-readable information, these messaging capabilities are also used as transport bearer services for applications that require very low data rates such as AVL. TETRA also supports a packet data (PD) service that provides a standard IP connectivity service that can be used to support other added-value applications with relatively higher bit rates (e.g. database lookup, picture messaging). However, the delivered data rates are peaked to 28.8 kb/s, achieved when the four slots of a 25-kHz carrier are combined and data is sent with the lowest protection for error control. To improve the support for data communication in TETRA, the ETSI Board in 2005 mandated the development of TETRA Release 2, commonly known as TEDS. TEDS offers new modulation options and wider radio channels (50, 100 and 150 kHz) to support data traffic rates of up to 500 kb/s, matching the speeds provided by cellular 2G GPRS/EDGE technology.

Other remarkable features supported by TETRA are DMO gateways, DMO repeaters and 'fallback' modes in TETRA base stations. A DMO gateway allows relaying the communication between a TETRA DMO terminal and a TETRA network that may not be directly reachable from that terminal. A DMO does a similar relying but now between two DMO terminals that cannot see each other. A typical application for the DMO repeater/gateway functionality is the following [19]. A DMO gateway is installed in a vehicle so that it automatically 'bridges' the network traffic to and from a person who left the vehicle and is working in the field, using a low-power handheld radio to talk to and from the TETRA network. With regard to the fallback mode in base stations, this allows a base station to continue to serve users if the backhaul link towards the rest of the infrastructure fails.

In terms of security, and similarly to all of the other digital PMR technologies used for PPDR communications, TETRA provides multiple security features for the verification of identities, protection of confidentiality and integrity and protection against lost or stolen terminals. In this regard, some of the main security features within the TETRA standard are the following [20]:

- Authentication. The authentication service allows a TETRA terminal and a TETRA network to prove each other's real identity, by proving to both parties the knowledge of a shared secret key (i.e. the authentication key K), unique for each terminal and only available in the terminal and a secured server of the home TETRA network.
- Air interface encryption (AIE). The AIE service allows encrypting the 'air interface' (AI), signalling and voice, between a TETRA terminal and its serving TETRA network. The TETRA standard supports a number of over-the-air TETRA Encryption Algorithms (TEAs), the differences being the types of users who are permitted to use them. Encryption keys can be static (e.g. usually preconfigured in the terminals) or dynamic (derived per connection as needed). Over-the-air rekeying (OTAR) methods are supported in TETRA to transfer in real-time secret keys securely to terminals.
- Enable and disable. This service provides a mechanism by which a terminal can be denied or allowed access to a TETRA system.
- Support for E2EE. It allows two parties to communicate in secure mode by ciphering the whole communication data at terminal level. In this case, a variety of other encryption algorithms can be used as deemed necessary by the national security organizations.

A schematic of a TETRA network architecture showing the standardized interfaces is depicted in Figure 1.4. The core of a TETRA network, named Switching and Management Infrastructure (SwMI), consists of a number of radio base stations (RBS) interconnected through one or a hierarchy of switches and/or routers (routers would be used in the case of TETRA over IP implementations). Typical components connected to the SwMI include the network management system, control room systems such as line dispatchers, PABX/PSTN/ISDN interconnections and IP gateways for PD networks. The gateway interfaces to PSTN, ISDN and PD networks have been standardized, in addition to an Inter-system Interface (ISI)

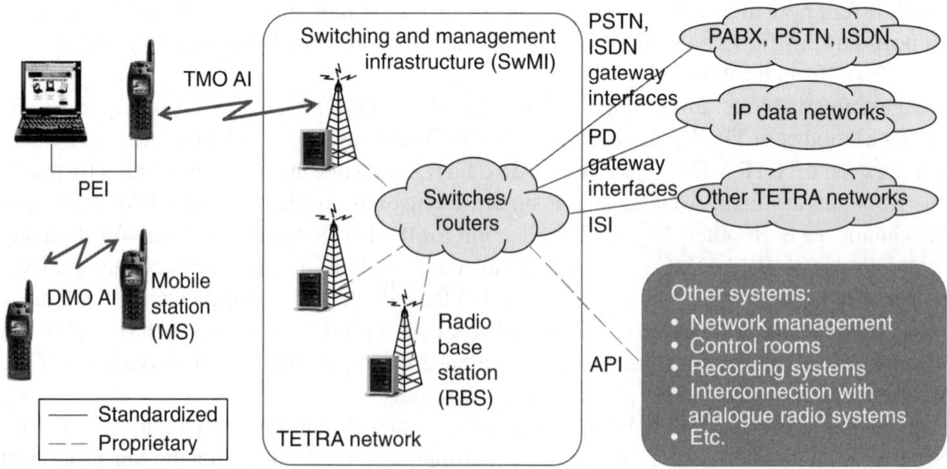

Figure 1.4 Network architecture and standardized interfaces of a TETRA system.

intended to interconnect TETRA networks. The remaining interfaces within the SwMI are proprietary implementations (vendor Application Programming Interfaces (APIs), which allow access to some of the SwMI services and functions). It is worth noting that the internal interface within the SwMI between the RBS and the switching/routing elements is not standardized either. On the radio side, two modes for the operation of the AI have been standardized: trunked mode operation (TMO) and DMO. A standardized Peripheral Equipment Interface (PEI) is also available. The PEI allows splitting the TETRA terminal in two separate devices: the terminal equipment (TE) and the mobile termination (MT). The equipment may be a PC or any other computing device, while the MT acts like a modem.

TETRA has been proven a suitable technology for large mission-critical networks. The ETSI TETRA standard has been deployed in over 120 countries worldwide, and, even though TETRA has been commonly adopted by many types of professional users, the emergency service sector remains the largest group of users using this standard. Indeed, most EU member states have rolled out national TETRA networks for PPDR (e.g. Belgium, Denmark, Germany, Finland, Sweden, etc.).

With regard to TEDS, equipment supporting the new data features entered the market in 2008–2009. However, the adoption of TEDS is still quite limited (TEDS is currently being deployed in some Northern countries). TETRA networks are also being deployed in North America, now that the Federal Communications Commission (FCC) allows the technology to be used there, but not for PS where the P25 standard is used.

1.4.3.2 TETRAPOL

TETRAPOL is another relevant NB technology in the European PPDR sector. TETRAPOL is a proprietary technology that was initially developed by a French company named MATRA in the 80s and gained a first-mover advantage over TETRA. Nowadays, the development of the technology remains in the hands of the French industrial group Airbus Defence and Space (which integrated the formerly known as EADS/Cassidian), which is the single provider worldwide. For the promotion of the technology, the TETRAPOL Forum was created, through which the specifications of the technology are available (in the form of Publicly Available Specifications (PAS)) to manufacturers who want to develop compatible products or solutions for TETRAPOL networks.

TETRAPOL is based on FDMA with 12.5-kHz carrier spacing, supporting one voice channel per carrier. This configuration gives some advantages in terms of coverage compared to TETRA systems, but at the expense of lower spectral efficiency. Like TETRA, TETRAPOL supports many features and functionalities for voice-centric services (multisite open channels, talk groups, direct mode, etc.). Fallback modes as well as DMO repeaters and direct mode to trunked mode (DMO–TMO) gateway repeaters are also supported in TETRAPOL systems. Another relevant characteristic of TETRAPOL is simulcast support (simultaneous transmission of the same information from multiple cell sites over the same RF channel). This allows systems to be rolled out even when few channels are available, as it is likely to be the case in most big cities. Nevertheless, data transmission capabilities of TETRAPOL are more limited than in TETRA.

TETRAPOL also includes the specification of some APIs, which enable third-party application developers to create complementary voice and data applications for TETRAPOL users. Applications developed using these APIs include command and control room solutions, AVL applications based on GPS location information, user management, network supervision and management applications.

Large PS networks based on TETRAPOL are nowadays deployed in France, Spain, Slovakia and Switzerland, along with smaller networks in the Czech Republic and Romania. While the largest market of the TETRAPOL technology has been Europe, the technology has also been deployed for nationwide or wide area PPDR in a few countries within Latin America, Africa and Asia, being one of the largest deployments the Mexican 'Red Iris' network with over 1 million terminals.

1.4.3.3 DMR

DMR is a European standard also developed under ETSI. The DMR standard, first ratified in 2005, was developed with the objective to create an affordable digital system with lower complexity than TETRA but able to satisfy the needs of many PMR users and easing the direct replacement of legacy analogue PMR [21]. The promotion and certification of the technology is conducted by the DMR Association.

The standard is designed to operate within the existing 12.5-kHz channel spacing in a wide range of frequency bands (i.e. from 66 to 960 MHz). DMR provides voice, data and other supplementary services. DMR comprises three substandards:

1. DMR Tier I is mainly designed for licence-free use in the 446-MHz band. Tier I provides for consumer applications and low-power commercial applications, using a maximum of 0.5-W RF power. With a limited number of channels and no use of repeaters and no use of telephone interconnects and fixed/integrated antennas, Tier I DMR devices are suited for personal use, recreation, small retail and other settings that do not require wide area coverage or advanced features. DMR Tier I is based on a FDMA AI.
2. DMR Tier II covers licenced conventional radio systems, mobiles and hand portables operating in PMR frequency bands from 66 to 960 MHz. This standard is targeted at users who need spectral efficiency, advanced voice features and integrated IP data services in licensed bands for high-power communications. It specifies two-slot TDMA in 12.5-kHz channels. Accordingly, DMR II provides the equivalent of two 6.25-kHz voice or data paths in a 12.5-kHz channel.
3. DMR Tier III covers trunking operation in frequency bands 66–960 MHz. Tier III supports voice and short messaging handling with built-in character status messaging and short messaging. It also supports PD service in a variety of formats, including support for IPv4 and IPv6. The Tier III standard is also based on a two-slot TDMA radio interface in 12.5-kHz channels.

DMR manufacturers have made a significant effort to guarantee that a certain number of DMR radio features are compatible across different manufacturers through interoperability tests. Some manufacturers have been certified as having that basic number of DMR features interoperable between vendors, and more manufacturers will pursue and obtain similar certifications in the future.

DRM equipment is sold today in all regions of the world, being actual applications mainly focused on the Tier II and III licensed categories. According to the DMR Association, while the DMR standard is being used in the PS market, the largest vertical market for this technology is actually the industrial sector.

1.4.3.4 Project 25

Project 25 (P25) is the standard for the design and manufacture of interoperable digital two-way wireless communications developed in North America under the auspices of the Telecommunications Industry Association (TIA). A P25 Steering Committee, collaborated by the Association of Public-Safety Communications Officials (APCO) International and other US national associations and federal agencies, was formed in 1989 to provide a set of industry standards for a PMR digital technology intended to replace the legacy analogue systems. The published P25 standards suite is administered by the TIA Mobile and Personal Private Radio Standards Committee (TIA/TR-8). Similar to TETRA TCCA and TETRAPOL Forum, the Project 25 Technology Interest Group (PTIG) was created to promote the success of Project 25 and educate interested parties on the benefits that the standard offers.

The initial specifications of P25 (P25 Phase 1) focused on the Common Air Interface (CAI) and vocoder as its baseline. Phase 1 included the specifications for 12.5-kHz FDMA equipment and systems that could interoperate with multiple vendors' radios in conventional or trunked mode, as well as legacy analogue FM radio systems. The CAI standard was completed in 1995. Since then, additional Phase 1 standards have been developed to address trunking; security services, including encryption and OTAR; network management and telephone interfaces; and the data network interface. Additionally, there have been ongoing maintenance revisions and updates to the existing standards.

P25 Phase 2 was designed to satisfy PS's need to transition to a 6.25-kHz or equivalent occupied channel bandwidth and maintain backward compatibility to Phase 1 technology, allowing for graceful migration towards greater spectrum efficiency. Although the need was identified for standards to address additional interfaces and testing procedures, the primary focus for the Phase 2 suite of standards was defined by a two-slot TDMA approach to spectrum efficiency as opposed to a 6.25-kHz FDMA technology. The Phase 2 suite of standards addressing TDMA trunking technology was completed and published in 2012, but the standards allowing initial product development were published back in 2010.

P25 supports both voice and data digital communications with data rates of 9.6 kb/s. A variety of system configurations are allowed, including direct mode, repeated, single site, multisite, voting, multicast and simulcast operation. Both conventional and trunked operations are supported in local and wide area configurations. P25 also offers high-power operation allowing large geographic areas to be covered with fewer sites than other technologies, making P25 technology an economical and efficient choice. The P25 standard itself is frequency agnostic. P25 equipment is available from numerous suppliers in VHF, UHF and 700-, 800- and 900-MHz frequency bands to meet the diverse frequency requirements of agencies around the world. The standard enables multiple frequency bands to be supported on one system. P25 supports secure communication through the use of federal government endorsed 256-bit key AES encryption, key management and equipment authentication.

In addition to the CAI, the P25 standard suite also enables interoperability for wire line interfaces. In this regard, a significant number of standards documents addressing the P25 Inter-RF Subsystem Interface (ISSI) have been developed and published. The standards for the Conventional Fixed Station Interface (CFSI) and Console Subsystem Interface (CSSI) have also been completed and deployed. Additionally, standards have been developed and published to address a number of interfaces relevant to security services, to include the Inter-Key Management Facility Interface (IKI). Interoperability testing is addressed through the P25 Compliance Assessment Program (CAP), a voluntary program that allows suppliers of P25 equipment to demonstrate that their respective products are compliant with P25 baseline requirements reflected in the suite of standards.

In the United States, P25 is widely adopted at different jurisdictional levels, from local to federal agencies, and has the support of the US Department of Homeland Security (DHS) that requires the implementation of this standard in the PMR equipment used for emergency communications [22]. P25 is also deployed in many other countries in the Americas, and there are some deployments throughout other regions in the world [31].

1.4.4 Main Limitations with Today's PPDR Communications Systems

Effective interoperable communications can mean the difference between life and death. Unfortunately, inadequate and unreliable communications have compromised emergency response operations for decades. Emergency responders need to share vital information via voice and data across services/disciplines and jurisdictions to successfully respond to day-to-day incidents and large-scale emergencies. Responders often have difficulty communicating when adjacent agencies are assigned to different radio bands, use incompatible proprietary systems and infrastructure and lack adequate standard operating procedures and effective multi-jurisdictional, multidisciplinary governance structures. This diversity is reflected in the different types of equipment and use of radio-frequency spectrum bands by PPDR organizations. Operational procedures are also quite different, which is a major problem for border security organizations.

The issues and problems identified with today's PPDR communications system can be well explained through the analysis of PPDR operations in a major incident.

PPDR scenarios have been described in a number of projects and initiatives (e.g. [5, 12, 23, 24]. The one described here was developed within the European research project HELP [25]. The scenario creates a hypothetical location, incident circumstances and response. However, the resources available at such a location are realistic as are the varied PPDR communications means available. For the purposes of emergency services operating methods and requirements, it is reasonable to suggest that, to a certain extent at least, 'an incident is an incident'. This means recognizing that, while there will be issues of scalability and certain incident-specific criteria (such as cross-border communication), it is both unnecessary and unrealistic to manufacture either large numbers of small-scale scenarios or one which is based on a colossal, continent-wide scale. It is far better to create a feasible, realistic incident that contains the relevant operational problems (in effect, a realistic, sequential series of events), from which a scalable technical solution can be developed. This approach is followed in the scenario developed in Project HELP, which concentrates on the early stages of a major incident (see Figure 1.5), rather than the incident as a whole. The scenario describes an incident which could happen in many locations and which, although having relatively small beginnings, expands considerably over a relatively short timeline. The reasoning behind this is that if a solution can be delivered where it matters, when operational resources are limited and operational intelligence

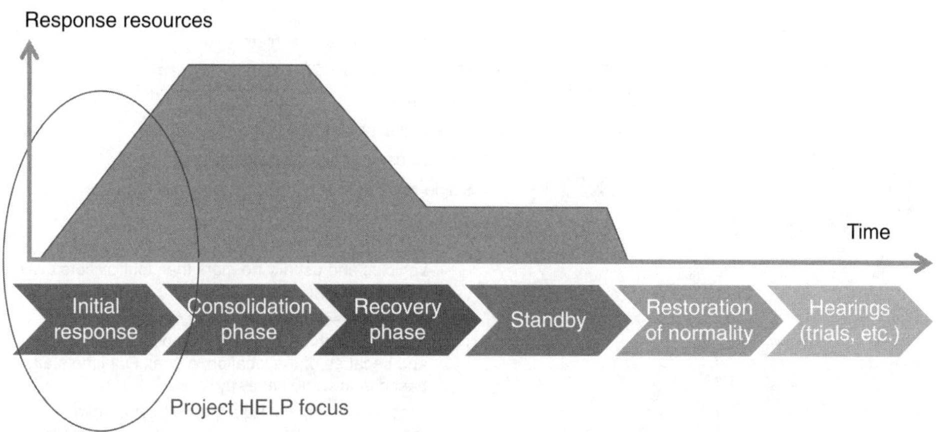

Figure 1.5 Typical stages of a major incident and focus of Project HELP scenario.

regarding the incident is still limited, then that solution can be extended into the longer term as the incident stabilizes and, ultimately, the affected area returns to a state of normality.

1.4.4.1 Description of the Area and Demographics

The location of this incident is on the coast and extending into a largely rural community of some 10 km^2. A coastal town of some 5000 inhabitants is divided by the mouth of a river. Beyond the town lies largely rural agricultural land that is sparsely populated.

The town contains rural police, ambulance and fire stations, all with limited capacity. In all cases, the services provide cover for the town itself and for predefined geographical areas beyond it. In the event of an incident outside this area, it is possible that some of these resources may be called to assist elsewhere. The police service provides a 24-h presence with two patrol vehicles and usually no more than four officers available at any one time. The fire and rescue service for the town is 'retained' (part-time), consisting of crews who are in other forms of full-time work and are called when required for fire and rescue duties by Short Message Service (SMS). The ambulance service for the area has one rapid-response vehicle. This is staffed by a skilled medical paramedic who can use specialist equipment such as defibrillators and administer controlled drugs. The vehicle is equipped accordingly. The safety of the people using the adjacent coastal area is primarily the responsibility of the local maritime rescue service. This area is serviced by an inshore patrol boat using a crew of three. The crew is provided using a 'retained' service and call-out system similar to (but discrete from) that of the fire and rescue service. There is a railway station in the town directly linked to the national rail network. This is policed by the national transport police. The area headquarters for each of the three principal services lie outside the scenario area. There is no hospital in the area, although there is a doctors' surgery in the town. It provides general medical services during office hours and has no emergency response structure.

A sketch map of the area and the emergency service resources available is shown in Figure 1.6. The map shows the area geography; the location of main road and rail systems; the location of police, ambulance and fire stations; and the key installations.

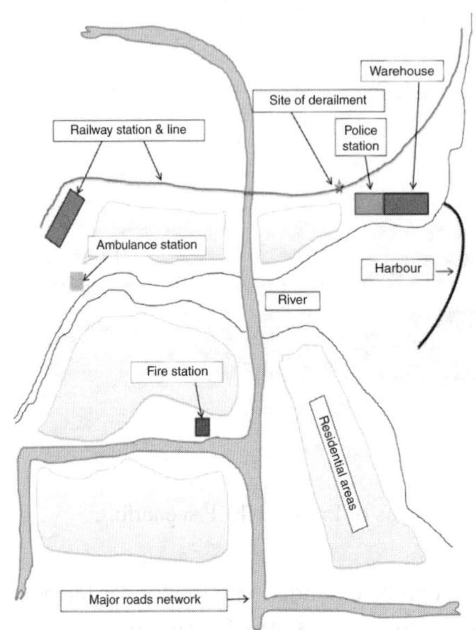

- Coastal town divided by river
- 5000 inhabitants + 25% summer tourism
- Rural area
- Railway line and station
  - Transport police responsibility
- Single-road bridge
- Emergency services:
  - Police service: 24-h presence with two patrol vehicles and usually no more than four officers available at any one time.
  - Fire and rescue service: Part-time personnel. Equipped with one general-purpose fire appliances and because of the location, a boat. Full-time staff based in town 30 km away.
  - Ambulance service: Equipped with one rapid response car staffed by an ambulance paramedic. Additional ambulances, technicians and paramedics are in a town 30 km away.
  - Maritime rescue service: Part-time personnel. Equipped with an inshore rescue boat (IRB) using a crew of three.

Figure 1.6 Scenario map of the area and the emergency service resources available.

1.4.4.2 Description of the Available Communications Means

The police and ambulance services each use their own area of a common, dedicated communications system for voice and data transmission (herein referred to as the public safety network (PSN)). In addition, they may use the commercial cellular telephone network in the area. A base station of the dedicated PSN is located on the roof of the police station. This base station has a diesel-powered generator as backup in the event of mains power failure. As the area is mainly rural in character, the PSN has limited capacity in comparison to an urban area.

The fire and rescue service uses a mixture of analogue and digital radio equipment on VHF frequencies. The nearest VHF base station used by the fire service is located on top of a mountain some 50 km distant.

The nearest VHF base station used by the transport police is located on the local railway station. This organization uses an autonomous VHF analogue radio system on a different frequency to that used by the fire and rescue service.

The local police, ambulance and fire facilities are connected to their control centres (which are not necessarily co-terminus with their headquarters and are outside the incident area) by the PSN (police and ambulance only), VHF analogue/digital system (fire services), internal secure computer networks (usually hardwired or microwave), landline telephone and fax.

Two public mobile networks (PMNs) are assumed to cover the area. Commercial telecom providers deliver voice and data communications (e.g. Internet access) over these infrastructures. BB access is provided although capacity and coverage are reduced in comparison to nearby urban areas.

The following communications equipment is available for local PS services:

- PSN handheld terminals for police and ambulance services that also support direct or back-to-back mode, in areas where the network coverage is poor or non-existent
- Dedicated VHF TE for the fire and rescue service and transport police
- COTS systems, used to supplement the above, particularly for data exchange (e.g. RIM Blackberry terminals)
- Commercial cellular telephony, a combination of terminals owned by the organization or owned privately by individuals within organizations

In addition, police patrol and ambulance vehicles are usually equipped with higher-power (e.g. 10 W) vehicle-mounted PSN terminals. In some instances, these vehicles may be equipped with gateway functionalities (e.g. handheld terminals can access the services of the PSN through this vehicle-mounted equipment). Moreover, out-of-area emergency units that arrive later to assist may also bring communications equipment as described previously as well as more specialized equipment for emergency management such as fast deployable network equipment (e.g. to establish an incident area network for video, data and sensor communications) and satellite communications equipment.

Table 1.6 summarizes the above-mentioned wireless communications means.

Table 1.6 Wireless communications means available for the PPDR services to assist in the emergency response.

Type of communications means	Available wireless communications means	Description
Permanent communications networks	Public safety network (PSN)	Dedicated, digital trunked communications system for voice and data transmission. The police and ambulance services regularly use this network for their internal communications
	VHF base stations for analogue/ digital public safety communications	Mixture of analogue and digital radio equipment on VHF frequencies regularly used by fire and rescue services and transport police
	Public mobile networks (PMNs)	Two mobile cellular networks in the area host a number of commercial telecom providers. These provide voice and data communications. Broadband access is provided although capacity and coverage are reduced in comparison to nearby urban areas
Handheld, portable or vehicle-mounted communications equipment available for local PPDR services or brought into the incident area	PSN and VHF handheld terminals Commercial cellular telephony and COTS systems for data exchange Vehicle-mounted PSN terminals and gateways	This is the equipment regularly available for local PPDR services and most out-of-area emergency support units
	Fast deployable network equipment to establish incident area networks Satellite communications equipment	Specialized equipment for emergency management that can be brought by out-of-area emergency support units

1.4.4.3 Brief Description of the Incident and Impact on the Communications Means

At some stage during normal working hours on a weekday, a train carrying liquid petroleum gas (LPG) is derailed close to the town. Two of the wagons are ruptured allowing LPG to leak out. This gathers around nearby houses. After some minutes, an explosion occurs, followed by a widespread fire. This fire attacks a building that contains paints and thinning chemicals in large quantities. The onshore wind will cause toxic fumes from the fire to move across a largely residential area. There are people trapped inside the building: staff in the basement and staff and customers on the ground floor. One of the adjacent buildings involved is home to local police station and has on its roof one of the local PSN base stations.

Initial calls are made to the emergency services from members of the public over the PMNs and from nearby premises over the landline telephone system. Local 'retained' fire services are called out via SMS, and the area police and ambulance services are tasked to attend. Information on the status and extent of the incident is being passed by the first resource at the scene via radio to their control centre and from there by telephone and data transfer to the other control centres, from where it is relayed back out to the respective resources at or attending the scene. The local retained fire service arrives on the scene. They also request out-of-area support from their own service: the fire is beyond their capacity to bring under control. Dense smoke and fumes are rising from the fire and drifting inland. As the local ambulance service arrives at scene, they also request additional support via their own communications channels as there are an unknown number of casualties inside the building, compounded by the fact that the fire is spreading and putting further lives at risk. Other local resources are mobilized to assist. For example, maritime rescue units are commanded to keep boats out of the harbour and provide information and assistance to those vessels out in the bay area.

This incident requires not only the existing local emergency services to respond and collaborate but also requires the attendance of additional resources from outside the area as a matter of urgency. Some of the additional teams are not authorized users on the PSN and brought some equipment that is not interoperable with the local communications system. The emergency service support personnel and equipment are moving into the area to be directed by operational-level commanders. Tactical-level command posts are established as close to the affected area as it is safe to do. Communications links between the different command levels must be established and maintained. These mobile control centres should have direct voice/data communications links back to their respective control centres.

Resources entering the building to tackle the fire and/or rescue people inside will require the use of direct mode (back-to-back) communication as radio coverage from base stations may not stretch to the interior of buildings and will certainly not deliver to subterranean levels. Moreover, the PSN is stretched to capacity simply by the attendance and requirements of the local emergency crews. When support emergency resources start to arrive from outside the area, the network capacity is exceeded. From the outset, some communications will take place across the mobile cellular network, and this will increase as it becomes more difficult to get through on the PSN. Things get much worse when the PSN base station on the roof of the police station building is destroyed by the fire.

1.4.4.4 Focus Areas

The full description of the Project HELP operational scenario [26] included a timeline of events, where the evolution of the incident raised the need for many different communications among many different actors and where a number of limitations in terms of communications needs became evident during the incident. An excerpt of the timeline of events is reproduced in Table 1.7. As shown in the table, for each event, the required communications activities are described along with the limitations faced in the considered scenario. As an illustrative example, Event 12 shows that the arrival of PPDR specialist teams and additional personnel to the incident area turns into network capacity limitations at scene that would put some limits on the utilization of advanced equipment for emergency response.

The overall timeline of events spanned over a time duration of 2 h. The limitations arisen were captured under the concept of 'focus area', which represents the key stress points (i.e. those where the PS services and/or technical systems would be under maximum strain) within the operational scenario. As a result of the analysis conducted in Project HELP, three focus areas were identified:

1. Providing enough communication capacity for PPDR units in the incident area. The high concentration of first responders in the incident area makes the PSN to be over capacity. If arriving out-of-area emergency support units do not bring additional on-board capacity (e.g. fast deployable base stations), then they will add significantly to the load already being placed on the networks. In addition, the capacity/coverage provided by infrastructure networks may not match the spatial/temporal capacity/coverage needs across the affected area. In turn, the limited available capacity and/or the lack of BB connectivity at scene prevent the utilization of advanced equipment and applications for emergency response that might be used by attending specialist emergency support units. Even in the case that priority access for PPDR users to PMNs is supported, organizational complications in managing the priority service frequently render it virtually useless.
2. Facilitating communications interoperability between PPDR units (local, support and command). In the considered scenario, supervisors at the scene are unable to communicate/coordinate between each other across emergency services due to interoperability issues between the dedicated PPDR systems (e.g. PSN and VHF systems). First responders arriving from other areas, even those equipped with compatible technology, are unable to use the existing PSN infrastructure. There is no capability to set up inter-agency channels when users are spread over several networks. Furthermore, there is a lack of coordination in the configuration of radio equipment brought to the incident area by different agencies.
3. Coping with sudden network base station failure during the incident response. It is highly probable that in a scenario like this, the PSN base station would have failed before tactical-level command units are properly established. Unless these units are able to move onto other networks, there is a very real risk that police and ambulance tactical-level control centres would be unable to carry out a major part of their role for some considerable time. Nevertheless, the usage of commercial network as alternative communications means when PSN base station fails is up to the emergency responders and not an automatic procedure at all. Relying only on back-to-back communications after PSN base station failure is likely to turn on some on-scene resources unable to communicate with their respective tactical-level control centres due to the limited transmission range of individual TE.

Table 1.7 Excerpt of the timeline of events in the Project HELP scenario.

Event number Time: (h:min)	Event description	Communications activity	Current limitations
1 0:00	Train carrying LPG derails LPG leaks out and disperses around nearby houses Mobile and fixed telephone calls reporting incident received by emergency control centres	Public telephone (fixed/mobile) calls directed to emergency services as requested by callers Mobile networks soon under strain as a result of calls reporting the incident, people trying to transmit pictures/video, etc. Note that many of these transmissions will not be to emergency services but to friends/family and the media	None to PPDR at this stage
2 +0:03	Local police and ambulance respond to calls from respective control centres and attend scene	Landline and cellular calls to emergency services continue Control centre calls to local police and ambulance resources over the PSN	Messages being received by disparate agencies/locations. Coordination and collation of information and response
		Incident detail transmitted to local fire station from control centres for staff information on arrival Fire call-out SMS system activated	Public calls routed based on their assessment – not necessarily correctly
…			
6 +0:12	Leaking LPG explodes causing massive fire	Reported to control centres by resources at scene	This mobile activity may well take mobile networks towards maximum capacity level
	Fire attacks the building containing paint and chemicals. People inside. Building structure is physically linked to local police station with PSN base station on roof	Significant increase in mobile calls from concerned members of the public and spectators Possibility that local TV or other media may be in the area and making demands on mobile network capacity	Significant threat now to PSN due to proximity of fire to the base station
…			
12 +1:10	PPDR specialist support units are moving into the area. Additional personnel to assist at the scene(s) are trying to get there. On arrival will be using existing communications networks	Increasing demand on mobile phone networks and PSN	Issues of congestion arising on the mobile networks due to significant use of the system by people in the area
			Communications equipment brought by support teams is not interoperable with local systems Some of the arriving support staff are not authorized subscribers on the existing PSN Available capacity at scene would put some limits on the utilization of advanced equipment for emergency response that might be used by attending specialized units
…			

1.4.4.5 Major Limitations

Based on the analysis of the previous scenario, the following major limitations associated with PPDR systems and arrangements used nowadays in emergency and DR scenarios can be drawn:

- **Lack of network capacity in emergency scenarios**. While the PPDR network operators have optimized the use of their communication systems in their day-to-day service, the situation changes dramatically when an emergency causes additional stress for the system (and the operators). Emergency scenarios usually lead to exceptionally high traffic loads, which a single wireless network may not be able to support. This situation can be worsened in scenarios with limited radio coverage (e.g. a traffic crash in a tunnel) or when parts of the communications infrastructure are damaged in the incident area.
- **Lack of interoperability**. The diversity of technologies used by PPDR organizations often inhibits cooperation between different agencies. Moreover, even when using the same technology, the networks cannot interoperate and the constraints on the security level constitute an additional barrier. As a result, first responders are frequently required to manage several disparate and often incompatible radiocommunication systems.
- **Lack of support for BB data rates in terms of both functionality and capacity**. The evolution of PPDR operations has created the need for applications where large amounts of data are exchanged between first responders or between the tactical-level front-line responders and multilevels of a hierarchical command structure. Data-intensive multimedia applications have a great potential to improve the efficiency of disaster recovery operations (e.g. real-time access to critical data such as high-resolution maps or floor plans; on-field live video transmission from helmet cameras to a central unit, telemedicine, etc.).

1.5 Regulatory and Standardization Framework

Governments and public administrations in each country are the ultimate responsible bodies for establishing the legal and regulatory provisions as well as determining the technical framework (e.g. frequencies, standards) for the delivery of PPDR communications, insofar as the PPDR sector is intimately connected to the public sector of society, either directly as part of the governmental structure or as a function with is outsourced under strict rules. Regulatory, organizational, operational and technical elements can vary substantially from country to country, since the PPDR structures are necessarily tuned to the country's specifics. However, national provisions shall be necessarily conjugated with international regulation and harmonization measures. In turn, international regulation and technical harmonization for PPDR communications is central to facilitate a successful cooperation in PPDR operations in international responses to disasters and cross-border cooperation among many nations. This requires the establishment of international legal instruments to provide guidance in these cases and make national legislations to be conformant with the applicable international law [27], while safeguarding any specific national interests. Indeed, complementing or superseding national regulations for some emergency communications matters, regional regulations are already in place in some parts of the world, like in Europe and the United States/Canada. Together with regulatory provisions, the adoption of global standards for PPDR communications systems and the use of harmonized frequencies are also central aspects towards more efficient and effective PPDR communications at a regional and even global scale. Besides the benefits associated with being able to cooperatively utilize the PPDR resources of different countries in an effective manner, harmonization is also crucial for the ordinary citizens. Citizens are increasingly mobile; they travel for business, for holidays, etc. In order to provide an optimum level of security and accessibility to these citizens in emergency situations, the emergency telecommunications services (ETS) also need harmonization. In this context, a central role in the regulation and standardization of PPDR communications at international level is played by ITU, which is the United Nations' specialized agency in the field of telecommunications and information and communications technologies (ICTs). ITU provides key definitions of PPDR communications concepts and solutions, including standards and spectrum harmonization.

A comprehensive collection of regulatory aspects and existing known emergency communication standards worldwide has been released by the Global Standards Collaboration (GSC) [31]. The GSC is an international initiative to enhance cooperation between standards development organizations (SDOs) from different regions of the world in order to facilitate exchange of information on standards development, build synergies and reduce duplication. The present membership is represented by the following organizations: Association of Radio Industries and Business (ARIB, Japan), Alliance for Telecommunications Industry (ATIS, United States), China Communications Standards Association (CCSA, China), ETSI (Europe), ICT Standards Advisory Council of Canada (ISACC, Canada), ITU (International), Telecommunication Technology Committee (TTC, Japan), TIA (United States) and Telecommunications Technology Association (TTA, Korea). Within the GSC, a task force on emergency communications was established to further encourage cooperation and the sharing of information on standardization activities relating to communications in emergency situations. The addressed work covers not only PPDR communications but also standardization in the areas of communications from individuals/organizations to authorities, communications from authorities to individuals/organizations and communications among affected individuals/organizations. The aim of the document elaborated by the GSC is to identify commonalities, gaps and possible overlaps of emergency communications-related standards in all regions. The document is concluded by looking to the future and examining what may exist tomorrow as well as making proposals to the GSC in order to ensure enhanced harmonization and cooperation.

With all the above, this section first provides an overview of the main activities at the global level conducted within ITU on emergency communications regulation and standardization. Following this, a description of some leading regulatory initiatives and standardization work related to PPDR communications in a number of countries across the North and Latin America, Asia and Pacific and European regions is provided. Complementing these descriptions, Chapter 3 provides further details on a number of relevant initiatives that are currently shaping the way forwards towards the actual delivery of PPDR BB communications based on the LTE technology ecosystem. Moreover, standardization activities concerning the LTE technology and related mission-critical BB applications are covered in more details in Chapter 4, and further details on spectrum regulation for PPDR communications are covered in Chapter 6.

1.5.1 ITU Work on Emergency Communications

The ITU was established last century as an impartial, international organization within which governments and the private sector can work together to coordinate the operation of telecommunications networks and services and advance the development of communications technology. Article 1, Section 2 of the ITU Constitution provides that ITU shall '*promote the adoption of measures for ensuring the safety of life through the cooperation of telecommunication services*'. This mandate has been further enhanced through resolutions and recommendations adopted by past and recent World Telecommunication Development Conferences (WTDC) and World Radiocommunication Conferences (WRC), and ITU's Plenipotentiary Conferences, as well as its active role in activities related to the Tampere Convention. The Tampere Convention calls on states to facilitate the provision of prompt telecommunications assistance to mitigate the impact of a disaster and covers both the installation and operation of reliable, flexible telecommunications services [32].

ITU is organized in three sectors: Telecommunication Development (ITU-D), Standardization (ITU-T) and Radiocommunication (ITU-R). The activities concerning different aspects of emergency communications are addressed within the three sectors [27].

The core mission of the Telecommunication Development (ITU-D) Sector is to foster international cooperation and solidarity in the delivery of technical assistance and in the creation, development and improvement of telecommunications/ICT equipment and networks in developing countries. ITU-D engagement in development support for disaster communications includes partnership for direct assistance to the disaster-prone countries with technical assistance and support for operational costs, deployment of donated satellite phones in disaster-stricken areas and projects on rehabilitation and reconstruction of telecommunications infrastructure in earthquake/tsunami-hit areas. ITU-D works in close cooperation with the United Nations Office for the Coordination of Humanitarian Affairs (OCHA) and is a member of the Working Group on Emergency Telecommunications (WGET), an open forum including all United Nations' entities and numerous international and national and governmental organizations and NGOs involved in disaster response as well as experts from the private sector and academia. The role of the ITU-D under the Tampere Convention and other related instruments is explained in Ref. [27]. Currently, 46 countries have ratified the Tampere Convention on the Provision of Telecommunication Resources for Disaster Mitigation and Relief Operations. Further information on the role of ITU-D in emergency communications can be found at the ITU-D website [28].

Through its work on standardization, the ITU Telecommunication Standardization (ITU-T) Sector develops technical standards that facilitate the use of public telecommunications services and systems for communications during emergency, DR and mitigation operations. Although ITU-T is not involved in emergency and DR operations *per se*, it develops recommendations that are fundamental to the implementation of interoperable systems and telecommunications facilities that will allow relief workers to smoothly deploy telecom equipment and services. The World Telecommunication Standardization Assembly (WTSA), which meets every 4 years, establishes the topics for study by the ITU-T study groups. In this context, ITU-T main activities are related to the provision of an ETS that is defined as a national service providing priority telecommunications to authorized users in times of disaster and emergencies. A number of recommendations have been developed for call priority schemes that ensure that relief workers can get communications lines when they need to. Supplement 62 to the ITU-T Q-series Recommendations [34] provides a convenient reference to assist ITU-T study groups and other national and international SDOs as they develop recommendations and standards for ETS. It identifies published ETS-related recommendations and standards as well as those currently in work programmes. Further information of the role of ITU-T in emergency communications can be found at the ITU-T website [29].

The ITU-R Sector is actually the sector more directly involved in PPDR radiocommunications regulation and standardization. The role of the ITU-R is to ensure the rational, equitable, efficient and economical use of the radio-frequency spectrum by all radiocommunication services, including satellite services. The regulatory and policy functions of ITU-R are performed by World and Regional Radiocommunication Conferences and Radiocommunication Assemblies supported by study groups. WRC are regularly held every 4–5 years, and the decisions adopted are incorporated in the Radio Regulations (RR) treaty [33].

The subject of frequency bands for PPDR communications was an important item on the agenda of the WRC held in 2003 (WRC-2003). Previously, WRC-2000 approved agenda item 1.3 for WRC-2003 to consider identification of globally/regionally harmonized bands, to the extent practicable, for the implementation of future advanced solutions to meet the needs of PP agencies, including those dealing with emergency situations and DR, and to make regulatory provisions, as necessary, taking into account Resolutions 644 (Rev. WRC-2000) and 645 (WRC-2000). These resolutions requested ITU-R study groups to pursue studies on the identification of suitable frequency bands, as well as on facilitating cross-border circulation of equipment intended for use in emergency and DR situations – the latter point reinforced by the Tampere Convention on the Provision of Telecommunication Resources for Disaster Mitigation and Relief Operations. The focus in 2003 was to identify bands for mission-critical voice and data for PPDR agencies. Resolution 646 was approved by WRC-2003 including the identified regional harmonized frequency bands. Indeed, report ITU-R M.2033 [1] was delivered in preparation for WRC-03 agenda item 1.3, which defined the PPDR objectives and requirements for the implementation of future advanced solutions to satisfy the operational needs of PPDR organizations around the year 2010. Specifically, ITU-R M.2033 identified objectives, applications, requirements, a methodology for spectrum calculations, spectrum requirements and solutions for interoperability. From 2003 till date, ITU has been continuously working on preparing reports and recommendations on PPDR, but the last adopted PPDR resolution is still Resolution 646. In the last WRC-2012, to account for the new PPDR scenarios offered by the evolution of BB technologies, it was agreed to review and revise Resolution 646 for BB PPDR under agenda item 1.3 in the forthcoming WRC-2015. In particular, Resolution 648 invited ITU-R to study technical and operational issues relating to BB PPDR and its further development and to develop recommendations, as required, on technical requirements for PPDR services and applications, the evolution of BB PPDR through advances in technology and the needs of developing countries. Further details on spectrum regulation and frequency arrangements for PPDR are provided in Chapter 6.

With respect to PPDR communications standards, ITU-R Recommendation M.2009 identifies a set of radio interface standards applicable for PPDR operations. These standards are not developed by ITU but based on common specifications issued by different SDOs. The recommendation is intended to be used by regulators, manufacturers and PPDR operators to determine the most suitable standards for their needs. However, as explicitly noted in the Recommendation, the inclusion of ITU-R M.2009 standards does not preclude the use of other standards, if so considered by the administration that is the ultimate responsible for determining which technologies to deploy for PPDR operations.

The NB standards (and respective responsible SDOs) for PPDR operations listed in ITU-R M.2009 are TETRA, P25 and DMR. In addition, the following BB technologies are also listed in ITU-R M.2009:

- IMT-2000 CDMA-MC technology, developed within 3rd Generation Partnership Project 2 (3GPP2). This is the technology used in commercial CDMA2000 networks.
- IMT-2000 CDMA-DS, specifically UTRA FDD, developed within 3rd Generation Partnership Project (3GPP). This is the technology used in commercial UMTS systems deployed over paired bands, which are the vast majority of UMTS systems.
- OFDMA TDD WMAN, developed within the IEEE 802.16. This is the technology more commonly known as WiMAX.
- TDMA-SC, developed within 3GPP. This is the technology behind the EDGE systems evolved from the 2G GSM radio interfaces.
- IMT-2000 CDMA TDD, specifically UTRA TDD, technology is developed within 3GPP. This is the technology developed for UMTS systems to be deployed in unpaired bands.
- E-UTRA (LTE) technology, developed within 3GPP.

Further information of the role of ITU-R in emergency communications can be found at the ITU-R website [30].

1.5.2 North and Latin America Regions

In the United States, the vast majority of PPDR networks currently utilized are NB systems that are governed by Part 90 of the FCC's rules. Part 90 consists of various services utilizing regularly interacting groups of base, mobile, portable and associated control and relay stations for private (non-profit) radiocommunications by eligible users. These systems use Project 25 (P25) suite of standards (briefly described in Section 1.4.3.4). Standardization activities related to P25 and other aspects of PPDR communications in the United States are addressed under the auspices of the TIA.

In February 2012, with the passage of the Middle Class Tax Relief and Job Creation Act, regulatory and financial provisions were established for the build out of a dedicated National Public Safety Broadband Network (NPSBN) in the United States. The law's governing framework for the deployment and operation of this network, which is to be based on a single, national network architecture, is the new First Responder Network Authority (or FirstNet), an independent authority within the National Telecommunications and Information Administration (NTIA). FirstNet counts with a spectrum allocation of 10+10MHz in the 700-MHz band and is charged with taking 'all actions necessary' to build, deploy and operate the network, in consultation with PS entities at all jurisdictional levels and other key stakeholders.

In 2009, the National Public Safety Telecommunications Council (NPSTC), organization that provides a collective voice on communications issues for PS first responders in the United States, endorsed LTE as the favoured technology standard most suited to the development of this anticipated nationwide interoperable BB network [35]. A partnership between the NPSTC and the APCO International, which is the world's largest organization of PS communications professionals, was established in 2013 to move forward on technical standards issues related to PS BB communications. Another leading effort from the United States towards the adoption and improvement of LTE standard capabilities for PPDR use is the Public Safety Communications Research (PSCR) program, a joint effort between the National Institute of Standards and Technologies (NIST) and NTIA, which coordinates interoperability testing and standards development for the nationwide PS LTE network.

In Canada, NB PPDR networks are using several LMR standards including P25 and ETSI DMR standards. Currently, the only bands where a given standard is mandated for PS spectrum in Canada is in the bands 769–775 and 799–805 MHz, where the P25 standard was selected for operation on the interoperability channels. Industry Canada (IC), a department of the Government of Canada, has also mandated that all mobiles and portables that provide voice services must be capable of operating on the interoperability channels.

The PPDR community in Canada is also engaged in the allocation of dedicated spectrum in the 700-MHz band to deploy a BB network for PPDR. The Communications Interoperability Strategy for Canada (CISC), which is the result of the collaborative efforts of leaders representing all levels of government and emergency response services from across Canada, developed an action plan that tasks national emergency management partners to develop the 700-MHz implementation strategy.

The PS community's intent is to harmonize Canadian and US PS BB networks in the 700-MHz spectrum to enable cross-border communications in these bands and establish mechanisms/protocols to avoid interference issues.

In the Latin America region, spectrum for mobile BB PPDR has also been allocated in Brazil in the 700 MHz. Indeed, LTE deployments in very specific areas have been operational by Brazil's army as part of the infrastructure deployed for the soccer's World Cup in 2014. Aligned with the Brazilian approach, the Organization of American States (CITEL) recommended its member states across North, Central and South America that PPDR to consider the 700-MHz band in possible BB PPDR spectrum allocations.

1.5.3 Asia and Pacific Region

In Japan, a NB PMR standard named Integrated Dispatch Radio (IDRA) was specified by the ARIB, which is an external Ministry of Post and Telecommunication (MPT) affiliate and recognized standardization organization. The IDRA system was developed for use mainly in business-oriented mobile communications applications, encompassing emergency services to commercial and industrial organizations. In 2011, the digitization of analogue TV broadcasting service using VHF/UHF bands resulted in the allocation of 32.5 MHz out of the newly available spectrum for BB wireless communications systems for PS. Technical specifications for those BB systems have been also standardized in ARIB.

In South Korea, ETSI TETRA standards have been adopted by TTA, a non-government and non-profit organization for ICT standardization, even though TTA has also produced its own standards for satellite infra on multimedia DR. The government planned to build a mission-critical nationwide PPDR network for sharing among PPDR agencies and designated the National Emergency Management Agency (NEMA) to lead the program. The Korea Communications Commission (KCC) allocated the frequency spectrum of 806–811 and 851–856 MHz for the nationwide PPDR system. NEMA started implementation of the nationwide system based upon NB technology in 2003–2007. Since then, the system has been in operation in major cities and major express ways. The program was planned for the second phase of PPDR system, which includes BB service.

In China, GoTa based on CDMA and GT800 based on GSM are the popular digital trunking mobile communications systems. These systems were specified by the CCSA, a non-profit legal organization established by enterprises and institutes in China for carrying out standardization activities in the field of ICT. The Emergency Communication Special Task Group (ST3) in CCSA is responsible to carry out studies on comprehensive, managerial and architectural standards of emergency communication, including policy, network and technology supportive standards. A project called broadband wireless trunking (BWT) was launched in 2011 to support research, development, standardization and applicable evaluation for BB wireless professional communications, with the major applications focused on future PPDR [39]. The BWT project covers the stages of research, standardization and industrialization and is planned to conclude in 2018. In terms of spectrum allocation and technology, China is planning to use TDD LTE for BB PPDR in 1.4 GHz. China has also reserved frequencies in the 350–370-MHz range for national security radio networks using TETRA.

In Australia, the Australian Communications and Media Authority (ACMA) is undertaking a number of initiatives to improve spectrum provisions for PPDR in Australia [40]. In particular, ACMA has already made provision for 10 MHz of spectrum from the 800-MHz band for the specific purpose of realizing a nationally interoperable cellular 4G data capability, though precise frequencies will be determined at a later stage. ACMA has also created a class licence that provides 50 MHz of spectrum in the 4.9-GHz band for PSAs to share Australia-wide. This is intended to provide very high-speed, short-range on-demand capacity to areas of high activity to support a wide range of uses.

Throughout this region, the Asia-Pacific Telecommunity (APT) members support regional harmonization of frequency bands/ranges for future deployment of BB PPDR. It is recognized that different amounts of available spectrum may be used within bands depending on their national circumstances. This will provide flexibility to decide the amount of spectrum and the frequency arrangement that best meets their overall national BB PPDR requirements.

1.5.4 Europe Region

A regional regulation exists in Europe, superseding the national regulations for some topics and therefore requesting coordination between countries. The European Commission (EC), being the executive body of the EU, is responsible for proposing legislation, adopting and implementing measures. The European Council and the European Parliament adopt directives that are implemented by member states of the EU in national laws. In this context, the provision of BB PPDR services has been identified as a policy objective in several EC provisions and reports [36, 37].

A number of European Standards Organizations (ESOs) assist the EC by producing standards and specifications supporting the EU policies. This is mainly achieved through mandates issued from the EC towards the ESOs. Mandates are statements of policy intent where the EC and the member states request the relevant ESOs and their members to develop standards (or a standardization work programme) in coordination with regulatory requirements or other policy initiatives. ETSI is the ESO more directly involved in the standardization of PPDR and emergency communications systems.

Within ETSI, an ETSI SC EMTEL was established. The primary responsibility ETSI SC EMTEL is to solicit and capture the requirements from the stakeholders (including national authorities responsible for provisioning emergency communications, end users, the EC, communications service providers, network operators, manufacturers and other interested parties) and coordinate the ETSI positions on emergency communications-related issues. ETSI SC EMTEL has produced several documents with requirements for emergency communications between individuals and authorities/organizations, between authorities/organizations, from authorities/organizations to the individuals and among individuals (e.g. [3]). ETSI SC EMTEL maintains a report [38] with the European regulatory texts and orientations applicable for the emergency communications (such as EC directives, commission decisions) and other information or references such as generally applicable regulatory principles.

Other central contributions from ETSI in the field of PPDR communications are the TETRA and DMR standards, both briefly described in previous Section 1.4.3. In addition to these PMR standards, ETSI is also active in emergency calling systems, GMDSS and satellite emergency communications. ETSI is currently working in a number of EC mandates that are linked to PPDR from different angles: Mandate M/284 related to the maintenance of the ETSI harmonized standards in the field of private/professional LMR systems and equipment; Mandate M/493 related to the support of the location-enhanced emergency call service; Mandate M/496 related to the development of standardization regarding space industry; and Mandate M/512 on Reconfigurable Radio System (RRS) related to the development and use of RRS technologies in Europe. Within this latter, there is Objective C that proposes to explore potential areas of synergy among commercial, civil security and military applications in terms of network interfaces and architectures for dynamic use of spectrum resources and of architectures and interfaces for reconfigurable mobile devices for commercial and civil security applications. Also recently, the technical committee (TC), in charge of the TETRA specifications, evolved into the now called TETRA and Critical Communications Evolution (TCCE), which is the TC with responsibility for the provision of user-driven standards for PPDR communications over both BB and NB AIs. Outside ETSI, TC TCCE close cooperates with the TETRA and TCCA, particularly with regard to the development of requirements, use cases and architectures for mission-critical communications standardization.

ETSI TC TCCE and TCCA, together with other relevant organizations such as the NIST from the United States, are working closely with the 3GPP to advance the LTE specifications to better support the needs of critical communications users. Indeed, the 3GPP body unites seven telecommunications SDOs from Asia, Europe and North America (ARIB, ATIS, CCSA, ETSI, TSDSI, TTA, TTC), known as 'Organizational Partners', together with market representatives and a huge number of companies within the telecom sector across the world to produce the reports and specifications that define the 3GPP technologies. In the context of PPDR communications, group call system enablers and off-network services are among the extensions being added to the LTE specifications. A new 3GPP working group (called SA6 – 'mission-critical applications') has been created for the development of applications for specialized communications. These extensions to the LTE standard are described in detail in Chapter 4.

As to radio spectrum matters, another key organization in Europe is the European Conference of Postal and Telecommunications Administrations (CEPT). CEPT, through its Electronic Communications Committee (ECC), brings together 48 countries to develop common policies and regulations in electronic communications and related applications for Europe. CEPT takes also an active role at the international level, preparing common European proposals to represent European interests in the ITU and other international organizations. Concerning PPDR communications, the band 380–385/390–395 MHz is so far the only harmonized band for permanent NB PPDR systems in Europe, as established in CEPT ECC Decision (08)05. Indeed, the introduction of digital radiocommunication, the TETRA standard and the harmonized frequencies for the national emergency services were developed in response to the requirements of the Schengen Treaty obligations. Nowadays, Frequency Management Project Team 49 (FM PT49) within CEPT ECC is currently working on radio spectrum issues concerning PPDR applications and scenarios, in particular concerning the BB high-speed communications as requested by PPDR organizations. FM PT49 is intended to identify and evaluate suitable bands for European-wide harmonization of spectrum (both below and above 1 GHz), by taking into account cross-border communication issues and PPDR application requirements. A further insight into the CEPT and other European bodies' activities concerning PPDR BB spectrum is addressed in Chapter 6.

In this context, a few European countries have already established the path to follow to eventually offer critical voice and BB data for their PPDR agencies. Notoriously, the UK Home Office has initiated the procurement process that is expected to lead to the replacement of its current NB TETRA system (called Airwave) for a new system, known as the Emergency Services Network (ESN), which will make use of 4G/LTE technology. Further details on this UK program, together with other proposals and initiatives towards the delivery of mobile BB PPDR across Europe (France, Finland, Belgium), are covered under Chapter 3.

References

[1] Report ITU-R M.2033, 'Radiocommunication objectives and requirements for public protection and disaster relief', 2003.

[2] ETSI TS 170 001 (V3.3.1), 'Project MESA; Service Specification Group – Services and Applications; Statement of Requirements (SoR)', March 2008.

[3] ETSI TR 102 181, 'Emergency Communications (EMTEL); Requirements for communication between authorities/organisations during emergencies', February 2008.

[4] C(2003)2657, Commission Recommendation of 25 July 2003: 'Recommendation on the processing of caller location information in electronic communications networks for the purpose of location-enhanced emergency call services', published on O.J.E.U. L 189/49 the 29 July 2003.

[5] SAFECOM, US communications program of the Department of Homeland Security (DHS), 'Public safety Statements of Requirements for communications and interoperability v I and II', January 2006.

[6] ETSI TR 102 745, 'User Requirements for Public Safety', October 2009.

[7] ETSI TR 102 182, 'Emergency Communications (EMTEL); Requirements for communications from authorities/organisations to the citizens during emergencies', July 2010.

[8] Major Incident Procedure Manual, 8th Edition: London Emergency Services Liaison Panel; 2012.

[9] EU FP7 ISITEP Project, 'D2.3.1 – End-user requirements document draft', January 2014. Project official website: http://isitep.eu/about/ (accessed 25 March 2015).

[10] CEPT ECC Report 102, 'Public protection and disaster relief spectrum requirements', Helsinki, January 2007.

[11] CEPT ECC Report 199, 'User requirements and spectrum needs for future European broadband PPDR systems (Wide Area Networks)', May 2013.

[12] National Public Safety Telecommunications Council (NPSTC), 'Public Safety Communications Assessment 2012–2022, Technology, Operations, & Spectrum Roadmap', Final Report, June 2012.

[13] Larry Irving, Final Report of the Public Safety Wireless Advisory Committee (PSWAC) to the Federal Communications Commission and the National Telecommunications and Information Administration, 11 September 1996.

[14] Simon Forge, Robert Horvitz and Colin Blackman, 'Study on use of commercial mobile networks and equipment for "mission-critical" high-speed broadband communications in specific sectors', Final Report, December 2014.

[15] National Public Safety Telecommunications Council (NPSTC), Broadband Working Group, 'Mission Critical Voice Communications Requirements for Public Safety', September 2011. Available online at http://npstc.org/download.jsp?tableId=37&column=217&id=1911&file=FunctionalDescripton (accessed 25 March 2015).

[16] George F. Elmasry, 'Tactical Wireless Communications and Networks: Design Concepts and Challenges', Hoboken, NJ: John Wiley & Sons, Inc., 2012.
[17] Bharat Bhatia, 'Wireless Technology Standards for Emergency Telecommunications', ITU Workshop on Emergency Communications and Information Management, February 2012.
[18] Report ITU R M.2014-2, 'Digital land mobile systems for dispatch traffic', December 2012.
[19] 'Digital Radio in the Americas: A Guide for New Deployments and System Upgrades', RadioResource Mission Critical Communications, Educational Series, 2014.
[20] ETSI EN 300 392-7 V3.3.1, 'Terrestrial Trunked Radio (TETRA); Voice plus Data (V+D); Part 7: Security', July 2012.
[21] TETRA + Critical Communications Association (TCCA), 'TETRA versus DMR', White Paper produced by the TETRA SME Forum, a sub group of the TCCA, October 2012.
[22] Office of Emergency Communications, US Department of Homeland Security (DHS), 'SAFECOM Guidance on Emergency Communications Grants'. Available online at http://www.dhs.gov/safecom-guidance-emergency-communications-grants (accessed 25 March 2015).
[23] Wireless Innovation Forum (WINF), 'Use Cases for Cognitive Applications in Public Safety Communications Systems – Volume 1: Review of the 7 July Bombing of the London Underground', November 2007.
[24] Wireless Innovation Forum (WINF), 'Use Cases for Cognitive Applications in Public Safety Communications Systems – Volume 2: Chemical Plant Explosion Scenario', January 2010.
[25] EU FP7 Project HELP on 'Enhanced Communications in Emergencies by Creating and Exploiting Synergies in Composite Radio Systems'. Available online at http://cordis.europa.eu/projects/rcn/97890_en.html (accessed 25 March 2015).
[26] EU FP7 Project HELP Deliverable D2.1 EU FP7 Project HELP, 'Description of operational user requirements and scenarios', Editor Paul Hirst (BAPCO), June 2011.
[27] ITU, 'Compendium of ITU'S Work on Emergency Telecommunications', Edition 2007.
[28] ITU-D website on Emergency Telecommunications. Available online at http://www.itu.int/ITU-D/emergencytelecoms/ (accessed 25 March 2015).
[29] ITU-T website on Emergency Telecoms. Available online at http://www.itu.int/ITU-T/emergencytelecoms/ (accessed 25 March 2015).
[30] ITU-R website on Emergency Radiocommunications. Available online at http://www.itu.int/ITU-R/index.asp?category=information&rlink=emergency&lang=en (accessed 25 March 2015).
[31] GSC-EM TF Report, 'Draft Report of the Global Standards Collaboration (GSC) Task Force on Emergency Communications', June 2014. Available online at http://www.itu.int/en/ITU-T/gsc/Documents/GSC-18/meeting-documents/GSC(14)18_003a1_GSC-EM_Task_%20Force_Report.doc (accessed 25 March 2015).
[32] Tampere Convention on the Provision of Telecommunication Resources for Disaster Mitigation and Relief Operations, Tampere, 18 June 1998. Available online at https://treaties.un.org/Pages/ViewDetails.aspx?src=TREATY&mtdsg_no=XXV-4&chapter=25&lang=en&clang=_en (accessed 25 March 2015).
[33] ITU-R Radio Regulations. Available online at http://www.itu.int/pub/R-REG-RR (accessed 25 March 2015).
[34] Supplement 62 to ITU-T Q series Recommendations, 'Overview of the work of standards development organizations and other organizations on emergency telecommunications service'. Available online at http://itu.int/rec/T-REC-Q.Sup62-201101-I (accessed 25 March 2015).
[35] National Public Safety Telecommunications Council (NPSTC), '700 MHz Public Safety Broadband Task Force Report and Recommendations', September 2009.
[36] EC Decision 243/2012/EU European Parliament and of the Council establishing a multi-annual radio spectrum policy programme (RSPP). Available online at http://eur-lex.europa.eu/legal-content/EN/ALL/?uri=CELEX:32012D0243 (accessed 25 March 2015).
[37] European Union Radio Spectrum Policy Group (RSPG), 'Report on Strategic Sectoral Spectrum Needs', Document RSPG13-540 (rev2), November 2013.
[38] ETSI TR 102 299, 'Emergency Communications (EMTEL); Collection of European Regulatory Texts and orientations', April 2008.
[39] Shao-Qian Li, Zhi Chen, Qi-Yue Yu, Wei-Xiao Meng and Xue-Zhi Tan, 'Toward Future Public Safety Communications: The Broadband Wireless Trunking Project in China,' Vehicular Technology Magazine, IEEE, vol. 8, no. 2, pp. 55, 63, June 2013.
[40] Australian Communications and Media Authority (ACMA), 'Five-year spectrum outlook 2014–18', September 2014.

2

Mobile Broadband Data Applications and Capacity Needs

2.1 Introduction

While current PPDR operational practices primarily rely on the use of voice-centric communications and messaging services (e.g. status messages, short data messages, location information), the advantages of having real-time access to information via a broadband connection are obvious and pave the way for a wide range of data-centric, multimedia applications that can greatly improve the operations of PPDR organizations. The ability to provide focused and detailed data on developing situations to PPDR officers in the field (e.g. building/floor plans, hazardous materials data) is clear as well as the ability to relay comprehensive information back to remote control centres (e.g. real-time streaming video and pictures from the incident). Mobile broadband enables new use cases such as body-worn cameras, advanced navigation, on-route mapping of deployed assets, automated case processing, augmented reality applications and many others, real-time streaming video being among the most sought-after applications. Just as broadband data has become essential to daily consumer activities, broadband data is also expected to become an increasingly routine aspect of PPDR activities, both on a day-to-day basis and in large-scale emergencies.

This increasing demand for more data-centric applications in the PPDR community can be primarily associated with a combination of two (interrelated) factors.

One factor is the progressive transition towards more information-driven PPDR working practices. Information-driven operations have many benefits in terms of, for example, more accurate information for decision-making on incident response, better mobilization of field teams and, ultimately, more timely and effective emergency response. Usage scenarios for how PPDR users work on a day-to-day basis while out on patrol or away from command centres shows that the usage is evolving towards information-centric operations with greater sharing of information from a variety of sources (voice, data and video) [1]. One key exponent

Mobile Broadband Communications for Public Safety: The Road Ahead Through LTE Technology, First Edition.
Ramon Ferrús and Oriol Sallent.

of these information-driven practices is the increased *situational awareness*: PPDR personnel need to report critical information to command or supervisory staff in need of detailed descriptions of the incident. This affects law enforcement, fire/rescue and EMS units on a regular basis. While as of today much of the incident command process and decision-making still revolves around situational awareness frequently reported as a voice message, mobile broadband access coupled with today's smart devices are increasingly being used by PPDR agencies to provide enhanced situational awareness to the first responders. For example, the use of video streams in PPDR operations could help experts in safe locations or command centres to manage and provide guidance to those in containment or hot zones. The overall purpose of this way of working is to establish a (so-called) common operating picture (COP) between participating PPDR agencies, officers at incidents and those in control centres. In addition to the transition towards improved situational awareness, there is also a trend towards mobile command and control, which can greatly enhance the effectiveness and efficiency of incident response. This is driving a demand for simultaneous access to a much wider range of data-centric applications from these temporary incident command locations. Moreover, more daily routines are also increasingly taking advantage of mobile offices. This requires access to the same range of applications while in the field as an officer would have while in a control centre or headquarters. After all, this transition towards information-driven PPDR working practices clearly connects with the more general trend apparent within the wider society for access to a wide range of information on the move.

The second key factor stems from the experience gained with the use of mobile data applications within current narrowband systems, which are already proving useful to improve PPDR operations (e.g. automatic vehicle location (AVL) and tracking, short and status data messaging, database queries/access and – limited – transfer of imagery and video). This provides solid foundations and creates new expectations for mobile data use, further increasing the demand for more sophisticated applications. Indeed, this trend is inherent to technology innovation: initial data transmission technology enables initial use; if proved useful, initial use turns into expanded use over time; next, expanded use requires and creates opportunities for enhanced technology for improved efficiency and effectiveness; and eventually, enhanced technology is more capable and enables new usages thus driving additional demand. Moreover, these expectations within the PPDR community for more sophisticated data applications are clearly amplified by the capabilities demonstrated by the rich mobile application ecosystem flourishing nowadays in the commercial domain. The variety of mobile computing devices with huge processing and multimedia capabilities (e.g. smartphones, tablets, connected car systems, etc.) alongside powerful and versatile application development technologies are in increasingly widespread use in the commercial domain, attracting not only ordinary consumers but also business and professionals. In this regard, being able to capitalize on the core technologies of this ecosystem and so being able to replicate it within the PPDR sector is of utmost importance for the PPDR community, keeping it aligned to mainstream technological evolution driven from the more large-scale markets of fixed/mobile broadband communications.

Overall, there is a huge potential for broadband capability to transform communications services, fostering the migration from the current dominant voice-only mode to multimedia applications. Some of these applications may result in a large amount of data being exchanged across the first responders or between the responders and the control rooms and central offices. Of note is that the true value with data-centric applications lies in the effective analysis of the data deluge that broadband brings. It is envisioned that, as technologies mature, PPDR should see more robust applications that require smaller amounts of network bandwidth. However, the growing demand for applications and image resolution are expected to offset such improvements.

2.2 Data-Centric, Multimedia Applications for PPDR

There are dozens of potential applications that could benefit PPDR operations. The need for and the various types of PPDR data-centric, multimedia applications deemed critical for on-scene operations have been widely analysed [1–7]. Table 2.1 lists the sets of mobile data and multimedia applications identified by organizations such as the National Public Safety Telecommunications Council (NPSTC) in the United States, the European Telecommunications Standards Institute (ETSI) and the Electronic Communications Committee (ECC) of the European Conference of Postal and Telecommunications Administrations (CEPT) in Europe and envisaged to be in widespread use within the PPDR sector over the short and medium term.

Even though the exact terminology and the specific focus of the PPDR applications in demand may differ slightly between the document references cited in Table 2.1, very similar ranges of applications are identified. Some of the applications are expected to be used only on the incident scene, while others require data from remote systems. In some cases, the data would be monitored by dedicated individuals within the intervention team, mostly, but not limited to, by the incident commander or officers in charge of specific operations. In some other cases, critical data should be moved off-site to a centralized location, where it could be more efficiently monitored by others. Many of these applications require interoperability among agencies to capitalize fully on their operational benefit. Based on the aforementioned references, a further insight in these current and future mobile data applications in PPDR is given in the following.

2.2.1 Video Transmission

Video transmission applications are identified as critical by most PPDR disciplines. Video information can greatly improve situational awareness as well as serve to enhance command and control. Examples of video applications are wireless video surveillance and remote monitoring, vehicle-mounted video, helmet camera video, aerial video feeds and use of third-party camera resources.

In wireless video surveillance and remote monitoring applications, a sensor (fixed or mobile) can record and distribute data in video streaming format, which is then collected and distributed to PPDR responders and command and control centres. This may include:

- High-resolution video communications from wireless clip-on cameras to a vehicle-mounted laptop computer, used during traffic stops or responses to other incidents
- Video surveillance of security entry points such as airports with automatic detection based on reference images, hazardous material or other relevant parameters
- High-resolution real-time video from, and remote monitoring of, firefighters in a burning building

Table 2.1 Examples of data-centric, multimedia applications in demand for enhanced PPDR operations.

Document reference	Identified applications	
NPSTC report on 'Public Safety Communications Report: "Public Safety Communications Assessment 2012–2022, Technology, Operations and Spectrum Roadmap"', June 2012 [4]	• Access to third-party video/cameras (private and governmental) • Automatic location (both vehicle and personnel location systems) • Biomedical telemetry (patient and firefighter) • Geographic Information Systems (GIS) • Incident command post-video conferencing	• Incident command white board • Message and file transfer • Mobile data computers application usage • Patient/evacuee/deceased tracking • Sensor technology • Vehicle telemetry • Video (aerial video feed, vehicle-mounted video and helmet camera video) • Voice over IP cell phone access • Weather tracking
ETSI TS 102 181, 'Emergency Communications (EMTEL): Requirements for Communication between Authorities/Organizations during Emergencies', February 2008 [5] ETSI TR 102 745 'Reconfigurable Radio Systems (RRS): User Requirements for Public Safety', October 2009 [6]	• Verification of biometric data • Wireless video surveillance and remote monitoring • Video (real time) • Video (slow scan) • Imaging • Automatic number plate recognition (ANPR) • Documents scan • Location/tracking for automatic vehicle/officer	• Location transmission of building/floor plans and chemical data • Personnel monitoring/monitoring of public safety officer • Remote emergency medical service • Sensor networks • E-mail • Digital mapping/geographical information services • Database access (remote) • Database replication
CEPT ECC Report 199, 'User Requirements and Spectrum Needs for Future European Broadband PPDR Systems (Wide Area Networks)', May 2013 [7]	• Automatic (vehicle) location systems data to command and control centre (CCC) • Video from/to CCC for following operations and intervention • Video for fixed observation • Video on location (disaster or event area) to and from control room and for local use • Video conferencing operations • Non-real-time recorded video transmission • Photo broadcast/photo to selected group (e.g. based on location) • Incident information download/upload (text + images) from CCC to field units • Command and control information including task management, briefings and status information • Download/upload maps/scanned documents/pictures • Automatic number plate recognition/speed control automatic upload to databases	• Patient monitoring • Monitoring status of security worker • Operational database search (own + external) • Remote medical database services • ANPR checking number plate live • Biometric (e.g. fingerprint) check • Cargo data • Crash recovery systems • PDA synchronization • Mobile workspace + (including public Internet) • Software update online • GIS maps updates • Automatic telemetry including remote-controlled devices + information from (static) sensors • Hotspot on disaster or event area (e.g. in mobile communication centre) • Front-office–back-office applications • Alarming/paging • Traffic management system: information on road situations to units • Connectivity of abroad assigned force to local CCC

Vehicle-mounted video cameras in fire/rescue and law enforcement vehicles on an as-needed basis (vs. a continual video feed) would allow command post and control centres personnel to visualize the incident scene in relation to damage and apparent needs when compared to other incident scenes. Vehicle-mounted video also enhances on-scene safety by allowing third parties to check on the incident scene, verifying that personnel are accounted for and monitoring the success or failure of the incident mitigation plan. Moreover, vehicle-mounted video allows the incident command team to 'see' the incident and develop a better perspective of the operational requirements. In the absence of video, the command staff must rely on a radio transmission description of the scene from first arriving units. It is a common practice in many countries to equip patrol cars with video technology to record any incidents. There are clear advantages in being able to transfer this information to control centres in real time, in order to keep command centre staff fully informed. Video can also facilitate the identification of individuals and vehicles on location, so that the officials may be given additional instructions on-site.

Helmet cameras can transmit live information to fixed or mobile control centres. Video from helmet cameras could enhance situational awareness and allow for better decision-making. For example, appropriate command staff and supervisors need access to helmet-mounted cameras to obtain real-time video images from firefighters inside a burning building to better assess needed resources and tactics. Video would also allow a subject matter expert to provide remote technical assistance. For example, a building engineer may provide advice to firefighters working inside a collapsed building on the status of a load-bearing wall that is in danger of collapse. A chemical plant engineer could look at helmet camera video of a damaged valve in a 'hot zone' at the factory and provide advice on how to best shut the leak down without creating more problems or damage. Trauma centre physicians and other critical care physicians have expressed the need to visualize the patient while the ambulance is on the way to the hospital. Certain low-volume, high-risk procedures could also be performed more safely under the video guidance of a physician who was monitoring and guiding the patient care team remotely. Law enforcement personnel could use video feeds to monitor the progress of officers conducting a sweep of a building for a dangerous suspect, could allow personnel at the command post to confirm the identity of a subject discovered by the sweep team and could provide expert support during an assessment of a bomb or explosive device. In a correctional facility, video feeds from officers provide an additional level of safety and security and accountability. These cameras would be capable of supporting high-, medium- and low-resolution 'situational awareness' mode of video display. The resolution needed by the incident commander would depend on the type of incident and the issue being addressed. Cameras can also be installed in robots. Robots with high-resolution cameras could investigate a building before human firefighters are committed, to see if there are additional hazardous, flammable or explosive materials present.

Live video transfer from cameras in helicopters already takes place in many PPDR scenarios. Airborne video could originate for a staffed unit, like a law enforcement helicopter, or from a non-staffed unit, such as a remote-controlled or radio-controlled aircraft. Access to aerial video feeds can help law enforcement command staff to develop appropriate situational awareness of the incident scene, plan evacuation routes, monitor crowd behaviour and movement and monitor the progression of the emergency. Likewise, fire/rescue command staff could use separate real-time video to view the incident scene, progression of the emergency, identification of adjoining building fire exposures and other risk factors and to observe movement of the fire or chemical cloud. An infrared video feed is especially important to determine fire spread inside a large warehouse structure, to look for injured persons in the dark who may not be visible to first responders and to see the liquid level of various storage tanks containing flammable materials. Rural areas also reported that video is a critical tool in the management of large wild land fires to monitor fire spread, to determine the viability of fire breaks, to plan and monitor evacuation routes and to identify homes and structures which may need immediate evacuation.

Command staff and supervisors need the ability to view third-party video feeds, including those from both public and private organizations. Many businesses, apartment complexes and industries use security video on a daily basis (e.g. CCTV systems). Appropriate command staff and supervisors should be able to view the security camera videos of an office building pointing at the specific place where, for example, the fire alarm is sounding. This situational awareness, including the presence or absence of smoke and fire, allows for appropriate decisions on the deployment of resources. Better management of available resources would allow one of the fire apparatus to respond to another emergency a few blocks away, instead of sending other, more distant, fire department resources to that scene. As well, law enforcement personnel should be able to view third-party security camera video while arriving on the scene of a robbery alarm, shooting or other violent crime to determine appropriate tactical actions that are needed to protect human life. Video is also extremely helpful when dealing with large crowds and large-scale events. The ability to track a fleeing suspect in a crowded mall (where airborne video is unavailable) would be greatly enhanced via the ability to monitor security video from the various exit points in the area. The recent phenomena involving flash mobs may also be better managed via the availability of video resources. Access to existing traffic camera systems is also critical to assess traffic flow and congestion when determining suitable evacuation routes or checking on the status of a dedicated evacuation route. These cameras also allow wide area access to monitor smoke plumes and chemical cloud releases.

2.2.2 Geographic Information Systems

A Geographic Information System (GIS) is a computer system that analyses, displays, edits, integrates, shares and/or stores multiple types of information items indexed by geolocation coordinates and time information (space–time location). Just as a relational database containing text or numbers can relate many different tables using common key index variables, GIS can relate unrelated information, typically organized in layers, by using the space–time location as the key index variable.GIS systems will play an important role in the incident management of the future: GIS and location intelligence systems can be the foundation for creating powerful decision support and collaborative applications for PPDR. For example, incident commanders may need to visualize street-level detail and use stored information from various GIS layers. In addition to viewing streets and landmarks, GIS may also provide aerial photography snapshots of an incident area. This information on the proximity of other buildings and exposures is critical when creating an incident action plan and when establishing a COP. GIS data may also display various utility layers including sewer, water, electric and gas lines and connections. Instantaneous access to site information is essential, for example, to help firefighting units execute plans more efficiently by allowing them to access and share relevant information quickly, without having to thumb through thick binders of paper information, about, for example, what's around and behind the place or building that they're about to enter. In this way, firefighters could determine what fire hydrants to connect to a single underground water supply line according to its dimension. It is also critical when determining where toxic chemicals may have travelled once they entered the storm water system.

Current existing applications for instantaneous access to site information can see significant evolution as PPDR broadband becomes prevalent and reliable, which can progressively leverage data from environmental sensors, biometric sensors and the firefighter location solutions, so that incident commanders could have the information needed to make better decisions than ever before. GIS data sources are becoming extremely robust including high-resolution aerial images and other data layers. The raw databases for a metropolitan area may contain several gigabytes (GB) of information, which may also be updated frequently. As the PPDR personnel on the scene need only a portion of that information in an incident response, the resulting operational model is one where GIS-based data and views are retrieved as needed for the incident. Therefore, the size of the incident area, the quantities of layers of data and the type of requested data determine the amount of data to be downloaded from the databases, which should be received typically within a few minutes. In this regard, mobile broadband capabilities are essential for the use of GIS applications by intervention teams on the field.

2.2.3 Location and Tracking

Related to the use of GIS applications to support PPDR operations, tracking of vehicles' and officers' positions is key for command staff and supervisors to be able to visualize personnel and vehicle resources on scene, including units responding and units in staging to make appropriate tactical decisions during an emergency. This is a common application already in use nowadays, which is typically implemented by means of GPS position localizers embedded in handheld and vehicular terminals, whose coordinates are sent periodically to the control centres through the narrowband PMR systems. Broadband transmission could further improve this application by allowing higher frequency updates as well as additional information that can be notified in addition to the spatial coordinates. AVL information should include all resources on scene such as EMS, fire/rescue, law enforcement and other public safety support units (i.e. mass care, public works, regional transit, etc.). For example, the decision to sustain the fire attack against a warehouse fire could be based on the number of fire trucks already in staging or in close enough proximity to the incident scene to be effective. A decision on managing the safe evacuation of a nursing home would be impacted by knowledge of the exact location and proximity of transit vehicles to the scene. Tracking the location of personnel and vehicles is also a critical piece of an agency's accountability and safety programme. Command staff and many supervisors also need to track individual public safety personnel, such as firefighters, who are on the scene of the incident. Automatic personnel location (APL) is especially critical for those employees who are conducting search and rescue operations inside the collapsed building and those who are working in the 'hot zone' with the gas leak. Law enforcement supervisors also need to track the location of on-scene deputies and police officers who leave their patrol car and are operating on foot in a hazardous situation (e.g. conducting a building search for an armed suspect). This type of location technology must support X, Y and Z coordinates, meaning that the incident commander must know if the injured public safety worker is in the basement or on the n-th floor. While this information may exist in multiple disparate systems, the incident command team needs to able to visualize the incident scene and the resources on a single display screen. This would require that AVL/APL data be collected from various agencies, consolidated and then distributed to the command team and appropriate supervisors.

2.2.4 *Electronic Conferencing and Coordination Tools for Incident Command*

Incident command requires a variety of applications to fulfil its mission. It is both a consumer and producer of substantial amounts of information. In addition to voice communications, electronic conferencing tools in PPDR can facilitate the sharing of information between the intervention team leaders at the emergency scene and local officials in the agency's (or other) control centre or headquarters facility in an interactive way. Video conferencing is one of the key features, allowing a number of 'distributed' meeting participants, who may be dispersed in several locations, to share video and audio signals.

PPDR incident command might also benefit from other features typically found in electronic conferencing tools, such as data conferencing (i.e. sharing of a common whiteboard among participants) and application sharing (e.g. participants can access a shared document or application from their respective computers simultaneously in real time). Hence, a common whiteboard application can greatly improve incident command coordination and response by easily sharing different and diverse data information such as notes, documentation and marked-up photographs. This information could be sent early in the incident to give all units on the scene an overview of the event and the operational plan. Furthermore, as the incident command system is more fully implemented, a formal incident action plan can also be distributed to all units on the scene in a more rapid and effective way. These action plans may comprise images, electrical plans, hazardous material information, building drawings, ingress/egress points and other contents that can require several megabytes (MB) of information to be distributed to dozens of personnel. As of today, incident action plan documents are frequently converted to PDF files and transmitted via e-mail communication or printed and distributed in hard copy [3]).

Incident command tools can also embrace support for collaborative management (coordination). Collaborative management tools are intended to facilitate and manage group activities. An example of collaborative management is a workflow system to manage tasks and documentation within a knowledge-based business process for PPDR operations.

2.2.5 *Remote Database Access and Information Transfer Applications*

In a general sense, this category of applications covers those electronic communications tools used to transfer messages, files, data or documents between PPDR officers on patrol or at the incident site and databases or information systems located in remote locations such as headquarters and control centres. Access to specific remote databases and applications to retrieve information while working on a routine operation or emergency scene can greatly improve PPDR working practices.

Remote database access and information transfer applications can benefit from reduced latencies and increased throughputs delivered through mobile broadband connectivity. Some examples are given in the following.

Law enforcement personnel must frequently query remote databases and be able to determine if a subject has a prior criminal history or an arrest warrant. In this kind of use, remote database access could be complemented with document scanning features. Hence, in patrolling or border security operations, PPDR officers can verify a document like a driving licence in a more efficient way. Documents scan is also useful in border security operations where people, who cross the borders, may have documents in bad condition or falsified. Automatic number plate recognition (ANPR) is also a typical use case within law enforcement activities, where a camera captures licence plates and transmits the image to headquarters or a centre with the plate data to verify that the vehicles have not been stolen or the owner is a crime offender.

Firefighters could be informed of the layout of a building by downloading images or video to a handheld device. In case of an emergency crisis or a natural disaster, PPDR responders may have the need to access the layout of the buildings where people may be trapped or where dangerous chemicals are kept. Fire rescue units need the ability to retrieve building pre-plan information, photographs and diagrams of hazardous materials storage, location of hydrants and water valves. Chemical data, building or floor plans can be requested to the headquarters and transmitted to the first responders.

EMS personnel also use data to review detailed drug information and poison control documentation and advise on various emerging health threats. Rescue ambulances could have access a web-based application to show hospital availability and also to track ambulance/patient destinations. In a similar way, a regional transit bus being used to move evacuated persons to a shelter should be able to access information on street closures, best routes and other information on their assignment.

Immediate access to real-time weather information might be essential at many emergency scenes. This is particularly true with wild land fires where changes in wind and humidity can cause significant changes in fire behaviour. Command staff and appropriate supervisors at the scene of a hazardous materials emergency also need to monitor wind speed and direction to determine evacuation areas and where 'shelter in place' orders should be given. While a single weather data station may be adequate for some incidents, the capability should exist for a consolidated view of multiple on-site weather reporting units.

2.2.6 PPDR Personnel Monitoring and Biomedical Telemetry

Vital signs of PPDR officers could be monitored in real time to verify their health conditions. This is particularly important for firefighters at fire ground incidents and officers involved in search and rescue operations. This involves monitoring, recording and measuring of basic physiological functions, such as heart rate, muscle activity and body temperature, of PPDR personnel, as well as other aspects such as air supply status, ambient temperature, presence of toxic gases, etc. A biomedical telemetry system would send these metrics from the individual first responder to an outside monitoring post and potentially move the data off-site to a central monitoring and recording station at their headquarters.

Besides constant monitoring of PPDR personnel, a biomedical telemetry system can also be central for remote emergency medical services (discussed in next section) as well as for other more specific applications such as identification of persons. For example, PPDR officers may check the biometric data of potential criminals (e.g. fingerprints, facial and iris recognition) during their patrolling duty and transmit it in real time to the headquarters or a centre with the biometric archives. This would be a positive method of identification during field interrogation stops.

2.2.7 *Remote Emergency Medical Services*

Through transmission of video and data, medical personnel may intervene or support the team in the field for an emergency patient. Biomedical telemetry can be used for sick and injured patients. In this regard, EMS personnel could be able to transmit a patient's heart rhythm, including a full 12-lead electrocardiogram (ECG), from scene to hospital for physician interpretation. A high-speed and reliable network would enable more complex types of biomedical telemetry monitoring and diagnostic applications. For example, certain types of testing to detect the presence of poisoning require interpretation by a specialist. Additionally, monitoring would be needed for serial blood glucose readings, oxygen saturation levels and carbon monoxide tracking.

Applications for tracking patients/evacuees/deceased are also crucial in disaster response and recovery operations. Command staff and supervisors need a mechanism that will track all persons on the scene and their eventual disposition. Paramedics currently attach a barcode or radio frequency identity (RFID) bracelet on each patient and then use a grocery store scanner-type device to enter in brief demographics before uploading a snapshot photograph and all information for centralized tracking and distribution to the receiving hospital. This same system would also track all evacuated persons, again using RFID or barcode scanner technology, allowing a snapshot photo of the citizen and demographics which are then uploaded to a server for distribution to the command post and public information centres and websites.

2.2.8 *Sensors and Remotely Controlled Devices*

Sensors networks could be deployed in a specific area and transmit images (thermal) or data to the PPDR responders operating in the area or to the command centre at the headquarters. Third-party sensors can also be leveraged. Hence, command staff and appropriate supervisors need to be able to 'connect' to various automated building systems to view alarm codes and conditions. For example, the fire department may need to determine or change the status of the air conditioning and ventilation system during a toxic gas leak. Law enforcement officials may need to access a specific building's security system to determine where a suspect may be based on a log of door activations. Likewise, there is a need to know the status of various hazardous sensors installed in many government buildings and large assembly areas, which are used to detect poisonous gases, radioactive materials, biological and other profiles. During an emergency inside an industrial plant, it would be critical to remotely monitor the status of various mechanical and automated systems, remotely turning on or off surveillance microphones or surveillance cameras (including remotely aiming or pointing the camera) and activating and deactivating alarms.

Vehicle telemetry is also of great relevance. A large number of vehicles can be typically present at the scene of almost all major emergencies. It is not uncommon in the fire service for dozens of fire trucks to be stationary and providing pump support. In some cases, particularly wild land fires, these engines may be pumping water for days as the firefighter crews rotate on and off duty. Public safety needs a system where the vehicle's health data is transmitted back to the incident command post for evaluation by personnel assigned to logistics. For example, attempting to determine which fire trucks need to be refuelled can be a daunting task. The ability to react to a report of falling oil pressure could prevent a major mechanical failure. Various telemetry systems also in use or envisaged within a range of PPDR usage scenarios include control of moving fixed assets (e.g. vehicles, equipment in hospitals, etc.)

Robotic devices can also be used to record images within badly damaged buildings that are too unstable for officers to enter or to operate within explosive areas or in underwater searches. There is likely to be increasing use of drone vehicles and aircraft over the next few years, mainly to obtain surveillance information without putting at risk the lives of the emergency services personnel (e.g. robotics-controlled bomb disposal). A drone vehicle might take the form of an unmarked car fitted with a number of concealed video cameras and a broadband wireless link. The vehicle's cameras will record all motion so as to enable the investigators to watch the footage in real time over the wireless link from the safety of a more distant location. A number of companies already market mobile CCTV systems that can relay real-time video via 3G mobile networks. Unmanned aeronautical vehicles (UAVs) are also increasingly deployed by the military, for example, to provide remote surveillance over wide areas. In this case, substantial bandwidth might be required, both to support surveillance video signals and the control and telemetry signals necessary to fly the UAV remotely.

2.2.9 *Mobile Office*

Access to common office applications such as e-mail and web browsing and access to intranets can be made available to PPDR officers also when patrolling or responding to an incident. E-mail and other office applications (e.g. contact databases, workflow applications, web browsing, etc.) could be used to improve the timely response of routine administrative processes as well as improve response in emergencies. For example, incident reporting can be performed via mobile devices, reducing the need to return to headquarters/control centres to access office applications. Incident-specific information could be exchanged using web applications. These applications would enable incident commanders to access their agency intranet to pull down various documents and templates. Of specific interest are building pre-plan documents that contain draft version of the floor layouts, access points, control rooms, etc.

Other data-centric applications such as business and data analytics are also increasingly being introduced in PPDR organizations, but they mainly remain accessible from desktop positions and cannot be used from officers in the field. For example, law enforcement personnel conducting research at their desks can benefit from systems designed to sort through information from multiple databases and provide officers with information relevant to the scene of an incident. While such tools are undoubtedly useful in its current form, the potential benefits will be realized fully when officers in the field are able to access this type of applications via handheld smart devices with mobile broadband connectivity.

2.3 Characterization of Broadband Data Applications for PPDR

A proper characterization of the enriched multimedia tools and applications that could benefit PPDR operations is central for a proper assessment of the technological needs as well as for communications resources dimensioning. The set of applications that could better serve the needs of a given PPDR agency and their specific characteristics and requirements (e.g. level of dependability, video resolution required, number of simultaneous users, deployment time in an incident, etc.) can be rather diverse among PPDR disciplines and even so among different PPDR agencies under the same discipline.

Mainly motivated by the need to estimate the amount of radio spectrum necessary for broadband PPDR, a number of PPDR organizations in Europe have created a list of PPDR applications based on their current operational experience and their vision of future working practices. The list of applications is largely based on PPDR studies from Germany, France, Finland, the United Kingdom, Belgium, the Netherlands and the European Telecommunications Standards Institute (ETSI). These studies involved governmental PPDR agencies and were led by the Radio Communication Experts Group (RCEG) within the auspices of the LEWP, a preparatory working group of the EU Council that participates in the EU legislation processes in the area of law enforcement.[1] After discussions with the EU member states' representatives, a list of applications was consolidated and adopted as the so-called LEWP/ETSI Matrix of Applications (referred to as the 'LEWP/ETSI Matrix' in the following). The LEWP/ETSI Matrix constitutes a toolbox of PPDR applications to be used either individually or in different combinations subject to the demands of the operational situation being attended. The scope of the LEWP/ETSI Matrix includes narrowband, wideband and broadband mobile data applications. The applications are not linked to any specific technology or implementation platform. The LEWP/ETSI Matrix was agreed between European PPDR organizations and is recognized by CEPT administrations as being representative in terms of future PPDR applications. ETSI took the lead in developing the technical parameters and definitions associated with each application. Furthermore, the LEWP/ETSI Matrix was later on complemented by ETSI with the addition of a spectrum calculation module for user-defined operational scenarios. The LEWP/ETSI Matrix is publicly available as an Excel tool in Ref. [7].

Table 2.2 describes the applications and services included in the LEWP/ETSI Matrix, which have been grouped into seven categories (i.e. location data, multimedia, office applications, download operational information, upload operational information, online database enquiry and miscellaneous).

For each application described in Table 2.2, the LEWP/ETSI Matrix captures the following requirements/characteristics:

- Throughput. Provides a rough, qualitative estimation of the relative throughput required for the application to provide the adequate quality of service, distinguishing between low (L)-, medium (M)- or high (H)-throughput levels.
- Use. Provides an estimate of the number of times that a given application is used per month and per user. 'L' indicates less than 10 times, 'M' between 10 and 30 times and 'H' higher than 30 times).
- Users. Estimation of the relative number of users of a given application in a typical PPDR operational scenario. Characterized as 'L' when used by less than 20% of the users, 'M' for a percentage between 20 and 70% and 'H' when used by more than 70% of the users.
- Mobility. Indicates if the application is used on the move or from fixed positions. Expressed in three levels of mobility: L, M and H.

[1] The main activities of LEWP are related to discussing the legislative proposals of the EU Commission in the area of law enforcement and creating means that would facilitate practical cooperation of law enforcement institutions at the EU level.

Table 2.2 Type of applications and services included in the 'LEWP/ETSI Matrix of Applications'.

Application/services	Description
Location data	
A(V)LS data to CCC (persons + vehicles positions)	Sending (automatically) location information from units to the control centre
A(V)LS data return	Sending (automatically) location information from the control centre (or software applications) to units (individual + groups)
Multimedia	
Video from/to CCC for following + intervention	Video information from and to special police units on suspects (e.g. hot pursuit)
Low-quality additional feeds	Extra cameras for observation with lower quality, which can be switched to higher quality when relevant
Video for fixed observation	Video information to control room or special observation room from a fixed location (most time building under observation)
Low-quality additional feeds	Extra cameras for observation with lower quality, which can be switched to higher quality when relevant
Video on location (disaster or event area) to and from control room – high quality	Video information to control room or special crisis centre from units on the location and to the units on what is happening; only a few high-quality video links
Video on location (disaster or event area) to and from control room – low quality	Video information to control room or special crisis centre from units on the location and to the units on what is happening; some more low-quality video links
Video on location (disaster or event area) for local use	Video information between the command unit on the location and the units on what is happening; some medium-quality video links which are only local.
Video conferencing operations	Video conferences between management, specialists, etc. (like in other businesses) + for coordination on the field
Non-real-time recorded video transmission	Sending a selected part from a recorded video in a later stage to control room or coordination centre
Photo broadcast	Picture (e.g. from wanted person) to a big group of officers
Photo to selected group (e.g. based on location)	Picture (e.g. from missed child) to those officers which are in the relevant search area
Office applications	
PDA personal information manager (PIM) synchronization	The 'normal' applications like mail, agenda search of the public Internet, etc.
Mobile workspace + (including public Internet)	The facility to do with a laptop 'on the street' the same as in the office (also the back-office applications, e.g. to fill in a file)
Download operational information	
Incident information download (text + images) from CCC to field units + net-centric working	Information regarding an incident from the control room to the field units. Can be text, pictures, images, maps, etc.
ANPR update hit list	Automatic update from the wanted cars (hit list) for the automatic number plate recognition application
Download maps with included information to field units	Sending maps with additional information (extra info on buildings, location of officers, routes, etc.) from the control room to the field units
Command and control information including task management + briefings	Sending all kind of briefing information from the control room to the relevant units

(*continued*)

Table 2.2 *(continued)*

Application/services	Description
Upload operational information	
Incident information upload (text+images) to CCC+net-centric working	Information regarding an incident from the field units to the control room. Can be text, pictures, images, maps, etc.
Status information+location	(Automatic) sending of status information (on route, arrived, incident closed, etc.)+location from field units to control room
ANPR or speed control automatic upload to database including pictures (temporary 'fixed' cameras+from vehicles)	ANPR/speed control application: automatic upload to database including pictures from relevant cars. Info is coming from temporary 'fixed' cameras+from vehicles equipped with ANPR or speed measurement equipment
Forward scanned documents	Making a scan from document(s) by field units and send that to control room or colleagues. Includes medical health-care information
Reporting including data files (e.g. pictures, maps)	Making a report (can be pictures, images or map info included) by field units and send that to control room or colleagues
Upload maps+schemes with included information	Sending maps with additional information (extra info on buildings, location of officers, routes, etc.) from the field units to the control room or other field units
Patient monitoring (ECC) snapshot to hospital	Sending patient information (e.g. ECC) from ambulance or from the field to hospital: only a limited snapshot
Patient monitoring (ECC) real-time monitoring to hospital	Sending patient information (e.g. ECC) from ambulance or from the field to hospital on real-time basis
Monitoring status of security worker	Specific fire application: firemen are equipped with safety measurement equipment which will send out a warning when there is a risk for the fireman; usually, fire brigades will send the information locally to a commander at the scene. Rescue services need to send the data back to a supervisor over the main network
Online database enquiry	
Operational database search (own+external)	Database enquiry by field units from all the back-office databases+relevant external databases
Remote medical database services	Database enquiry by medical field units from the relevant (external) medical databases
ANPR checking number plate live on demand	On-the-spot number plate control by field units via connection to the car registration database
Biometric (e.g. fingerprint) check	With special equipment checking biometrics and sending this info to the relevant database to check (hit check)
Cargo data	Database enquiry by field units from the relevant external databases with cargo information (by logic cargo numbers)
Crash recovery system (asking information on the spot)	On-the-spot control by fire units where they use the hydraulic scissor for cutting a car open to rescue people (via the car registration database)
Crash recovery system (update to vehicles from database)	From the most common cars, the car drawings are stored in the fire truck to save data communication. Updates are then needed

Table 2.2 *(continued)*

Application/services	Description
Miscellaneous	
Software update online	Online software updates for the terminals in use
GIS maps updates	Updates from geographical maps which are stored on the terminals
Automatic telemetry including remote-controlled devices + information from (static) sensors	All kind of telemetric information: from and to remote control devices + information from (static) sensors (e.g. observation)
Hotspot on disaster or event area (e.g. in mobile communication centre)	A temporally hotspot for local broadband data on a crisis/disaster/investigation area or at a big planned event
Front-office–back-office applications	The possibility to work 'on the street' with the normal 'in the office' front and back-office applications
Alarming/paging	Paging function to alarm PSS people (e.g. fire people to go to the fire centre)
Traffic management system: information on road situations to units	Information to the field units on which roads to used, blockages, etc.
Connectivity of abroad assigned force to local CCC	Availability for forces from other countries to get in contact with the local control room via data communication

- Quality of experience (QoE). Indicates whether short interruptions/impairments can be tolerated in the connection. Expressed as L, M and H levels.
- Start-up availability. Indicates how soon the application should be ready for use by PPDR agencies. Expressed mainly as two levels: 'ready', if the application needs to be used with zero or almost zero start-up time, and 'low', when there is enough time to switch on.
- Delay. Indicates if there are delay restrictions in data delivery. Expressed in three levels (L, M and H), being H associated with the most demanding timeliness.
- Continuous operational availability. Indicates the mission-critical level of the application. Expressed as L, M and H levels of criticality.
- Peripherals. Indicates whether the application needs some sort of peripherals for field units such as smartphones, PDA, pagers, modem, routers, satellite communications equipment or any special equipment.
- Screen. Indicates whether the application needs screens/data displays for field units.
- Security. Indicates the relative level of security requirements (confidentiality and integrity). Expressed as L, M and H levels of security.
- Multicast/broadcast. Indicates whether the application needs to support group calls and broadcast delivery. If broadcast/multicast delivery is needed, the number of intended recipients is estimated in three levels as L, M and H.
- Timeliness. Indicates the urgency to introduce a certain application. Expressed as 'now', which includes applications that are partly already in use as of today; 'short', for applications required in less than 2-year time; 'medium', for a period between 2 and 5 years; and 'long', when the real need is not foreseen before 5 years or more.

Moreover, as the relevance of the aforementioned requirements/characteristics can differ per application, the LEWP/ETSI Matrix indicates which are the three most relevant parameters for each application. An excerpt of the LEWP/ETSI Matrix is illustrated in Table 2.3, which

Table 2.3 Characterization of the 'multimedia' applications.

Type of application + services	Throughput	Use	Users	Mobility	QoE	Start-up availability
Video from/to CCC for following + intervention	**H**	L	M (emergency vehicles)	*H*	L	Ready when vehicle is ready
Low-quality additional feeds	**L–M** (depending on quality)	L (but more than above)	M (emergency vehicles)	*H*	L	Ready when vehicle is ready
Video for fixed observation	**M** (H on HD)	H	L	L	L	L (mostly enough time to switch on)
Low-quality additional feeds	**L–M** (depending on quality)	H	L (but more than above)	L	L	L (mostly enough time to switch on)
Video on location (disaster or event area) to and from control room – high quality	***H***	L	L	L	M	Take along on ad hoc basis
Video on location (disaster or event area) to and from control room – low quality	***M***	L	L	L	M	Take along on ad hoc basis
Video on location (disaster or event area) for local use	***M–H***	L	L	L	M	Take along on ad hoc basis
Video conferencing operations	***M*** (H on HD)/priority needed	L	L	L	L	L
Non-real-time recorded video transmission	**H**	L	L	L (but H when using in cars)	*M*	L (mostly enough time to switch on)
Photo broadcast	M	L	M	*H*	L	Ready
Photo to selected group (e.g. based on location)	M	L	L	***H***	L	Ready

Delay	Continuous operational availability	Peripherals	Screens	Security	Broadcast/ multicast	Timeliness
M (has to be in line with speech; max 1 or 2 seconds)	***H***	Modem/ router	Yes, if receiving	L	M	M
M (has to be in line with speech; max 1 or 2 seconds)	***H***	Modem/ router	Yes, if receiving	M	M	M
M (has to be in line with speech; max 1 or 2 seconds)	*M*	Modem/ router	None	H	L	M (partly already in use)
M (has to be in line with speech; max 1 or 2 seconds)	*M*	Modem/ router	None	H	L	M
M	**H** (availability at Golden Hour essential)	Modem/ router	Yes, if receiving	L	L	Now (partly already in use)
M	**H** (availability at Golden Hour essential)	Modem/ router	Yes, if receiving	L	M	Now (partly already in use)
M	**H** (availability at Golden Hour essential)	Modem/ router	Yes, if receiving	L	M	Now (partly already in use)
M (has to be in line with speech; max 1 or 2 seconds)	**L** (but M/H if used in crises)	Special equipment	Graphic	M	*L*	L
M (has to be in line with speech; max 1 or 2 seconds)	M	Modem/ router	None	M	No	M
L	**M**	PDA/ smartphone	Graphic	M	***H***	Now (partly already in use)
L	**M**	PDA/ smartphone	Graphic	M	*M*	Now (partly already in use)

shows the characteristics of the applications under the 'multimedia' category. The three most important parameters per application are marked off using, in this order, (i) boldfaced, (ii) both boldfaced and italicized and (iii) italicized letters. For example, for the 'video to/from CCC for following + intervention' application shown in Table 2.3, it can be observed that the most important parameter is the throughput, estimated as 'high'. The second and third parameters in relevance are the continuous operational availability and the mobility level, respectively, both also estimated as 'high'.

2.4 Assessment of the Data Capacity Needs in Various Operational Scenarios

The amount of data capacity needed is strongly dependent on the type of operational scenario. Several works (e.g. [3, 4, 7–14]) have described detailed usage scenarios within the PPDR sector, illustrating both the range of applications that might be used within daily operations and the applications used to respond to specific incident types. Sustained in these previous works, this section firstly provides some estimates of the throughput requirements for different types of PPDR data-centric, multimedia applications. Then, a quantitative assessment of the required data capacity for a number of 'typical' PPDR operational scenarios within the categories of (i) day-to-day operations, (ii) large emergency/public events and (iii) disaster scenarios is presented.

2.4.1 Throughput Requirements of PPDR Applications

PPDR data-centric applications can have very different rate requirements that place varying levels of demand on the capacity of a broadband mobile network. The estimation of the load on the network for each of these applications is necessary for the dimensioning of the infrastructure and spectrum assets.

A quantitative estimate of the throughput requirement per application is included within the LEWP/ETSI Matrix. For each of the applications listed in Table 2.2, the details of a single data transaction are characterized, distinguishing between transactional- and streaming-based applications and between uplink (end user to the fixed network) and downlink (fixed network to end user) transmissions when appropriate. For transaction-based applications, the size of a data transaction is estimated as the number of bytes of information required in each single transaction. For streaming-based applications, the peak bit data rates and duration (in minutes) are given. In addition, for each application either transactional- or streaming-based, the average number of transactions/sessions per user in a peak busy hour in normal conditions is provided according to PPDR end users' experience. Based on this characterization of the transaction details, estimates for the bit rate requirements per hour per user can be easily derived. Table 2.4 shows the resulting throughput estimates of the PPDR applications according to the application characterization provided in the LEWP/ETSI Matrix.

As observed from Table 2.4, video streaming applications are expected to be by far the largest contributors to the peak hour throughput, both in uplink and downlink directions. The considered peak data rates for video applications vary from 768 to 64 kb/s because of the different uses of streamed video that can have very different quality attributes (e.g. the required video quality differs considerably between having only some level of situational awareness and using the video source as focal point to follow the details of a fast and complex PPDR intervention). Hence, video throughput is highly dependent on a required video quality, which largely depends on the resolution of the image and the frame rate (number of images

Table 2.4 Throughput estimates for PPDR data applications (based on the LEWP/ETSI Matrix).

Application	Transaction details				Peak hour throughput (kb/s)	
	Peak bit rate UL/DL (kb/s)	Data per trans. UL/DL (in blocks of 1000 bytes)	Duration (min)	Transaction per hour	UL	DL
Location data						
A(V)LS data to CCC (persons + vehicles positions)		0.08/–		240	0.04	0.00
A(V)LS data return		–/1		60	0.00	0.13
Multimedia						
Video from/to CCC for following + intervention	768/768		60	1	768.00	768.00
Low-quality additional feeds	64/64		60	1	64.00	64.00
Video for fixed observation	384/–		20	1	128.00	0.00
Low-quality additional feeds	64/–		20	1	21.33	0.00
Video on location (disaster or event area) to and from control room – high quality	768/768		60	1	768.00	768.00
Video on location (disaster or event area) to and from control room – low quality	64/–		60	1	64.00	0.00
Video on location (disaster or event area) for local use	192/192		60	1	192.00	192.00
Video conferencing operations	256/256		10	1	42.67	42.67
Non-real-time recorded video transmission		2000/2000		1	4.44	4.44
Photo broadcast		–/50		2	0.00	0.22
Photo to selected group (e.g. based on location)		–/50		2	0.00	0.22
Office applications						
PDA PIMsync		5/5		2	0.02	0.02
Mobile workspace + (including public Internet)		100/100		5	1.11	1.11
Download operational information						
Incident information download (text + images) from CCC to field units + net-centric working		–/50		2	0.00	0.22
ANPR update hit list		–/8		1	0.00	0.02
Download maps with included information to field units		–/50		1	0.00	0.11
Command and control information including task management + briefings		–/50		1	0.00	0.11

(*continued*)

Table 2.4 (*continued*)

Application	Transaction details				Peak hour throughput (kb/s)	
	Peak bit rate UL/DL (kb/s)	Data per trans. UL/DL (in blocks of 1000 bytes)	Duration (min)	Transaction per hour	UL	DL
Upload operational information						
Incident information upload (text + images) to CCC + net-centric working		50/–		1	0.11	0.00
Status information + location		0.1/–		5	0.00	0.00
ANPR or speed control automatic upload to database including pictures (temporary 'fixed' camera's + from vehicles)		40/–		50	4.44	0.00
Forward scanned documents		100/–		0.1	0.02	0.00
Reporting including data files		1000/–		1	2.22	0.00
Upload maps + schemes with included information		50/–		1	0.11	0.00
Patient monitoring (ECC) snapshot to hospital		50/–		1	0.11	0.00
Patient monitoring (ECC) real-time monitoring to hospital	15/–		15	1	3.75	0.00
Monitoring status of security worker		1/–		120	0.27	0.00
Online database enquiry						
Operational database search (own + external)		1/50		2	0.00	0.22
Remote medical database services		1/50		2	0.00	0.22
ANPR checking number plate live on demand		0.1/2		5	0.00	0.02
Biometric (e.g. fingerprint) check		20/2		1	0.04	0.00
Cargo data		0.1/2		1	0.00	0.00
Crash recovery system (asking information on the spot)		0.2/50		1	0.00	0.11
Crash recovery system (update to vehicles from database)		–/50		0.1	0.00	0.01
Miscellaneous						
Automatic telemetrics including remote-controlled devices + information from static sensors		0.1/0.1		60	0.01	0.00
Front-office–back-office applications (e.g. online form filling)		10/10		3	0.07	0.07
Alarming/paging		0.1/1		1	0.00	0.00
Traffic management system: information on road situations to units		–/10		4	0.00	0.09

per second) as well as a variety of factors such as target size, motion level and lighting level. In recent years, the coding engines that convert the raw video information into a highly compressed data stream have dramatically improved, and this trend is likely to continue. The state-of-the-art commercial video coders are capable of sustaining very high-quality video with a relatively low bit rate and requiring relatively low computing resources.

Following the streamed video, recorded video transmission in both uplink and downlink and reporting of operational information such as pictures in uplink are the applications that require larger volumes of data per transaction. These applications are estimated to send data volumes in the order of a few MB per transaction. It is worth noting that while these applications may only represent a few kilobit per second on average in the busy hour, the instantaneous data rate achieved when sending this data can be quite relevant from a quality of service point of view, since it will determine the amount of time needed for data delivery (e.g. delivery of a 1 MB picture from the incident in 5 s may be a reasonable target).

Typical data rate values for data and video PPDR applications have been also reported by the NPSTC in Ref. [4]. As the figures from the LEWP/ETSI Matrix, NPSTC's reported data rate values are intended to represent typical data rates necessary to sustain sufficient quality of service. Table 2.5 shows the resulting throughput characterization of the PPDR applications according to NPSTC's studies, which represent the usage impact of a single user of each application. In this case, streamed video applications (referred to as 'incident video' by NPSTC) also represent the most important source of traffic, distinguishing in this case three general video categories:

1. High quality. This represents 1 Mb/s throughput requirement and provides high resolution (e.g. standard-definition television) and high frame rate communications. High quality is capable of high motion in a highly dynamic range of light, with small target size, and discrimination capable of facial recognition.
2. Medium quality. Related to 512 kb/s throughput requirement, medium resolution and high frame rate communications. Medium quality is capable of high motion in high dynamic range of light, with small target sizes, and discrimination capable of licence plate recognition. It provides an overview of an incident scene and enables visualization of broad elements of action.
3. Low quality. Representing video feeds with 256 kb/s throughput requirement, low resolution and high frame rate communications. Low quality is capable of low motion, with large target size, high dynamic light range and object identification. Low speed is capable of providing situational awareness with some level of perspective of each video source. It provides a large area tactical view but little specific details.

Video throughput numbers are experienced in both uplink and downlink directions depending on the source and destination of each video stream. For example, a low-speed situational awareness stream may be transmitted from the incident location on the uplink and then back down to incident command on the downlink. Therefore, these rates are applied to the uplink and downlink separately. A 'return path' data speed is also included in the NPSTC analysis to accommodate acknowledgement or other traffic associated with two-way communications where appropriate.

Following streamed video, the use of GIS tools and the tracking of civilians are the next two applications considered in NPSTC analysis that could consume the largest bandwidth. In the case of GIS, as described in Section 2.2.2, a variety of geospatial-based data may be required by incident commanders and other personnel at the incident. GIS data sources are becoming extremely robust including high-resolution aerial images and other data layers. For the throughput characterization, NPSTC assumes the downloaded data for each GIS view is 350 KB and that data

Table 2.5 Throughput estimates for PPDR data applications (based on the NPSTC report [4]).

Application	Transaction details			Peak hour throughput (kb/s)	
	Peak bit rate UL/DL (kb/s)	Duration (s)	Transaction per hour	UL	DL
Incident video – high quality (DL) (aircraft)	16/1024	3600	1	16	1024
Incident video – medium-quality (DL) traffic camera	16/512	3600	1	16	512
Incident video – low quality (DL), situational	16/256	3600	1	16	256
Incident video – low quality (UL), situational	256/16	3600	1	256	16
Incident video – high-quality (DL) helmet/vehicle	16/1024	3600	1	16	1024
Incident video – high-quality (UL) helmet/vehicle	1024/16	3600	1	1024	16
Incident video – medium-quality (DL) helmet/vehicle	16/512	3600	1	16	512
Incident video – medium-quality (UL) helmet/vehicle	512/16	3600	1	512	16
Incident video – medium-quality (UL) video conference	512/16	3600	1	512	16
Incident video – medium-quality (DL) video conference	16/512	3600	1	16	512
Automatic location (UL+DL) vehicles	0.04/0.04	1	240	0	0
Automatic location (UL+DL) personnel	0.04/4.00	1	240	0	0.27
Geographic Information Systems (GIS) – street view	16/160	1	5	0.02	0.22
GIS detailed view	68/683	1	60	1.13	11.38
File and message transfer UL	0.02/0	1	4	0	0
File and message transfer DL	0/0.02	1	4	0	0
Patient and evacuee and deceased tracking	13/5	60	60	13	5
Biotelemetry – first responder (UL+DL)	0.13/0.13	30	120	0.13	0.13
Biotelemetry – patient	2.70/0.03	300	50	11.25	0.11
Vehicle telemetry	2.70/0.03	300	4	0.90	0.01
Third-party sensors	0/0.03	30	2	0	0
Weather tracking	13.30/13.30	60	12	2.66	2.66
Voice over IP cell phone access	10/10	3600	1	10	10

must be transmitted over 4–5 s in order to achieve a reasonable quality of service. This results in a requirement of around 700 kb/s in downlink. With regard to the application for tracking of civilians, it is considered that PPDR personnel collect a 100 KB dataset that includes patient demographic information, images and medical information. The data is not extremely time sensitive and could be transmitted over 1 min. Considering that one person can process one patient per minute, the impact on the system is estimated as 13.3 kb/s in uplink.

2.4.2 Day-to-Day Operations Scenarios

Day-to-day operations scenarios can be considered as the minimum requirement for PPDR activity. In this context, two everyday life situations where it is considered that mobile broadband communications are in place are described in the following [7]: a road accident and a traffic stop police operation. In addition, some estimates for the background traffic load that could be expected as a result of routine PPDR operations are also provided.

The road accident scenario illustrates the response to a car crash where the occupants are assisted by police and EMS personnel. An ambulance and a helicopter are involved. There is exchange of data information with control rooms (e.g. patient's profile for remote diagnosis), and a cardiologist from a hospital joins the incident group call (voice and video) providing advice. The simplified timeline of the scenario is as follows (a more detailed description can be found in Ref. [7]). At 8:40, the emergency control centre (ECC) in the area is informed of a car crash. Information received by the ECC from people who called consists of voice descriptions together with photos and videos sent though public networks (assuming PSAP support multimedia capabilities, commonly referred to as next generation 911/112). Information provided by the sensors in the car, which has access to public networks, can also be used. Emergency and medical services as well as police are dispatched immediately. The crews are given the best route to reach the accident location and all information available about the car trash through the PPDR communications services. An alert is also sent to place a helicopter on standby. Paramedics arrive at the car crash site at 8:52. On their arrival and having quickly assessed the situation, they ask for the helicopter and an ambulance. Thanks to the images and the situation description (location), the helicopter can land very close to the accident with very little help from personnel on the ground. The police also arrive and secure the area to ensure authorized personnel only to enter the area. The police start to gather evidence to help to establish what exactly happened. There are two injured people, one still trapped within the car. The first patient is a male that responds to voice and has no visible injuries but complains of shoulder pain. The paramedic uses several devices with radio interfaces to assist him (e.g. ECG, vital measurements). The patient's profile is sent to the ECC and from there to the patient's hospital. The male patient has a medallion providing medical data. Based on this data, the doctor in the ECC makes a diagnosis of a heart attack and a decision to take him to another hospital where specialist facilities exists is made. Data needed for initial treatment to stabilize the patient is transmitted to the ambulance. The second patient is female. She is unresponsive but breathing, with an open head injury. At 9:02, a heavy rescue vehicle arrives and the car roof that was collapsed is cut and removed. At 9:12, a cardiologist from the hospital joins the incident group call (voice and video), providing advice to the male patient in the ambulance. At the same time, the female patient is taken care of by the doctor. All medical data is sent to the control room to be put in the database. The medical crew monitors the female patient while she is transported in the helicopter.

Based on the above description, the following main communications requirements are identified:

- All information about the car crash is communicated to emergency and medical services (location, pictures).
- Patient information is sent to the ECC and then sent to the ambulance.
- Images and video link sent to the helicopter (considering that low-flying helicopters may use terrestrial networks).
- Video of patient on accident site to hospital.

Table 2.6 Data capacity required in some illustrative day-to-day scenarios.

Scenario	UL/DL	Application/use	Data rate per application (kb/s)	Total data rate (peak traffic) (kb/s)
Road accident	UL	Incident video (768 kb/s peak rate, 1 user)	768	1300
		Data transfer	512	
	DL	Incident video (768 kb/s peak rate, 1 user)	768	1300
		Data transfer	512	
Traffic stop police operation	UL	Incident video (768 kb/s peak rate, 1 user)	768	1300
		Data transfer	512	
	DL	Incident video (768 kb/s peak rate, 1 user)	768	1300
		Data transfer	512	
Background traffic	UL	Incident video – low-quality additional feeds (64 kb/s on average per user, 10 simultaneous users)	640	1500
		Fixed video (64 kb/s on average per user, 5 simultaneous users)	320	
		Fixed video – low-quality additional feeds (11 kb/s on average per user, 20 simultaneous users)	220	
		Other applications (location, patient monitoring)	320	
	DL	Incident video – low-quality additional feeds (64 kb/s on average per user, 9 simultaneous users)	576	876
		Other applications (photos, download maps, etc.)	300	

Considering that not all of these transmissions would be simultaneous, the peak use is estimated as the aggregation of two video streams (768 kb/s in downlink and the same in uplink) together with a simultaneous data transfer need of approximately 512 kb/s in both directions (voice traffic is not considered in these data rate computations). This totals 1300 kb/s both in uplink and in downlink. A summary of these estimations is captured in Table 2.6.

In the traffic stop police operation, two police officers on routine traffic patrol observe a car running through a red traffic light at an intersection. The patrol signals to the control room through predefined data message that a pursuit is beginning. The camera in the patrol's vehicle starts to record the offending vehicle, whose plate number is automatically used to query a remote database. The video is made available to the control room and authorized personnel connected through the police information system. As a response to the database query, the police patrol is informed that the car is not stolen and additional information about the registered owner. The offending vehicle stops. The camera is still recording and the video feed remains available on demand to the dispatch centre. The two police officers approach the car and note that there is only one driver. They request driver's licence, but no documentation is provided. One of the officers observes what he believes to be the remains of drugs in the ashtray. He decides to search the suspect vehicle and contacts dispatch to request a backup unit. A second unit is so forwarded to the incident place, and a specific group call is created for the incident to share voice and data information. In this way, the second unit can access all the incident data (video and database) while on the move to the incident site. The supervisor and the second unit bring up the real-time video of the event in the control room and in the second unit vehicle. The backup unit arrives on scene. The suspect is ordered to get out of the vehicle. A white substance is found that appears to be cocaine. The suspect is put under arrest and a transport vehicle is dispatched from the control room upon request. The transport unit also joins the group call to access all information. After the arrest, one police officer takes the driver's biometric sample and checks it against a database, which returns name, photo and specific information about the driver. He has been previously arrested by drug possession. The suspect is taken to jail by the transport unit. After the suspect has left, the police officers take images of the car and the drugs and complete all the steps and data forms needed to make the car to be taken by a tow truck and to complete the report of the incident. Therefore, when the driver arrives at jail, all data and forms are ready.

Based on the above description, the following main communications requirements are identified:

- Video to control room (camera of patrol vehicle) and database query (licence plate)
- Video feed available on demand (uplink to control room and downlink to backup vehicle near site)
- Access to databases (return with information and possible photo)

Like in the road accident scenario, considering that not all of these transmissions would be simultaneous, the peak use is also estimated as the aggregation of two video streams (incident video at 768 kb/s in uplink and video feed available in downlink) together with a simultaneous transfer need of approximately 512 kb/s associated with database queries and responses in both directions. This also totals 1300 kb/s in both uplink and downlink. A summary of these estimations is captured in Table 2.6.

In addition to the specific needs that may arise due to the occurrence of common incidents as those described previously, CEPT Report 199 [7] also characterizes what it could be the overall background traffic generated by a number of routine activities whose communications needs would be typically served by a single cell site. For the estimation of the background traffic, the applications and their data rates averages are taken from the LEWP/ETSI Matrix. Among the assumptions considered for the computation of the background load, it is considered that high-quality video feeds are not used. Moreover, the number of simultaneous applications is chosen independently of the size of the cell (which is clearly a rough approximation). Under these assumptions, the estimated load of a cell site sector due to routine traffic is provided in Table 2.6.

2.4.3 Large Emergency/Public Events

While recognizing that the size and preparation time for large emergencies and/or public events may vary a lot between distinct scenarios, two illustrative scenarios are discussed in the following [7]: the Royal Wedding in London in April 2011 and the London riots that occurred in August of the same year. The former scenario is a pre-planned event, while the latter is unplanned. The description and capacity estimates of both scenarios are developed considering how PPDR broadband communications might have been used had they been available.

The Royal Wedding can be considered as a high-profile, high-security event that draws vast crowds to get a close look and to be part of a historic occasion. Security operations however have to strike a balance between allowing the crowds to get close and have sight of the couple and keeping the royal family and other very important people (VIP) attending the ceremony safe. The security-optimized route planned for the royal procession varies from wide tree-lined roads to narrow streets overlooked by tall mainly office accommodation providing an opportunity for, for example, a terrorist attack. On the wedding day, the police and other security organizations perform their last-minute security checks and report to control. A crowd of well-wishers estimated to be around 1 million is lining the short wedding procession route between Westminster Abbey and Buckingham Palace. Covert and overt teams, totalling a force of 5000 officers, are deployed to mingle with the crowd and to look for suspicious activity. Trained officers and observers are stationed at key rooftop vantage points along the route. More than 80 VIPs are given close-protection bodyguards. A trained assassin or terrorist is not normally distinguishable from others in a crowded situation, being their behaviour what might alert security officials. In many cases, however, what is seen as suspicious behaviour turns out to be unrelated to a security threat. A judgement between acting too soon, and causing significant disruption and embarrassment to the proceedings, and acting too late is a fine line. Intelligence delivered through fast and good-quality communications is vital in such situations to help with the split-second decisions PPDR officials have to make.

Policing of the Royal Wedding route is managed in sectors. From a communications perspective, each sector is managed identically. Subject to how the radio infrastructure overlays the Royal Wedding route, it is considered that there is a strong possibility of one site having to carry the traffic from two adjacent sectors. The estimates below are applicable to one sector only and are in addition to the routine traffic that would be associated with general crowd control:

- One video stream from the Royal Coach.
- Two video streams from each side of the road from close-protection officers lining the route, that is four streams per sector in total.
- One high-resolution picture sent per minute from the helicopter to the coach and each of the bronze commanders managing each section. Frequency of updates would increase in the event of an incident.
- Selected still pictures from the helicopter and fixed cameras along the route would be sent to the two covert teams mingling with the crowd as and when felt necessary.
- The 60 officers of the two covert teams in each sector provide GPS-based location updates every 5 s.

Therefore, this scenario considers up to five simultaneous active cameras in total, four along the path and one on the coach. On this basis, capacity estimates are captured in Table 2.7, totalling around 5 Mb/s in uplink without accounting for other routine communications that could take place simultaneously in the surroundings of the Royal Wedding parade and so be served from the same cell site. Of note is that the estimates in Ref. [7] are for communications using a permanent wide area network. Additional capacity deployed on-site may also be provided by other means such as wireless local area networks. This additional capacity could be used for local data exchanges between responders on the site and for specific applications (e.g. if a robot is deployed, the necessary command and control links would be provided locally). This additional local capacity is not accounted in the capacity assessment of the Royal Wedding scenario.

The second analysed large event scenario is based on the riots occurred in London in August 2011. The rioting was triggered by a fatal shooting of a 29-year-old man by police. The riots started in the Tottenham area of London where the shooting occurred and then rapidly spread to neighbouring areas around Tottenham and areas further afield both within in London and outside of London. Rioting occurred over 4 consecutive days. In this kind of scenarios, a typical Bronze commander (i.e. level of management corresponding to the operational level within the typical command structure hierarchy – see Chapter 1) would manage locally up to 300 police officers. Due to the number of rioting focal points in close proximity to each other, two sub-Bronze commanders were deployed, one for each rioting focal point reporting to a common Bronze commander. The geographic spread and scale of rioting in London required 16 000 additional police officers being drafted in from surrounding forces into London to assist the Metropolitan police contain the situation (through an arrangement known as mutual aid). In this kind of scenarios, the immediate issue for police is the protection of the public and property. Identifying and arresting the perpetrators of criminality would not be an immediate priority. Analysis of photographs and video captured at the time would be used post riots to identify culprits for later questioning.

The main communications requirements for this unplanned emergency can be summarized as below:

- Two high-quality video streams, one from each sub-Bronze command area, being fed back to Gold and Silver command. The video would be used to help in managing the situation and would be recorded remotely for evidential purposes.
- Bronze and sub-Bronze commanders receive regular GPS-based location updates of the officers under their command.
- Interactive maps pushed to officers on the ground to help them to navigate to where they needed to be and to show which areas/streets they should avoid. Particularly important as many of the officers deployed to contain and quell the rioting were not familiar with the area in which they were working.
- An infrared video feed from a helicopter fed to firefighters on the ground to help them tactically fight large fires.
- A video feed from a helicopter fed back to Gold and Silver command.
- Numerous still pictures captured by police officers transmitted back to Gold and Silver command to help manage the situation and to be recorded remotely for evidential purposes.

The above communications needs exclude ambulance, fire (except for the downlink infrared video feed) and routine traffic in the surrounding area. On this basis, capacity estimates are captured in Table 2.7, totalling around 4 Mb/s in downlink and 1.8 Mb/s in uplink. As in the Royal Wedding scenario, the estimates in Table 2.7 are for communications using the permanent wide area network, thus not considering additional on-site capacity that may be

Table 2.7 Data capacity required in a large emergency and a massive public event scenarios.

Scenario	UL/DL	One video stream on coach	Data rate per application	Total data rate (peak traffic) (kb/s)
Royal Wedding in London in April 2011	UL	One video stream on coach	768 kb/s	4590–4840
		Four video streams along coach path (768 kb/s per stream)	3072 kb/s	
		One high-resolution picture from helicopter to control centre every minute (some MB per picture every minute)	250 kb/s (average) – 500 kb/s (peak to increase delivery speed)	
		Other communications (including GPS updates)	500 kb/s	
	DL	Selected still pictures are sent to the covert teams. Resulting traffic amount not specified	Not estimated	Not estimated
London Riots in August 2011	UL	Two video streams from sub-Bronze command areas (768 kb/s per stream)	1536 kb/s	4072
		Infrared video from helicopter	768 kb/s	
		Video from a helicopter to central command	768 kb/s	
		Pictures from officers transmitted back to central command (and GPS information from officers)	1000 kb/s	
	DL	Infrared video to firefighters	768 kb/s	1768
		Interactive maps	1000 kb/s	

provided by other means such as wireless local area networks and used for local data exchanges between responders on the site or for specific applications.

2.4.4 Disaster Scenarios

In unplanned mass events and major incidents, especially natural disasters where the location and requirements are not known in advance, significantly higher communications needs at very short notice are likely to occur. A detailed analysis of the broadband communications needs in four major incidents has been addressed by NPSTC [3] as part of an assessment conducted to identify PPDR spectrum and technology requirements in the United States for the 10-year period, from 2012 to 2022. The disaster scenarios covered are:

- **Hurricane**. Scenario based on Hurricane Charley, which hit the Central Florida area on 13 August 2004. The operations characterized focus on the collapse of an apartment complex building with dozens of injured and trapped citizens. Another 500 residents in the area needed to be evacuated from damaged buildings. There were reports of a natural gas leak near the scene, and law enforcement agencies were notified of looters moving into the area as night fell. EMS and fire/rescue agencies arrived at the scene, established an incident command system and started an assessment of damage and injuries. Law enforcement units arrived and started to secure the area and move crowds of displaced citizens out of the danger zone to a common holding area. The incident command immediately assigned crews to start searching the collapsed apartment building and assigned additional crews to locate and secure the broken gas line and to begin treating injured persons. EMS personnel were organizing the treatment of multiple personnel via use of the area's mass casualty incident plan. As the scope of the incident was fully identified, the incident command requested additional units to respond to assist those already on scene. A secondary staging area was identified several blocks from the incident where incoming fire trucks and ambulances would park until they were given a specific assignment and were called in to the scene. The analysis of the incident focuses on the operations conducted in an area of 1 square mile, involving around 220 responders and 60 vehicles.

- **Chemical plant explosion**. Scenario based on a chemical plant explosion occurred in the large industrial corridor between the City of Houston and the City of Pasadena, in Texas, along the Houston Ship Channel. A large explosion and fire were reported along with the presence of a chemical gas cloud moving across an interstate highway. Dozens of workers at the chemical plant were reported to be injured or missing. The cause of the explosion was not known at the beginning, but it was later revealed that an employee had been fired several days earlier and made threats during his departure. Fire and EMS personnel reported they would arrive near the scene and set up an incident command post approximately 2 miles upwind from the incident. Additional units would be sent directly to the scene to meet with plant officials and to determine the extent of the incident. Law enforcement units were arriving at the command post, while other units were sealing off the area and starting to evacuate nearby businesses in the 'safe zone'. Law enforcement personnel were also working in conjunction with Department of Transportation staff to close the interstate in the area of the explosion. Additional fire and EMS personnel were deployed directly to the scene in protective gear to locate and remove injured workers and to ensure a complete evacuation of the staff. Hazmat crews were on scene performing a detailed assessment of the damage that triggered the chemical cloud. Specialized sensors were used to test the level of toxicity of the fumes. A representative from the chemical plant was also sent to the command post to provide assistance in the decision-making process. The incident involves 200 responders and 50 vehicles, deployed across an area of 5 square miles.
- **Major wild land fire**. Scenario centred on a large wild land fire that occurred in 2003 in Southern California called 'The Old Fire'. This fire burned an area of more than 35 square miles while destroying 993 homes and causing six fatalities. Fanned by high winds and fuelled by abundant dried vegetation, this incident grew quickly and taxed the ability of local emergency officials to manage evacuations and road closures ahead of the fast-moving firestorm. At the incident peak, more than 1000 vehicles were on scene providing firefighting, security and support functions. Unlike the other three disaster incidents considered by NPSTC in which the peak activity period occurred in the first 90 min, this wild land fire incident experienced a peak activity period at around the 4-h mark. Fire and rescue units would arrive in the area of the reported fire and immediately conduct an assessment to determine the size of the fire, how quickly it was spreading, how quickly it was growing in size and intensity, what exposures were immediately threatened, and what exposures were soon to be threatened. They would make rapid decisions regarding the need for additional resources and start to implement an attack strategy. Law enforcement units would arrive at the command post and would be briefed on which areas and neighbourhoods were in immediate danger. Law enforcement representatives would start

directing the closure of certain roadways and initiate evacuation of targeted neighbourhoods. The incident involves over 2000 responders and 1000 vehicles, deployed across an area of 35 square miles.
- **Toxic gas leak**. This scenario was patterned after a report of a toxic gas leak in a large public assembly building near the National Mall in Washington, DC. While this type of incident has not actually occurred in the DC metro area, it is in their domestic security threat profile, and public safety response options have been practiced. Reports to the emergency call service (i.e. 911) indicate that dozens of citizens have collapsed inside the building, while hundreds of others are fleeing out into the streets. Citizens in the area are flooding 911 service with calls reporting some type of unknown emergency is occurring at the building. Additional calls are coming from inside the building reporting the location of downed persons. Responding units are receiving conflicting information on the type and extent of the emergency. PPDR units from multiple jurisdictions would arrive on the scene almost simultaneously given the nature of compact and overlapping jurisdictions and the distributed way in which the emergency would be reported. The first wave of units has arrived before any clear operational picture has been established. Fire, EMS and law enforcement representatives would establish an incident command post a safe distance from the scene of the emergency. Fire personnel in protective gear would move directly to the scene to start removing injured civilians. Hazmat personnel would be conducting a rapid assessment of the situation while also using sensor sniffer technology to identify the type of chemical involved. Law enforcement personnel would be working to seal off the area, preventing additional access by citizens into the danger area. Law enforcement personnel would also be interviewing those who had fled the building in an attempt to determine what happened. EMS personnel would be setting up triage and treatment areas, requesting additional transport ambulances and alerting area hospitals of the incident. The incident involves over 300 responders and more than 120 vehicles in 1 square mile area.

The list of applications identified as critical by the PPDR agencies was very similar across the four types of disaster scenarios. Remarkably, all agencies reported that access to GIS files was critical as well as the need to access real-time video feeds from the incident scene back to the incident commander. The list of data applications identified as being essential to emergency response and management was[2]:

- Access to third-party video/cameras (private and governmental)
- Automatic location (both vehicle and personnel location systems)
- Biomedical telemetry (patient and firefighter)
- GIS
- Incident command post-video conferencing
- Incident command white board
- Message and file transfer
- Mobile data computers application usage
- Patient/evacuee/deceased tracking

[2] It is worth noting that PPDR agencies involved in the NPSTC assessment indicated that the majority of these applications were needed for day-to-day operations in addition to becoming critical at the scene of a major incident.

Table 2.8 Data capacity required in disaster scenarios.

Scenario	UL/DL	Most capacity demanding applications.	Total data rate (peak traffic) (Mb/s)
Hurricane	UL	Incident video – medium-quality (UL) helmet/vehicle (38%) Incident video – high-quality (UL) helmet/vehicle (25%) Incident video – low quality (UL), situational (13%) Incident video – medium-quality (UL) video conference (13%) Patient and evacuee and deceased tracking (3%)	4.7
	DL	Incident video – high quality (DL) (aircraft) (29%) Incident video – medium-quality (DL) helmet/vehicle (22%) Incident video – medium-quality (DL) helmet/vehicle (15%) Incident video – high-quality (DL) helmet/vehicle (15%) Incident video – low quality (DL), situational (7%)	8.1
Chemical explosion	UL	Incident video – medium-quality (UL) helmet/vehicle (34%) Incident video – high-quality (UL) helmet/vehicle (22%) Incident video – low quality (UL), situational (22%) Incident video – medium-quality (UL) video conference (11%) Voice over IP cell phone access (2%)	5.2
	DL	Incident video – high quality (DL) (aircraft) (27%) Incident video – medium-quality (DL) helmet/vehicle (20%) Incident video – medium-quality (DL) helmet/vehicle (14%) Incident video – low quality (DL), situational (14%) Incident video – high-quality (DL) helmet/vehicle (14%)	8.6
Wildfire	UL	Incident video – medium-quality (UL) helmet/vehicle (29%) Incident video – low quality (UL), situational (25%) Incident video – high-quality (UL) helmet/vehicle (20%) Incident video – medium-quality (UL) video conference (10%) Voice over IP cell phone access (5%)	12.0
	DL	Incident video – medium-quality (DL) helmet/vehicle (23%) Incident video – low quality (DL), situational (19%) Incident video – high quality (DL) (aircraft) (16%) Incident video – high-quality (DL) helmet/vehicle (16%) Voice over IP cell phone access (4%)	15.2
Toxic gas leak	UL	Incident video – medium-quality (UL) helmet/vehicle (41%) Incident video – high-quality (UL) helmet/vehicle (27%) Incident video – low quality (UL), situational (14%) Incident video – medium-quality (UL) video conference (7%) Voice over IP cell phone access (3%)	8.6
	DL	Incident video – medium-quality (DL) helmet/vehicle (30%) Incident video – high quality (DL) (aircraft) (20%) Incident video – high-quality (DL) helmet/vehicle (20%) Incident video – medium-quality (DL) traffic camera (10%) Incident video – low quality (DL), situational (10%)	11.9

- Sensor technology
- Vehicle telemetry
- Video (aerial video feed, vehicle-mounted video and helmet camera video)
- Voice over IP cell phone access
- Weather tracking

Based on the reference values established for the peak throughput, session duration and sessions per hour to characterize the usage of each application (see Table 2.5), the number of users for each application was determined for each of the four incidents. Based on this characterization, the capacity requirements for video and data applications traffic rate estimated to serve PPDR operational needs in the four scenarios are summarized in Table 2.8, which provides the averaged traffic generated by all users during the busy hour both in uplink and downlink along with the five most capacity demanding applications in each scenario. The applications and usage considered in the NPSTC analysis reflect the applications that are used while personnel are physically located at incident command and while the incident is at its peak communications activity, being so representative of the busy hour usage, specifically within the first 2 h of units arriving on scene. As shown in the table, the highest aggregated traffic is originated within the wildfire response (15.2 Mb/s in downlink), with around 75% of the traffic associated with incident video.

References

[1] Analysis Mason, 'Report for the TETRA Association: Public safety mobile broadband and spectrum needs', Report no. 16395-94, March 2010.

[2] WIK Consulting and Aegis Systems, 'PPDR Spectrum Harmonisation in Germany, Europe and Globally', December 2010.

[3] ETSI TS 170 001 v3.3.1, 'Project MESA; Service Specification Group – Services and Applications; Statement of Requirements (SoR)', March 2008.

[4] National Public Safety Telecommunications Council, 'Public Safety Communications Assessment 2012–2022, Technology, Operations, & Spectrum Roadmap', Final Report, 5 June 2012.

[5] ETSI TS 102 181 V1.2.1, 'Emergency Communications (EMTEL); Requirements for communication between authorities/organisations during emergencies', February 2008.

[6] ETSI TR 102 745 V1.1.1, 'Reconfigurable Radio Systems (RRS); User Requirements for Public Safety', October 2009.

[7] CEPT ECC Report 199, 'User requirements and spectrum needs for future European broadband PPDR systems (Wide Area Networks)', May 2013.

[8] US NYC Study, '700 MHz Broadband Public Safety Applications and Spectrum Requirements', February 2010.

[9] US FCC White Paper, 'The Public Safety Nationwide Interoperable Broadband Network: A New Model for Capacity, Performance and Cost', June 2010.

[10] APT Report on 'PPDR Applications Using IMT-based Technologies and Networks', Report no. APT/AWG/REP-27, Edition: April 2012.

[11] IABG, 'Study of the mid- and long-term capacity requirements for wireless communication of German PPDR agencies', June 2011.

[12] ETSI TR 102 485 V1.1.1 (2006–07), 'Technical characteristics for Broadband Disaster Relief applications (BB-DR) for emergency services in disaster situations; System Reference Document', July 2006.

[13] Wireless Innovation Forum, 'Use Cases for Cognitive Applications in Public Safety Communications Systems – Volume 1: Review of the 7 July Bombing of the London Underground', November 2007.

[14] Wireless Innovation Forum, 'Use Cases for Cognitive Applications in Public Safety Communications Systems – Volume 2: Chemical Plant Explosion Scenario', February 2010.

3

Future Mobile Broadband PPDR Communications Systems

3.1 Paradigm Change for the Delivery of PPDR Broadband Communications

In most parts of the world, the prevailing model being used nowadays for the delivery of mission-critical narrowband PPDR communications (mainly voice-centric and low data rate services) can be well described by the following principles:

- **Use of dedicated technologies**. As discussed in Chapter 1, PPDR communications today rely mostly on the use of PMR technologies (e.g. TETRA, TETRAPOL, DMR, P25), most of them conceived in the 1990s in parallel with the second generation (2G) of mobile communications systems (e.g. GSM).
- **Use of dedicated networks**. Networks deployed for PPDR use are mainly private, dedicated networks that are built and operated with the specific purpose to serve the PPDR communications needs of a single agency or a number of them. Use of these systems by other non-PPDR users is rather limited or even not allowed in most cases.
- **Use of dedicated spectrum**. The operation of current narrowband PPDR networks is based on the use of dedicated spectrum, which is specifically designated for PPDR use.

This delivery model, while it has proven to be able to provide the PPDR users with the levels of control and high availability that they require, has resulted in a niche market with far less innovation and higher price points for communications equipment in comparison to the commercial wireless communications domain. It is worth noting that the adoption of such a delivery model was in practice the only way forwards at a time that digital technologies such as GSM had just started to boost the commercial mobile communications market. Clearly, GSM had not been conceived to fulfil mission-critical voice's needs (e.g. PPT, off-network

Mobile Broadband Communications for Public Safety: The Road Ahead Through LTE Technology, First Edition.
Ramon Ferrús and Oriol Sallent.

operation), nor were the commercial deployments in the 1990s thought to be able to provide the levels of coverage and availability, among other requirements, necessary for emergency communications. Therefore, the case for separate systems for commercial and emergency communications was considered a must at that time. Alongside the huge market size differences, innovation in the PPDR domain has also been constrained by limited, and sometimes fragmented, funding within the PPDR community. At the end, investment in PPDR communications systems is greatly dependent on public authorities' budgets (i.e. taxpayer's money) insofar as the PPDR sector is intimately connected to the public sector of society, either directly as part of the governmental structure or as a function that is outsourced under strict rules. In this context, funding for PPDR organizations is usually decided at political/government level, and budget for new radio equipment may be limited or approved in specific time frames. Furthermore, the budget is often allocated to different public safety organizations with little or no coordination to pool the demand and benefit from a higher buying power in the communications equipment procurement processes. Some consequences of this limited and fragmented budget have been the delayed network buildouts (some countries in Europe are nowadays still deploying TETRA networks) and long equipment life cycles (e.g. 10–15 years or more), clearly increasing the risk of technology obsolescence.

Connected with the funding problem is also the issue of designating dedicated spectrum for PPDR radiocommunications. The PPDR community usually expects the government to grant priority resources, such as exclusive spectrum. However, the allocation of additional dedicated spectrum to cope with the increasing PPDR traffic demand is certainly a challenging question for public administrations since the most suitable spectrum bands needed to build cost-effective PPDR networks with broadband capabilities are the same highly valued bands demanded by the market to provide commercial wireless communications.

Last but not least, this delivery model based on dedicated products, systems and spectrum has not contributed to override technical interoperability barriers among multiple separate dedicated systems that have been built just to serve the needs of a given agency or jurisdiction. Indeed, even in Europe where there has been a clear push towards the deployment of large-scale dedicated PMR networks shared by multiple PPDR agencies, interoperability problems are still a challenge to address, both at the national level (e.g. some countries have deployed regional networks that are not interoperable) and at the European level among PPDR organizations from different nations.

In this context, the following trends are driving today the evolution of PPDR communications in Europe [1], most of them being also applicable to other regions across the world:

- Voice communications has always been the main mission-critical application, but data communications are increasingly being used to support a number of PPDR data-rich applications. Furthermore, new PPDR applications require new uses and approaches for telecommunications. Some examples are ad hoc networks, sensor networks and support for high data rate ground–air links.
- Security challenges like terrorism and environment disasters have raised public awareness and increased the political support to enhance the capability and efficiency of PPDR organizations.
- Government entities, industry and regulators are advocating for a closer integration between public safety and commercial network infrastructures.

- The progress of the European integration is a driving force for a closer cooperation among PPDR organizations across Europe. As a consequence, there is an increasing support at the political level to remove (technical but also operational) interoperability barriers among national organizations or among European member states.

On the other side, conservative forces may obstacle the evolution of PPDR communications:

- PPDR organizations have already made large investment in dedicated networks. It is unlikely that these infrastructures are replaced with new technologies in the near future.
- Security and data protection are essential requirements in the PPDR domain. PPDR organizations have the concern that their data are safely protected from unwanted access by outsiders. Solutions to provide full interoperability may not be accepted if they do not provide adequate security.
- Radio-frequency (RF) spectrum is increasingly congested for an increasing number of services, and it may not be available for future technical solutions.

From the previous discussion, it becomes evident that new approaches to PPDR communications delivery are needed to be able to cope with the increasing demand for broadband and interoperable PPDR communications. Therefore, the 'dedicated technology/network/spectrum' paradigm shall be revisited and evolved towards new approaches, able to conjugate the following principles:

- **To pursue higher imbrication with the commercial wireless industry**. This is essential for the PPDR sector to achieve economies of scale and keep pace with technology evolution and innovation swiftly evidenced by the much wider commercial wireless industry.
- **To embed business sustainability criteria in the delivery of PPDR communications**. Synergies and cost-sharing approaches have to be found to leverage investments and combined efforts of the public sector for the delivery of PPDR services and the commercial and private sectors. In this respect, opportunities for network and spectrum sharing principles must be seized.
- **To achieve higher harmonization and cooperation among countries** for PPDR communications matters, facilitating cross-border operations as well as disaster responses requiring international assistance and security operations involving officials from a number of countries.
- All of the aforementioned, not compromising the **high control**, **security and resilience standards** required by the PPDR community, as currently being delivered through the dedicated narrowband communications systems in use nowadays.

3.2 Techno-economic Aspects Driving the Paradigm Change

As discussed in the previous section, the PPDR communications delivery model needs to evolve towards more effective and cost-efficient solutions while maintaining the high reliability standards required in PPDR communications. Changes in the way that PPDR communications systems are designed, managed and funded will follow. Changes in the market structure and roles of diverse market players can also be anticipated.

This section identifies and discusses the main techno-economic drivers that are anticipated to be pivotal in this evolution [2]. Remarkably, leveraging mainstream cellular technology is one of the key foundations for the next generation of PPDR mobile broadband communications systems. No less important for an effective and cost-efficient deployment of the PPDR mobile broadband service is seizing synergies among the PPDR and other sectors with regard to the provisioning and use of the two key assets needed to deliver mobile broadband PPDR services: the network infrastructure and the associated radio spectrum. The key techno-economic drivers across these three dimensions – technology, network and spectrum – are compiled in Figure 3.1. A detailed discussion on these drivers follows in the next subsections, providing some illustrative estimates from multiple studies that help to add perspective into the potential economic benefits that could be achieved.

3.2.1 Technology Dimension

Technological advances in the commercial domain have led to top-of-the-line radio technologies able to achieve performance levels close to Shannon's bound. The state of the art of commercial wireless technology evolution is Long-Term Evolution (LTE) mobile broadband technology, which has emerged as the leading standard in 4G technology. LTE networks first became publicly accessible in 2009 in Oslo (Norway) and Stockholm (Sweden). Nowadays, LTE networks are extensively deployed by most of the largest communications service providers worldwide for consumer-based data and information services. For the first time in the

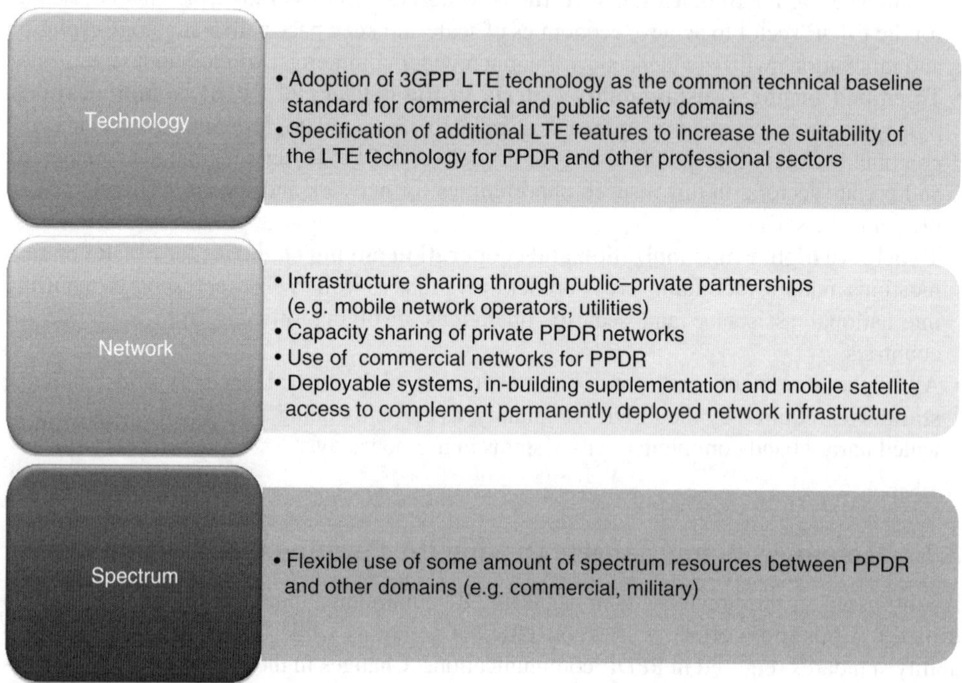

Figure 3.1 Techno-economic drivers for future PPDR communications.

cellular communications history, LTE has brought the entire mobile industry to a single technology footprint resulting in unprecedented economies of scale.

LTE technology provides a high bit rate, low-latency IP connectivity service that could be readily used to deliver many of the new demanded PPDR video and data-rich services.[1] With such a powerful technology already in place, the development of a completely new mobile broadband communications standard specifically conceived for PPDR data-centric communications could hardly be justified and would require far too much resources and time due to the complexity of modern communications technologies. A specialized public safety and critical communications technology cannot attract the level of investment and global R&D that goes into commercial cellular networks.

In this context, the adoption of LTE technology as the global standard for next-generation emergency communications broadband networks is gaining strong momentum among the PPDR community worldwide. Becoming part of the global LTE ecosystem is seen by the PPDR community as crucial to gain several advantages such as more choice of terminals, lower prices, possibility to roam to commercial networks and benefit in the long term with the adoption of further developments. Many organizations within the PPDR and critical communications community, such as the Association of Public-Safety Communications Officials (APCO) Global Alliance [3], the National Public Safety Telecommunications Council (NPSTC) [4] and the TETRA and Critical Communications Association (TCCA) [5], have explicitly endorsed the LTE standard as the baseline technology for the delivery of future critical communications broadband services. Remarkably, these efforts are not isolated from each other but seek to benefit from wide consensus building and coordination. In June 2012, the TCCA and the NPSTC announced that they had signed a memorandum of agreement (MOA) to underscore their joint commitment to the need to develop mission-critical public safety communications standards for LTE-based technology. Alignment of the critical communications and PPDR community to a common global standard is expected to create a rich ecosystem of devices and applications spurred by the standard-based designs, open intellectual property environments, commitments from chipset manufacturers, large communities of developers and interest from consumer electronics manufacturers.

A key milestone in this process was the involvement of the 3rd Generation Partnership Project (3GPP),[2] which embraced the initiatives coming from the PPDR community and committed to deliver the necessary standard enhancements to make the LTE system more suitable for this purpose. Specific standardization work related to PPDR communications started under the Release 12 of the LTE specifications with wide support from the mobile industry and involvement of the PPDR stakeholders. Standards for the first batch of PPDR requested features are expected to be completed under Release 13, which is planned to be frozen by March 2016. Among the key new features incorporated into the standard are the support for device-to-device (D2D) communications and group communications enablers. It is worth highlighting that these extensions of the LTE specifications are not only relevant for the PPDR sector but also important to raise new business opportunities in the commercial and other professional sectors (e.g. transportation, utilities, government). As expressed by 3GPP officials [6], there is a need to strike a balance between more and less customization, to make use of commercial products while meeting the specific requirements for PPDR and critical

[1] See Chapter 2 for a description of the data-rich applications in demand by the PPDR community.

[2] 3GPP is the organization in charge of developing the LTE standard. The official website is at www.3gpp.org.

communications. Therefore, the approach being followed by 3GPP is to preserve the strengths of LTE in the commercial domain while adding the features needed to support critical communications and so seeking to maximize the technical commonality between commercial and critical communications aspects. A detailed description of the LTE features that are being working out within 3GPP to address the specific PPDR communications needs is covered in Chapter 4.

The alignment to commercial technologies offers huge opportunities for creating and exploiting synergies between these two worlds, which have remained virtually separated to date. Remarkably, the use of common technical standards for commercial cellular and PPDR offers advantages to both communities:

- PPDR community gets access to the economic and technical advantages generated by the scale of commercial cellular networks. Using equipment developed for the mass market instead of niche products, the PPDR community will profit from the economies of scale, faster innovation and high competition between vendors. The same applies for the market of end-user devices and dedicated software, where even stronger competition should be expected.
- The commercial cellular community gets the opportunity to address parts of the PPDR market as well as gaining enhancements to their systems that have interesting applications to consumers and businesses.

LTE devices supporting critical emergency services will need to support many of the features and design considerations used today in PMR products, including high-performance batteries, radio, antenna and audio; rugged components and enclosures and ergonomics based on 'high-velocity human factor' industrial design. Despite this necessary level of customization with respect to consumer devices, economies of scales are expected to bring down the cost of those parts of the devices and network equipment that add the most to the bill of materials (BoM). Table 3.1 shows the degree of commonality that the components of a PPDR device are expected to have with respect to commercial devices [7]. As reflected in the table, the components that add the highest customization costs (operating system, baseband chipset and RF chipset) are anticipated to be fully leveraged. Nevertheless, these expected benefits are still to be proven in practice, and they have been put into question by some within the PPDR community [8]. Indeed, similar claims were made for the GSM-Railway (GSM-R) technology,

Table 3.1 Components and effect of customization on cost of LTE devices for PPDR use.

Component	Hardware	Software/ middleware	Operating system	Baseband chipset	RF chipset	RF front end
Degree of commonality to commercial devices	Medium	Medium	100%	100%	100%	Low
Effect of customization on cost	Low	Medium	High	High	High	Low

which was built upon the successful commercial GSM standard and aimed at being a cost-efficient digital replacement for previously existing incompatible in-track cable and analogue railway radio networks in Europe. As a matter of fact, the requirements for GSM-R terminals were sufficiently different from standard terminals, and the volumes sold so small that they have eventually worked out much more expensive, even exceeding the price points of comparable terminals designed for the TETRA market. This is because the niche GSM-R market has far fewer terminals and many fewer competing suppliers than that of the TETRA market. Therefore, lessons learnt from the GSM-R case should not be disregarded and be used to achieve a truly cost advantage in future PPDR-grade LTE equipment. In this respect, some manufacturers have started unveiling LTE-capable, mission-critical handheld devices (mainly for the US market) with announced price points around $1000, which is above high-end commercial smartphones but lower than typical high-end PMR devices that are in the range of €2–4K.

3.2.2 Network Dimension

A wide consensus exists among PPDR users on the need of dedicated network infrastructures for mission-critical PPDR communications. This is the main approach followed so far with most of the current PMR networks that serve the PPDR community worldwide. Nevertheless, given that the support of data services dramatically increases the number of required cell sites in comparison to current narrowband network footprints, huge investments are required to roll out dedicated mobile broadband infrastructure, which may not be deemed convenient or even affordable to some public administrations.

Generally, the so-called total cost of ownership (TCO) of a mobile cellular radio network over the long term includes both the cost to build the network (capital expenditures (CAPEX)) plus the cost of keeping the network up and running (operational expenditures (OPEX)) over 10–20 years as well as the depreciation of assets. That may be taken over 3–10-year amortization, depending on the item. The major cost elements to be taken into account in any TCO analysis include [9]:

- CAPEX elements:
  Network sites to host the base stations (BSs), transmission and switching equipment, network operation centres, etc., with backup sites as needed
  Network elements: radio and transmission equipment, gateways, internal cabling, etc., with backup dual units
  Civil works and cable laying for backhaul and core network
  Civil works for site building, mast erection, etc.
  Power supply infrastructure and heating, ventilation and air conditioning (HVAC) systems
  Mobile terminals (handsets) and vehicle terminals (specialized or generic)
  Wayleaves for backhaul and core network ducts, with alternative routing
  Data centre with infrastructure, equipment and software licence
  Spectrum licences (if applicable; if payable with annual fees, then under OPEX, as well)
  Operational support systems for network management and telecommunications management network
  Business support systems
  Back-office and front-office centres and equipment (legal, accounting)

- OPEX elements:
  Operational, maintenance, maintenance teams 24×7, design and development staff/high-resilience (HR) cost
  Power supply infrastructure and HVAC with uninterruptible power supply (UPS) lasting for several days and operation with main supply charges
  Data centre infrastructure and operations, equipment, software maintenance, energy costs and annual software licences
  Operate administration, back office (payroll, legal, accounts) and front office ('sales', etc.) with all staff costs
  Operational costs of hardening, including extra security, power and equipment maintenance and site protection
  Cost of capital

Approximately, (up to) 70% of any mobile cellular network cost is the radio access network (RAN). Much of the RAN cost is not the radio and transmission equipment but the site real estate (either rented or purchased). In addition to that is the cost of wayleaves for passing backhaul ducts and cabling into the core network. Another reason for the high cost associated with a PPDR network is the large amount of required redundancy: extra switches maintained in 'standby' mode, extra transceivers at BSs in key locations, multiple backhaul paths to bypass link failures and batteries and generators to provide electricity when main power is lost. Another form of redundancy (used in TETRA) is that BSs have overlapping coverage to ensure continuity of service if a BS fails. All these redundancies are intended to make the networks highly resilient, to achieve high availability and sustain PPDR services when other communications systems fail. For LTE to provide PPDR users with a similar level of network resilience, similar measures will be needed.

The quantification of the costs of a nationwide mobile broadband PPDR network has been the subject of several studies, most from the United States. A possible model to estimate the cost for such a network is developed in Ref. [10]. The model is based on the computation of the required number of cells in the network, since the overall network cost is considered to be roughly proportional to the number of cell sites. The number of cells in a network depends on the maximum cell size that is, in turn, dependent on many factors such as the terrain characteristics, subscriber density in the served area, capacity requirements per subscriber for routine PPDR operations, aggregate capacity required in an emergency response, minimum data rate requirements at cell edge and other important link budget parameters (e.g. terminal's maximum transmit power, coverage reliability margins). The authors in Ref. [10] apply the model to estimate the cost of a network that would cover 99.998% of the US population (equivalent to roughly 83% of the US geographic area) under three scenarios whose distinguishing factors are the frequency band and the amount of available spectrum. The three considered network scenarios are (i) 10 MHz of spectrum in the 700-MHz band, (ii) 7.5 MHz of spectrum in the 168-MHz band and (iii) 7.5 MHz of spectrum in the 414-MHz band. Each scenario is analysed under three different traffic configurations: (i) voice only, (ii) data only and (iii) data and voice. The data-only scenario would be appropriate if PPDR agencies continue to rely on their existing wireless systems for voice communications, while the voice and data scenario would eventually allow PPDR to phase out existing systems and rely on a single network to support all the communications. Table 3.2

provides a summary of the total number of cell sites required, the upfront deployment costs, the recurring annual costs and the total costs calculated over a 10-year period for each of the three network scenarios and traffic targets.

The costs in Table 3.2 are calculated considering an estimate of \$500K per site in upfront deployment cost (i.e. CAPEX) and \$75K per site in annual operating costs (i.e. OPEX). Cost estimates only consider the costs associated with the installation and operation of cell sites and not the costs of the backbone network components or the costs of network planning and administration. Also, handset costs are not part of the infrastructure and therefore are not included. Cost estimates in Table 3.2 are mainly indicative values and can change dramatically by adjusting a few critical input parameters such as the coverage area's signal reliability, building penetration margin, aggregate capacity required in emergencies, highest user data rate required and population/area buildout requirements. In fact, there is no widely accepted model to assess the capacity requirements for a PPDR network, which is essential for developing these cost estimates. Therefore, more work is required in this area to fix these critical input parameters. In fact, the estimated total network costs in Table 3.2 are roughly 30–50% less than other estimations [10, 11], but the differences cannot be explained because many assumptions done are not publicly available. In any case, a conclusion that can be reached from Table 3.2 is the significant effect that the selection of the frequency band has on the overall number of sites and so in the network cost. As drawn from results in Ref. [10], the network at 168 MHz requires roughly 30–50% fewer cell sites and costs roughly 30–50% less than a comparable network operating at 414 MHz, with all other factors held constant. Besides, the results from Table 3.2 also indicate that moving from a voice-only PPDR system to a data-only system dramatically increases the number of cell sites required. However, moving from data only to both data and voice has a much smaller impact. This can facilitate the convergence of legacy PMR services and emerging data-intensive (e.g. PPDR/multimedia)

Table 3.2 Cost analysis of building a nationwide network for PPDR in the United States.

Frequency band (MHz)/ bandwidth (MHz)	776/10 MHz	168/7.5 MHz	414/7.5 MHz
Voice traffic only			
Total number of cells required	3 700	1 000	1 900
Upfront deployment cost (M\$)	1 900	500	950
Operating costs (M\$)	280	75	140
Ten-year total cost (M\$)	3 400	910	1 700
Data traffic only			
Total number of cells required	18 200	6 200	10 700
Upfront deployment cost (M\$)	9 100	3 100	5 400
Operating costs (M\$)	1 400	470	800
Ten-year total cost (M\$)	16 600	5 700	9 800
Data and voice traffic			
Total number of cells required	22 200	12 300	18 400
Upfront deployment cost (M\$)	11 100	6 200	9 200
Operating costs (M\$)	1 700	900	1 400
Ten-year total cost (M\$)	20 300	11 200	16 800

services into the same mobile broadband network infrastructure in a long term and so avoid duplicated infrastructures for voice on one side and data on the other.

According to the previous cost estimates, the TCO per user and per year would fluctuate for a network in the 700-MHz band in the range of $600–2000 considering a PPDR subscriber base between 1 and 3 million [11]. Just for illustrative purposes, it's worth noting that these values are roughly higher than the current TCO per user per year figures for nowadays TETRA networks in Europe, as captured in Table 3.3. Costs estimations given in this section shall be considered only as indicative values of the order of magnitude of the investment required to deploy new dedicated broadband networks both in absolute and relative terms compared to current narrowband deployments. As reported in a recent study for the European Commission [9], accurate costs of current PPDR narrowband networks vary considerably, and sometimes, they cannot be properly estimated from the published accounts from governments or PPDR network operators. An indicative measure provided in this study is that EU member states plus Norway have spent over €14.6B deploying TETRA and TETRAPOL networks for PPDR, plus almost €4B on mobile and portable terminals, and they spend an additional €1.35B each year operating these networks (possibly a low estimate). About 23 450 BSs serve over 1.5 million users nowadays in Europe, for an average of 64 users per BS. However, since there are almost 5 million police, fire, EMS and rescue workers, there must be a great deal of equipment sharing (three shifts in 24 h) or continuing use of other mobile networks.

While the market has witnessed these levels of investment in the past due to the imperative need to provide the PPDR community with mission-critical voice communications, budget constraints faced by many public administrations in the current (and foreseeable future) economic climate make unlikely to see a generalized adoption of a delivery model based on the deployment of new 'stand-alone' LTE-based dedicated networks to cope with the demand for data-intensive applications. In this context, four key cost-saving dimensions are expected to be properly seized when deploying dedicated capacity for PPDR: (i) infrastructure sharing through public–private partnerships, which can allow the deployment of the PPDR dedicated network based on marginal costs of adding or contributing new equipment into an existing infrastructure as opposed to the deployment of a stand-alone network; (ii) capacity sharing of private PPDR networks, so that users other than PPDR agencies (e.g. utilities, transportation) can be served and charged for the use of the network; (iii) use of commercial networks' capacity as an integral part of the PPDR communications solution, which can contribute to alleviate coverage and capacity requirements in the deployment of dedicated capacity; and (iv) use of transportable/ fast deployable equipment, which can be central to lowering the amount of permanently deployed network infrastructure and provide a cost-efficient solution to face localized capacity

Table 3.3 Costs of current dedicated TETRA networks.

Network	CAPEX (M€)	OPEX (M€)	TCO (M€)	Users	TCO/user per year (€)
Virve (Finland)	134	222	356	50 000	475
ASTRID (Belgium)	99	259	358	40 000	596
Airwave (United Kingdom)	952	2649	3601	200 000	1200

Reproduced from Ref. [12].

surges, improved coverage in underserved areas and increased redundancy. A further insight into each of these four cost-saving dimensions is addressed in the following.

3.2.2.1 Infrastructure Sharing through Public–Private Partnerships

Infrastructure sharing with private partners (e.g. mobile network operators (MNOs), utilities) arises as a central cost-saving dimension, especially by bringing down the site acquisition cost that is one of the biggest cost contributors. Examples of infrastructure sharing in the commercial mobile communications domain can be found in both mature and developing markets and can constitute a solid starting point for the development of new public–private collaboration models with the PPDR domain [13]. Network sharing may take many forms, ranging from passive sharing of cell sites and masts to sharing of the RAN and other active elements of the mobile core network.[3] MNOs, who already have a massive infrastructure deployed, are clear candidates for infrastructure sharing. In addition, utilities also are considered a promising partner for a shared deployment, and use, of a mission-critical network because they have a lot of infrastructure in place (e.g. towers, power, communications backhauling facilities) that can be leveraged to add the dedicated network equipment [14]. Thus, infrastructure sharing that can allow the deployment of the PPDR dedicated network based on the marginal cost of adding a new RAN for PPDR access to an existing tower or site (which already has backhaul to a functioning core network) and hardening the site can be a cost-effective option as opposed to the deployment of a fully stand-alone network.

An estimation of the cost savings that could be attained through infrastructure sharing with private entities was conducted by the Federal Communications Commission (FCC) [11]. An incentive-based partnership model is assumed for the estimates, under which public safety network operators will partner with commercial operators or system integrators to construct and operate the network using dedicated public safety broadband spectrum. Under this model, the vast majority of sites are built by a commercial partner, either a wireless operator, an equipment vendor or a system integrator. The model assumes a 700-MHz LTE network. Costs include installing and operating the dedicated 700-MHz RAN and sharing backhaul and IP core transport systems, including ancillary and support systems and services. The IP network architecture enables public safety agencies to have their own dedicated servers for applications and services requiring high levels of security and privacy. The costs to deploy the public safety network following the incentive-based partnership model are compared to the cost of building a stand-alone public safety network. The technical requirements and capabilities under both approaches are identical, considering in both cases a total number of cell sites close to 45 000, 80% of them being new builds. On this basis, the incentive-based partnership model considers the marginal cost of adding a new RAN for public safety to an existing tower or site, which already has backhaul to a functioning core network and hardening costs of the tower or site. In contrast, for the stand-alone network, the full cost for public safety capabilities is accounted. Therefore, the main differences between the two models emerge in the cost per cell site in both CAPEX and OPEX, the costs in zoning and site acquisition (because of the need for many more new cell sites beyond the base required for public safety narrowband PMR networks), the costs of backhaul from the cell sites and the costs for a core network. Table 3.4 compares the costs of these two approaches in terms of CAPEX. Overall, the FCC's analysis yields cost-saving estimates for a 10-year period by at least 60% considering both CAPEX and OPEX.

---

[3] The technical capabilities built in the LTE standard for RAN sharing are described in Chapter 4.

Table 3.4 Comparison costs of the incentive-based partnership and stand-alone network deployment models.

Network	Incentive-based partnership	Stand-alone
Number of sites	44 800	44 800
Cost of an urban upgraded site (K$)	95	164
Cost of an urban new site (K$)	N/A	223
Cost of an suburban upgraded site (K$)	95	213
Cost of an suburban new site (K$)	N/A	288
Cost of an rural upgraded site (K$)	216	247
Cost of an rural new site (K$)	363	394
Total CAPEX for sites including hardening (B$)	6.3	12.6
Backhaul – installation to core fibre ring, non-rural sites (B$)	0	2.1
IP core equipment, network operation centres (B$)	0	1.0
Total CAPEX	$6.3B	$15.7B

Reproduced from Ref. [11].

A similar study addressed by Bell Labs research [14] also points out considerable cost savings in the range of 40–50% achievable through infrastructure sharing. In this line, the First Responder Network Authority (FirstNet), who is backed with $7B in federal grants for the deployment of the LTE-based nationwide network in the United States, early pointed out that building a stand-alone network is likely to be impractical from a cost standpoint, and rather, ways to leverage the existing US wireless mobile communications infrastructure through partnerships with network operators shall be explored (for comparative purposes, cumulative US wireless network investments are estimated over $350B) [15]. According to FirstNet officials, up to 70% of the cost of the network could be in cell site locations.

3.2.2.2 Capacity Sharing of Private PPDR Networks

Building a dedicated network for PPDR while allowing excess capacity to be used by users other than PPDR agencies (e.g. utilities, transportation) can result in further cost savings to public budget in terms of new revenue streams from the private sector. This is considered a feasible approach owing to the fact that PPDR users are not expected to use all of the available capacity in the dedicated network (PPDR communications systems are typically designed under worst-case capacity assumptions, and fortunately, most of the time, these large-scale incidents that would put the capacity under stress are not taking place).

A joint-use network for PPDR and other mission-critical users is a possible realization of this type of capacity sharing [9, 16]. Clearly, deploying a joint-use network can result in further cost savings as opposed to rolling out separate networks for different types of users. Nevertheless, being able to balance the likely misalignment of requirements that different types of users may have (PPDR, utilities, transportation) and guaranteeing fair competition with private mobile operators that can also be competing for some of these professional/business users are hurdles for this approach. A clear example is electrical utilities in their efforts to transition towards smart grids, encompassing an upgraded energy network to which two-way digital communication between supplier and consumer, intelligent metering and monitoring systems have been added. A consolidated approach to support the critical communications services in smart grids is still an open point, and a PMR type of network implemented on a shared platform basis is among the potential solutions (as pointed out in the public consultation launched by EC in February 2012 on the use of spectrum for more efficient energy production and distribution [17]).

Economic estimates for the deployment of a joint-use network are available for the case of network capacity sharing between PPDR and commercial traffic, though this approach shows even higher difficulties when it comes to balance misaligned requirements and conflicting interests of PPDR and commercial traffic. In such a joint commercial PPDR network, most of the time, the overall capacity will be mostly available for commercial traffic (excepting that capacity needed for PPDR routine operations), while PPDR users will be allowed to use a higher fraction of the overall capacity to satisfy increased capacity requirements that might arise in an emergency response. A first attempt to promote the deployment of a network to serve both PPDR personnel and commercial subscribers on the same spectrum and infrastructure was carried out by FCC in the United States. Back in 2007, when the FCC designated the first 10-MHz portion of the 700-MHz spectrum band specifically for public safety broadband use, the FCC also created a 10-MHz commercial licence for the spectrum adjacent to the public safety allocation, the so-called D-Block. The D-Block was auctioned in 2008 under the condition that the auction winner would have been obligated to build a nationwide public safety-grade network on the 20 MHz of combined spectrum to be shared by public safety and commercial users. This was done in an attempt to have a commercial entity fund and build out a public safety-grade network in exchange for discounted access to spectrum. This auction concluded without a winning bidder emerging, a fact that has been widely attributed to the considerable uncertainty about the requirements that would be placed on such a PPDR-grade network. In this context, the cost of such type of a joint-use network that uses 20- of 700-MHz spectrum to serve both commercial subscribers and PPDR personnel was analysed in Ref. [10] and reproduced in Table 3.5. In this analysis, the design requirements assumed for the joint-use network were the same as those considered for a PPDR-only network option built to the PPDR standards. The evaluations conducted focused on the estimation of the number of cells and network costs of a PPDR-only network and a joint-use network, carrying both data and voice traffic. Considering a market penetration of 10%, the number of sites of the joint-use network is shown to be 12% less than that of a PPDR-only network. The reason is due to the higher aggregated capacity that can be supported by the cells of the joint-use network, thanks to the availability of the additional spectrum. Therefore, authors in Ref. [10] claim that the total cost can be brought down by close to 15% (from $20.3B to $17.7B), from which PPDR user should only cover a fraction and the rest will be borne by the commercial provider. The business case of this joint-use network that would result from a kind of public–private partnership between PPDR agencies and a commercial network operator has been analysed by the same authors in Ref. [18].

3.2.2.3 Use of Commercial Networks' Capacity

While the cost-saving approaches discussed in the previous two subsections are sustained in the principle that PPDR broadband communications are solely supported on a private dedicated network (potentially sharing infrastructure and/or enabling access to other type of users),

Table 3.5 Number of cells and network costs of a PPDR-only network versus a joint-use network.

Type of network Frequency band (MHz)/bandwidth (MHz)	PPDR-only network 776/10 MHz	Joint-use network 776/20 MHz
Total number of cells required	22 200	19 400
Upfront deployment cost (M$)	11 100	9 700
Operating costs (M$)	1 700	1 500
Ten-year total cost (M$)	20 300	17 700

Reproduced with permission from Ref. [10]. © Elsevier.

a radically different approach consists of directly using the capacity offered by the commercial mobile broadband networks for the provisioning of PPDR services. Though conceptually very different, using commercial and dedicated networks is not mutually exclusive but complementary. Indeed, relying on commercial network capacity can be the first step for a reduced time to market and investment, while a dedicated network can be progressively deployed in specific areas and be used together with the commercial capacity in a longer term [19].

This approach can be realized through the introduction of business agreements and technical solutions able to satisfy the more stringent requirements in terms of control and reliability demanded by the PPDR community. In this respect, the adoption of a mobile virtual network operator (MVNO) model by a PPDR service provider where the critical control functions remain in the MVNO's hands (e.g. PPDR subscriber management, security, policy control, etc.) constitutes a plausible solution for the exploitation of the commercial capacity, as developed in EC FP7 HELP Project [20, 21]. A key constituent element of this solution is the support of prioritization services for PPDR applications in the commercial network. Indeed, prioritization capabilities already specified in LTE technology constitute a powerful framework to manage capacity allocation when congestion arises [22], as could be the case in large crisis in populated places where commercial capacity can become saturated. A more detailed description of the MVNO model is addressed in Chapter 5.

The cost of provisioning PPDR services through commercial networks' capacity will depend on a large extent on the service-level agreements (SLAs) established between the MNO and the PPDR organizations (or a PPDR service provider on their behalf). SLAs will define, among others, the functional and technical aspects (e.g. service availability, prioritization capabilities, etc.) that the MNO must satisfy and the cost for the service. If the agreed SLAs are similar to those being provided by the MNO to its commercial users, provisioning of capacity for PPDR services mainly constitutes a new stream of revenue for the MNO as those with other business users (e.g. enterprises, transportation, etc.). On the other hand, if SLAs target to increase the degree to which PPDR users can rely on commercial networks (which PPDR will certainly ask for), this will turn into an impact on both CAPEX and OPEX of the MNOs. For example, the capabilities needed to implement priority access are optional features within LTE-based equipment so that their deployment will bring additional costs for the acquisition/upgrading of network equipment as well as increased costs for the operational management of these capabilities.

The costs of implementing prioritization need to be considered in the MNO's business model and eventually be transferred to PPDR users (e.g. accounted in the service fees) and/or government (e.g. public funding for the deployment of prioritization capabilities in the commercial networks). Beyond the consideration that the solutions for the delivery of priority access services should not bring financial risks to MNOs, it is worth noting that the activation of priority access in an emergency response can reduce the amount of capacity available to support citizen's communications in a moment that basic mobile communications services are most valuable to the citizens. Therefore, the addition of specific clauses in customers' contracts to describe the acceptable level of service degradation in emergency conditions might be required. In addition, MNO's potential loss of revenue from citizens due to the activation of priority access in a crisis situation might be covered by means of insurances in order to offset economic impacts on an MNO providing priority access to PPDR users.

Estimations of the costs of deploying priority access in commercial networks have been addressed in Ref. [23]. The analysis considered Spain as a reference country and assumed the deployment of priority access to voice calls (not mobile broadband) in the four Spanish commercial deployed networks. Estimations were in the range of €50M for CAPEX and €2–5M for operating costs. Regarding the service charges, illustrative values considered in the Spanish case were an annual subscription fees of around €50 per user and a flat fee per call of less than €0.20. Indeed, different business models were analysed in Ref. [24] depending on the involvement of the government, PPDR users and MNOs in the delivery of priority access:

- 'OnlyOp' model. In this model, it is considered that countries' governments should be responsible for providing the needed infrastructure for priority access, covering the associated CAPEX, while network operators should take care of OPEX based on their market position. All MNOs in a country implement the service and compete to gain market share. The revenue sources for MNOs from PPDR users are the activation/subscription fee, call-based cost and feature charge (per month per user). PPDR agencies can select the most advantageous operator since all options are available at the same time.
- '3Shared' model. As mentioned previously, all MNOs implement the service and compete to gain market share. However, CAPEX and OPEX are now to be shared among country governments, PPDR users and MNOs (e.g. government covers 50% of CAPEX and no OPEX, PPDR organizations cover 50% of OPEX and no CAPEX, and MNOs cover the remaining 50% of CAPEX and OPEX). In this case, since PPDR users participate in the cost distribution, activation/subscription fees might not apply and PPDR agencies are supposed to be paying lower service rates with respect to the 'OnlyOp' model.
- 'Exc' model. This model considers that public open contests are used to select the MNO(s) that implements the priority access service. Hence, only those MNOs interested in deploying the service apply to the call, and the decision award criteria have then to consider the distribution of the costs among the three parties and also the associated operational gain. This model would substantially decrease the amount of money to be spent by the government for enabling the service.

The most suitable model to be adopted depends on the specific conditions of the country under consideration and the different revenue sources that can be applied (subscription fees, call per minute costs, flat rates, etc.). In the case that public administration opt to assign public funds for priority access service provisioning, public administration would have the obligation to avoid any situation that could infringe free competence in the market.

Possible pricing options for priority roaming of PPDR users on commercial networks have been analysed in Ref. [25]. As argued by the authors in Ref. [25], enabling priority roaming could potentially lead to problems if PPDR agencies have no incentive to use commercial capacity efficiently or if roaming during unexpected events leads to costs that well exceed annual budgets or if PPDR roaming traffic reduces commercial revenue by displacing commercial subscriber usage and/or leading to increased subscriber churn. In this context, the analysis conducted in Ref. [25] shows that these risks are small or can be mitigated by the choice of the right pricing scheme. A key conclusion of the analysis is that a hybrid pricing scheme resulting from the combination of usage-based and flat-rate pricing is seen as the most suitable approach. In particular, the envisioned hybrid scheme would employ usage-based pricing as the default (i.e. for normal operation and during localized emergencies) and a flat-fee scheme invoked when a serious disaster occurs. This could mitigate the potential harm from a usage-based pricing scheme (i.e. harmful rationing during large-scale disasters) while still preserving the incentives that the usage-based pricing provides during more routine use and mitigating the risk of lost usage-based revenue for commercial carriers.

Schemes for privileged access to commercial cellular networks have already been adopted in some countries, though limited to voice communications. Examples of these systems are the Wireless Priority Service (WPS) used in the United States and the Mobile Telecommunication Privileged Access Scheme (MTPAS) deployed in the United Kingdom. Some technical details regarding these systems are addressed in Chapter 5. With respect to the business models currently adopted, their main characteristics are:

- US WPS:
  Government pays all infrastructure costs needed for the implementation of the service in the networks.
  Participation of MNOs is voluntary.
  Users have to pay for the priority service (upper and lower bounds of service costs established by the government). The WPS charges include an activation fee of up to \$10 and a monthly feature cost of no more than \$4.5, plus a usage fee of no more than \$0.75/min when WPS is invoked [26].

- UK MTPAS:
  Government pays all infrastructure costs needed for the implementation of the service in the networks.
  Participation of MNOs is mandatory.
  Users do not have to pay any additional cost. Mobile operators do not profit from providing the service.

3.2.2.4 Use of Transportable/Fast Deployable Equipment

Deployable systems (e.g. transportable radio BSs and network equipment) and in-building supplementation (e.g. distributed antenna systems) are additional complementary approaches that can help lowering the amount of permanently deployed network infrastructure and provide a cost-efficient solution to face localized capacity surges, improved coverage in underserved areas and increased redundancy.

A PPDR network dimensioned for a given coverage and capacity with the ability for ad hoc deployment of new network elements could be able to handle the traffic increase due to a temporary concentration of users or just to cover underserved areas that are seldom attended by users. Thanks to solutions based on mobile ad hoc deployable network elements, the network operator does not have to keep on a fully operating network for all time in the areas of no or very low utilization but extend capacity or coverage on demand to cover PPDR needs. This approach is likely to significantly reduce network CAPEX as well as result in operational savings since backhaul expenses may only be incurred on an ongoing basis. Furthermore, deployable systems are likely to play a central role during the early days of a wide area network (WAN) buildout, when fixed infrastructure coverage is expected to be spotty. A description of the types and key technical features of fast deployable equipment from a network architecture point of view is addressed under Chapter 5.

An analysis that quantifies the reduction in terms of the total number of BSs needed for building an economic nationwide public safety broadband network based on a combination of fixed and mobile BSs is provided in Ref. [27]. The wireless access points of the network consist of sparsely deployed stationary BSs for supporting light routine traffic and a distributed set of mobile BSs ready to be deployed quickly to any incident scene by vehicle or helicopter. A premise of the architecture is that a mobile BS can be dispatched to the incident scene as quickly as a large number of personnel and can be set up quickly to provide the wireless services needed. This imposes requirements on the density and placement of mobile BSs as well as the technologies that link the mobile BSs to the fixed infrastructure through, for example, a wireless backhaul. The proposed architecture in Ref. [27] is compared to a conventional architecture, in which the cell sites are designed to satisfy the throughput requirement due to the sum of both light routine traffic and heavy incident scene traffic. The analysis shows that the proposed architecture can potentially offer over 75% reduction in terms of the total number of BSs needed. In this respect, the advances in the so-called LTE small cells will undoubtedly benefit the development of this sort of mobile BS solutions. Indeed, small cells are increasingly seen as a key enabler for bringing LTE to the public safety sector. Due to their portability and the availability of virtualized/embedded mobile core solutions, small cells can help to establish and maintain communications even when there is no access to the core infrastructure. It is estimated in Ref. [28] that military, tactical and public safety LTE small cell shipments will account for over $350M in revenue by the end of 2020, following a compound annual growth rate of 45% over the 6-year period between 2014 and 2020.

Another type of fast deployable equipment is the relay node (RNs). An RN works as a low-power BS that can be deployed under the coverage of another BS (e.g. a high-power macrocell BS) to extend the coverage at cell edge and/or increase capacity at localized areas. An RN is connected to a BS (called the donor BS) via the radio interface and then provides access to user devices as usual (i.e. devices 'see' the RN as a normal BS). RNs are expected to be a cost-efficient way to fulfil requirements on high data rate coverage in next-generation cellular networks, like LTE networks. From a cost point of view, the main differences between an RN and a BS are that equipment costs for RNs as well as site costs (sites for RN can be, e.g. lamp posts) are in general less expensive and that RNs do not generate additional backhaul costs (backhauling use the air interface resources of the donor BS). This advantage of RNs is partly compensated by the fact that several RNs are needed to achieve the same service level that can be achieved with one conventional BS. In the context of a commercial LTE network, the business case of network scenarios with and without RNs characterized by the same service level (also referred to *isoperformance* scenarios) has been addressed in Ref. [29]. The methodology chosen was to compare the TCO of different deployments with and without LTE RNs in a coverage-limited scenario. The results provided show that the use of mid- and high-power LTE RNs (33 and 38 dBm) can yield to the operator a cost saving of 30% and more. The reasons for the cost benefit are the site-related costs, which are dominated by civil work costs associated with the opening up of new sites in the case of deploying new conventional BS.

3.2.3 Spectrum Dimension

Clearly, the deployment of LTE-based dedicated systems raises the issue of identifying the spectrum band(s) and spectrum management model(s) on which these systems can be deployed and operated. Even though the inherent spectrum flexibility associated with LTE technology (i.e. support of different frequency bands, transmission bandwidths from 1.4 to 20 MHz, carrier aggregation and both frequency and time division duplexing arrangements) will be a facilitator, political, regulatory and economic facets will have greater influence on the final solutions to be adopted.

The allocation of dedicated, exclusive-use spectrum has been so far the traditional approach to support PPDR communications. From a purely technical and operational perspective, an exclusive allocation of spectrum for PPDR is the preferred option of the PPDR community because it provides them with full control over the resource. Nevertheless, the allocation of enough dedicated spectrum for PPDR radiocommunications is a challenging issue for public administrations: suitable spectrum bands needed to support cost-effective PPDR communications with broadband capabilities are the same highly valued bands demanded by the market to provide commercial services. This creates the necessity of having a proper economic valuation of the spectrum.

As for any resource, including radio spectrum, the primary economic objective is to maximize the net benefits to the society that can be generated from that resource. Prices are used as an important mechanism to ensure users use the spectrum resources efficiently. The broad goals and objectives associated with spectrum pricing are [30]:

- Covering the costs of spectrum management activity borne by the spectrum management authority or regulators
- Ensuring the efficient use of the spectrum management resource by ensuring that sufficient incentives are in place
- Maximizing the economic benefits to the country obtained from the use of the spectrum resource
- Ensuring that users benefiting from the use of the spectrum resource pay for the cost of using spectrum
- Providing revenue to the government or to the spectrum regulator

The allocation of spectrum for PPDR services is expected to improve the overall effectiveness of PPDR organizations and is a key stimulus for economic growth, innovation and productivity within the PPDR communications industry. It is recognized that the estimation of the socio-economic benefits associated with the improvement of the overall effectiveness of PPDR response due to the allocation of additional dedicated or shared spectrum is hard to undertake: the economic value placed on PPDR spectrum is not readily quantifiable in pure market terms, mainly because it has to do with citizens' lives [31]. Some estimates of the socio-economic benefits in the United Kingdom, and further extended to other EU countries, have been reported in Refs [32, 33]. According to this work, an annual consolidated socio-economic value of around €34B is estimated for a set of 10 European countries assessed which represent a total population of approximately 300 million people. This figure is derived taking into account improvements in safety (e.g. reduction of the number of incidents and their impact in lives and properties) and efficiency of the PPDR forces (e.g. improvement in productivity).

A more tangible economic valuation of the spectrum is the so-called opportunity cost. One way to assess the opportunity cost is to estimate how much a buyer would have been willing to pay to use that spectrum for its most promising alternative use [34]. Therefore, considering that the most promising spectrum to be used for broadband PPDR is the same spectrum that commercial mobile operators are willing to use, a good estimate of the opportunity cost can be derived from the monetary sums offered by mobile operators in the auctions to assign spectrum for mobile communications. Table 3.6 shows the prices paid by mobile operators in the German and Spanish auctions, held in May 2011 and July 2011, respectively. Prices are provided per MHz of spectrum as well as per MHz and per head of population (MHz/pop), and they correspond to the winning bid for the most valued spectrum block in each auctioned band. As shown in the table, the highest prices were paid for the 800-MHz band, reaching €0.5/MHz of spectrum per head of population (€/MHz/pop) and €0.73/MHz/pop for 800-MHz spectrum in Spain and Germany, respectively. Remarkably, the value of spectrum below 1 GHz is between 33 and 10 times higher than spectrum in 2.6 GHz due to the better propagation characteristics that facilitate wider geographic coverage outside urban areas and better in-building penetration in dense urban areas.

Figure 3.2 provides additional data intended to compare the prices paid by MNOs for mobile spectrum in different countries and considering previous spectrum allocations for 2G and 3G [35]. The values are given in dollars per MHz and head of population ($/MHz/pop). As noted from the figure, the top position is for the US auction of 700-MHz spectrum held in 2008 where prices of $4.17/MHz/pop were paid in the top 20 areas, namely, the largest cities. Nevertheless, the US average at the time was $1.18/MHz/pop. The next highest paid prices were achieved at some auctions for 3G spectrum held in early 2000s. In the case of Germany, 3G spectrum in 2 GHz was worth at that time as more than fourfold of the value of the recently paid for 800-MHz spectrum.

In turn, the work in Refs [32, 33] compares side by side the opportunity cost of the alternative sale of 2 × 10 MHz in 700-MHz spectrum with the derived estimates for the socio-economic benefits. In particular, the one-off economic gain from spectrum auctions for the governments

Table 3.6 Price paid in German and Spanish auctions for 4G spectrum.

Band (MHz)	Price per MHz (value in M€/MHz – value in €/MHz/pop)	
	Germany, auctioned May 2010	Spain, auctioned July 2011
800	€59.6M/MHz – €0.73/MHz/pop	€23.0M/MHz – €0.5/MHz/pop
900	—	€16.9M/MHz – €0.367/MHz/pop
1800	€2.1M/MHz – €0.026/MHz/pop)	—
2000	€8.8M/MHz – €0.108/MHz/pop	—
2600 paired	€1.8M/MHz – €0.022/MHz/pop	€2.3M/MHz – €0.05/MHz/pop
2600 unpaired	€1.7M/MHz – €0.021/MHz/pop	Offered but no bids for it

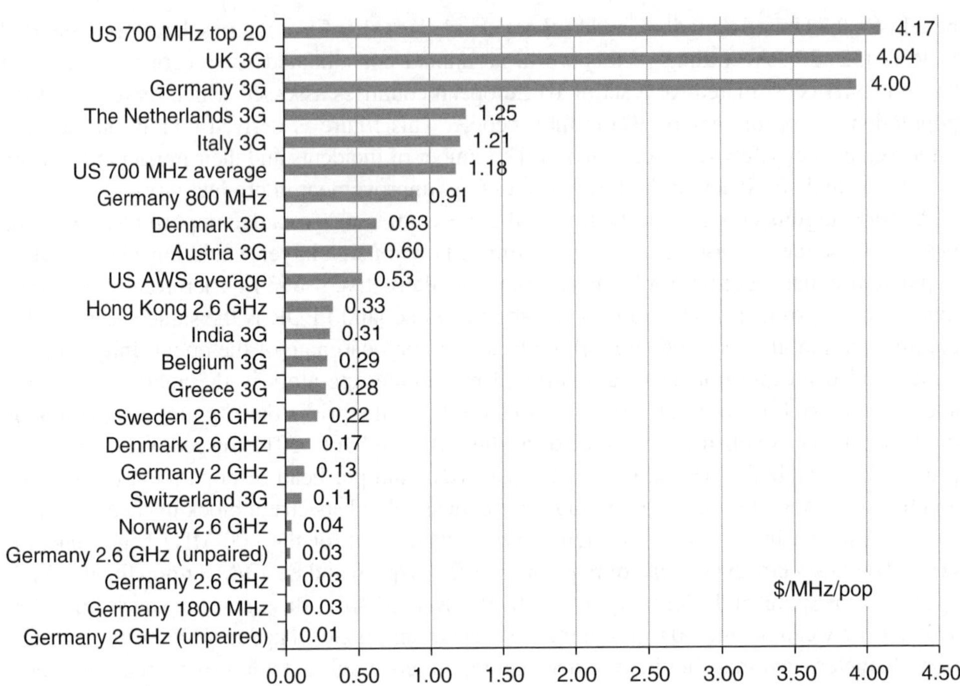

Figure 3.2 Comparative prices for mobile spectrum in different countries and bands. Reproduced from Ref. [35].

of the 10 countries assessed is estimated as €3.7B (equivalent to €0.61/MHz/pop for a total population of around 300 million people), significantly lower than the socio-economic benefits, estimated around €34B.

Overall, the proper economic valuation of the spectrum can be central to justify in economic terms the granting at no direct cost of some amount of spectrum to the PPDR community. Also, this valuation could be necessary in case that PPDR users are expected to pay some fees for the spectrum designation (e.g. this approach is used in some European countries for the use of GSM-R spectrum in the railway sector for its mission-critical communications [9]). It is worth noting that, while not being yet by far the prevailing option in most countries, the UK government's policy with regard to the delivery of next-generation emergency communications services is to divest itself of dedicated spectrum and for all users (including government) to pay market rates for the spectrum [36].

In this context of growing competition for spectrum, together with the general requirement that further progress is needed towards more efficient spectrum utilization [37], the introduction of flexible spectrum use models based on spectrum sharing principles is gaining momentum among regulators and industry [38] and might become instrumental in finding practical solutions for PPDR spectrum allocation and management. Indeed, the need of spectrum for PPDR communications shows a high fluctuation between the amount of spectrum needed in major incidents/events and that used for daily routine tasks. The obvious risk with the allocation of dedicated PPDR spectrum is not using such spectrum efficiently all the time or in all the locations. Therefore, in addition to any incentives in place for an efficient spectrum use within the public sector (e.g. administered incentive pricing practices [30]), sharing some amount of spectrum between PPDR and other uses can contribute to (i) guarantee a peak spectrum availability to satisfy exceptional spectrum needs in major emergencies and (ii) avoid having a large assignation of PPDR spectrum (e.g. allocated to face spectrum demands in worst-case incident scenarios) lying unused when not required for routine PPDR tasks.

The potential introduction of spectrum sharing approaches between the PPDR and other domains such as commercial and military brings new elements that impact on the economic and business dimensions. As discussed in Chapter 6 from a technical perspective, there are two main approaches that deserve close consideration: allowing secondary access to TV white spaces (WS) in the UHF bands for PPDR use and deploying a sort of Licenced Shared Access (LSA) regime able to ensure certain quality of service (QoS) guarantees in terms of spectrum access and protection against harmful interference for both all sharers.

In the case of shared access to TV WS spectrum [39, 40], PPDR industry can catch up and leverage the technology underdevelopment in the commercial domain to exploit the unused spectrum within the TV UHF bands. Good propagation conditions and the likely high availability of TV WS in low populated areas (e.g. rural areas) make this spectrum a valuable (at no acquisition cost) asset for PPDR communications. Besides, further regulatory/technical extensions could be conceived to increase the degree of reliability of this spectrum for PPDR use (e.g. higher authorized maximum transmission power for PPDR equipment and/or introduction of a priority access scheme to TV WS with preferential treatment for PPDR applications in emergency situations).

In the case of sharing with QoS guarantees, military bands can represent also an important asset to leverage (e.g. part of the NATO band within 225–380 MHz). This spectrum is key for military operations, but a significant amount of it is not used in a permanent and/or geographical uniform manner. Therefore, net gains can be obtained for PPDR if (unused) military spectrum can be temporary exploited by PPDR applications in a limited geographical area and when not needed by military users. With regard to sharing with commercial providers, this is not a new concept. Indeed, this type of sharing is allowed for emergency services in a few European countries as described in Ref. [41] and within the policy goals pursued by the European Commission [42]. A shared spectrum only available for PPDR under a major emergency is likely to be available for commercial use all the time and in all locations. The key point to assess here is how much a commercial operator would be willing to pay for that spectrum. Initiatives such as the implementation of an LSA regime in the 2.3-GHz band in Europe [43, 44] are expected to shed light on the economic valuation that this type of spectrum could have for the mobile operators. As stated by GSMA officials [45], mobile operators in general are not fundamentally opposed to the notion of shared access spectrum, but governments should continue to consider exclusive licenced spectrum as the primary source of spectrum for mobile broadband.

3.3 System View of Future Mobile Broadband PPDR Communications

The paradigm change fuelled by the techno-economic drivers discussed in the previous section clearly advocates for the gradual introduction of the LTE technology into the PPDR ecosystem, letting the PPDR community benefit from the synergies with commercial domain and from emerging wireless communications technologies and concepts. On this basis, the challenge for PPDR agencies and governments around the world is finding the right deployment scenario and associated business model for the delivery of mobile broadband PPDR communications. First moves towards the conception and/or provision of reliable broadband communications for emergency services are already underway across the globe, though there are yet significant differences in starting points and focus area. For example, the United States enacted legislation in February 2012 that established a single, nationwide governance body (the FirstNet), assigned 10+10 MHz of 700-MHz spectrum, and provided up to $7B funding for the development of a nationwide, interoperable public safety broadband network based on LTE technology. In contrast, the Electronic Communications Committee (ECC) in Europe is still working on the identification and evaluation of suitable bands for European-wide harmonization of spectrum. All of this is a context where the United Kingdom has already initiated the process to replace voice services currently provided through a TETRA network operated by Airwave by a new national voice and broadband mobile communications service, while other European countries are still finishing the deployment nationwide narrowband networks. A more detailed overview of the current major initiatives that are paving the way towards mobile broadband PPDR is covered in Section 3.4.

Certainly, the most suitable mobile broadband PPDR delivery models may differ among countries and regions to cope with specifics related to different geography and population distribution, different levels of reliance on public networks, different budgets and private players to be involved, etc. Nevertheless, the analysis of the techno-economic drivers identified in the previous section together with the approaches being adopted under the different ongoing initiatives allows us to draw a comprehensive system view of the high-level building blocks that are likely to form part in some way or another of most future broadband communications solutions for emergency services. The envisioned system view is depicted in Figure 3.3, and

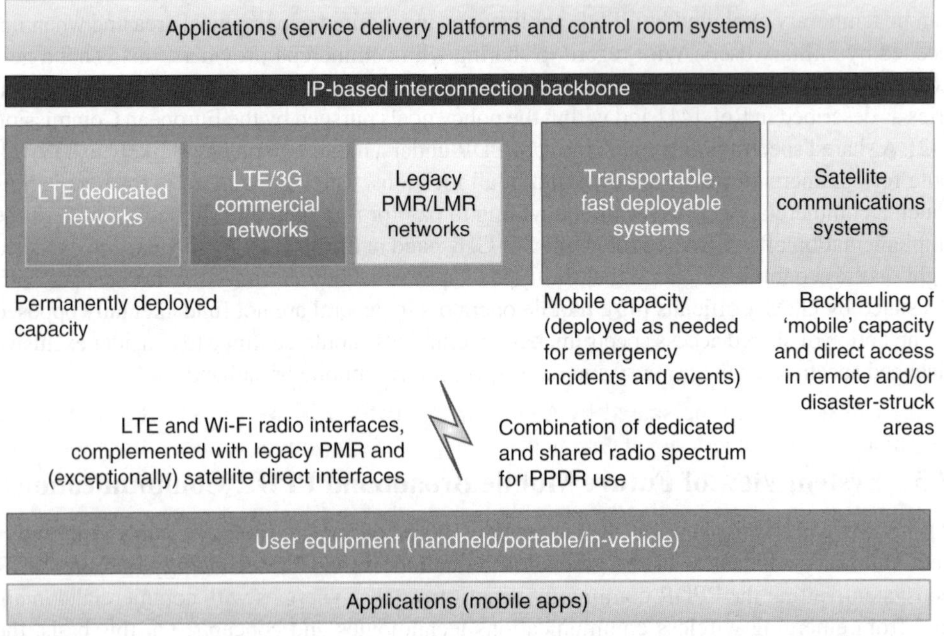

Figure 3.3 High-level building blocks in future mobile broadband PPDR communications systems.

its underlying principles and building blocks are explained in the following. Of note is to remark that, while implementations may differ, it is believed that the overall hierarchy should be similar, even if not all the pieces are included. As shown in Figure 3.3, a multilayered communications approach is envisioned:

- Permanently deployed WAN, consisting of a combination of private/dedicated and commercial LTE-based networks, extending or complementing today's narrowband PPDR networks
- Transportable, fast deployable infrastructure to provide extra capacity and/or coverage in the form of ad hoc local area networks (LANs) or as an extension of the permanent infrastructure
- Satellite access to support the deployment of the mobile capacity as well as possible direct access from portable PPDR user equipment in remote areas in which there is no infrastructure available or it cannot be deployed in an affordable manner

Interoperability among the multiple layers is essential so that PPDR users can get access to their services and interact with each other irrespective of the network they are connected to. In this regard, standardized service delivery platforms and applications are also central elements of the overall architecture. Remarkably, these service delivery platforms and applications shall leverage the functional division established in next-generation networks (NGNs) between transport and service layers (i.e. the underlying network mainly provides IP-based connectivity, while services and applications are implemented on top of this IP connectivity). This functional separation allows services and network layers to be provisioned and offered separately and, importantly, to evolve independently. Moreover, secure and reliable IP-based interconnection backbones should be in place to allow for the interconnection of the different access networks and the data and control centres.

Radio interfaces are expected to be mostly based on LTE and Wi-Fi,[4] complemented with legacy PMR technologies (e.g. TETRA, P25) and, for special user equipment, direct satellite interfaces. Both dedicated (i.e. assigned for exclusive use to PPDR) and shared spectrum components are envisioned. A brief insight into the components illustrated in Figure 3.3 follows, together with pointers to the sections and chapters of this book where more extended and detailed descriptions are given.

3.3.1 LTE Dedicated Networks

Assuming that spectrum and funding are available, the deployment of dedicated PPDR LTE network infrastructure can offer the ultimate availability, control and security features that can best satisfy the PPDR community. For economic sustainability reasons, dedicated LTE networks will most likely be shared by a number of PPDR agencies (police, fire, EMS) and potentially open to other critical communications user organizations (utilities, transportation, etc.). Dedicated networks will be built to meet the required coverage and availability criteria, and the users will have absolute control of the network. Higher standards of availability and resilience can be applied (e.g. hardening the network with backup generators, duplication of

[4] Though based on non-protected frequency bands, Wi-Fi is as of today the main technology used by professionals to exchange data in many industrial or logistical environments. Wi-Fi is a technology that has proven its worth, thanks to its performance, its reasonable cost and its simplicity of implementation.

key components, equipment and communications links and more robust installation) so that the dedicated network is able to withstand high levels of physical disruption caused by, for instance, strong winds and low-level earthquakes. Conventional wisdom is that dedicated infrastructure makes economic sense in metropolitan, urban and even some suburban areas, where capacity demands support the establishment of many cell sites located relatively close together. A detailed description of the LTE technological features that are being worked out by 3GPP to address PPDR communications needs is covered in Chapter 4. Moreover, the definition of 'public safety-grade' requirements as well as the description of challenges and solutions in building out and running dedicated LTE networks for PPDR is addressed in Chapter 5.

3.3.2 LTE Commercial Networks

Using commercial networks for the delivery of mobile broadband PPDR is believed to be a complementary rather than a mutually exclusive approach to the deployment of dedicated infrastructures, at least in the short and medium term. Indeed, public mobile networks are already being used nowadays by some professional users, including PPDR agencies, for non-mission-critical data applications. It is generally agreed that public networks can properly handle most routine traffic, whenever the public network is working under normal conditions. Moreover, even when dedicated LTE-based networks can be rolled out, the unpredictable nature of the time, place and scale of an incident renders it virtually impossible to ensure that the first responders will have proper support only from dedicated infrastructures during the emergency (e.g. due to lack of coverage, capacity or damaged infrastructure). In this context, the consideration of public mobile networks as an integral component for the delivery of PPDR services is anticipated to produce a number of benefits, including increased aggregate capacity, improved resiliency and enhanced radio coverage. However, it is a fact that, in emergencies, public cellular networks are likely to suffer from congestion and may eventually fail more easily than the PMR networks in use today for mission-critical voice [46, 47]. In any case, as commercial broadband networks are becoming an important part of a society's infrastructure, there is also increasing consensus that these infrastructures undoubtedly have a role to play in many critical communications solutions and can enable users to experience the benefits of enriched multimedia tools in PPDR operations in a relatively short time [15, 48, 49]. Consequently, the use of the public mobile broadband networks is anticipated to be a cornerstone for the provisioning of emerging data-intensive/multimedia PPDR services, yet the level of dependability on dedicated and/or commercial networks and their use can be quite varied across countries and regions. From a network operator's perspective, this approach enables different business opportunities to provide different grades of services for the public safety segment. Challenges and solutions in using public mobile broadband networks for PPDR are covered in depth in Chapter 5, including hybrid solutions that enable mobile broadband PPDR service delivery through both dedicated and commercial networks in a consistent manner.

3.3.3 Legacy PMR/LMR Networks

The introduction of LTE for PPDR is expected to complement, not to replace, the existing legacy PMR networks (e.g. TETRA/TETRAPOL/P25/analogue PMR), which will likely continue to be the best choice for mission-critical voice service in the near future. One obvious reason is that the key capabilities for mission-critical voice such as group communications and direct mode are still being introduced in the LTE standard and it could take several years before these features are fully developed and tested to meet the stringent requirements of PPDR. Moreover, until a new mobile broadband network is built and able to provide coverage equal to or better than the coverage currently provided by PMR systems, PPDR users cannot abandon their legacy systems. Therefore, before a mobile broadband solution can effectively replace current PMR systems, the LTE network and associated applications must be able to meet all of the requirements currently satisfied by the existing systems (both functionalities and coverage). In this context, the delivery of mission-critical voice over broadband can be set out as a long-term objective, without hindering or holding up the short-term benefits associated with the deployment of a mobile broadband solution initially mainly intended for the delivery of data-centric applications. Hence, the public safety community should create parallel paths to accomplish both long-term and short-term objectives. This view is sustained by relevant organizations such as NPSTC, APCO and TCCA [4, 5, 50]. In this context, interworking services with legacy systems and the adoption of PMR/LTE multimode user equipment are expected to be fundamental to PPDR users. In this respect, some interworking solutions between LTE and PMR systems such as TETRA and P25 are described in Chapter 5.

3.3.4 Transportable Systems and Satellite Communications

Transportable systems allow PPDR responders to bring the network with them for those events that occur in areas where it does not make sense to have a site around the clock (e.g. rural and wilderness environments). So instead of a permanent, fixed installation infrastructure, a bubble of coverage is deployed where and when needed. The use of transportable systems is anticipated to be central for network restoration, network extension and remote incident response. The use of transportable systems is not limited to PPDR network operators. Public MNOs can also contribute to disaster relief operations via transportable BSs.

There are different types of transportable systems, which are usually classified under the categories of cell on wheels (COWs) and system on wheels (SOWs). On one hand, COWs typically include a BS (e.g. LTE eNodeB) along with one or more backhaul transports (such as microwave or satellite). COWs require connectivity to a core network (e.g. LTE evolved packet core) to support application functionality. On the other hand, SOWs are fully functional systems that can act without backhaul connectivity, though these are likely to be more expensive systems than COWs. As a general approach, the use of COWs could make sense in dense, urban environments where connectivity to the core network can be guaranteed, while SOWs are more appropriate in rural environments and in disaster areas, where broadband backhaul connectivity is an issue. Deployable systems can also leverage the capability of both Wi-Fi and LTE technology to support hybrid approaches (e.g. transportable system that uses Wi-Fi interface to create a hot spot for local access by PPDR responder equipment and then relies on the use of LTE to provide the remote connectivity).

Together with deployables, satellite communications provide a unique and important method for PPDR to plan around the hazards of earth-based infrastructures that can be susceptible to all manners of natural and man-made catastrophes. This turns satellite communications platforms into important components within the complete tool kit of PPDR communications means. Satellite service can be offered in areas where there is no terrestrial infrastructure and the costs of deploying a fibre or microwave network are prohibitive. It can also support services in areas where existing infrastructure is outdated, insufficient or damaged.

In particular, very small aperture terminal (VSAT) solutions can be used to provide the backhaul connectivity to the deployable solutions (e.g. backhaul connectivity for COWs). A typical VSAT may have full two-way connectivity up to several Mbps for any desired combination of voice, data, video and Internet service capability. Communications on-the-move (COTM) solutions are also important for PPDR, enabling applications such as mobile command and control where a vehicle can serve as a mobile command post while in-route and as a fixed command access point for personnel upon arrival at the designated location when local terrestrial and wireless infrastructures are not available. Moreover, mobile satellite service (MSS) solutions can also be in place for PPDR users, allowing the use of portable satellite phones and terminals. MSS terminals may be mounted on a ship, an airplane, a truck or an automobile. MSS terminals may even be carried by an individual. The most promising applications are portable satellite telephones and broadband terminals that enable global service. In addition, solutions that integrate satellite and cellular technologies are also appealing for PPDR use (e.g. satellite chips inserted in handheld devices or adaptors to turn the cellular device into a satellite device). Chapter 5 outlines some further details on the use of satellite communications for the interconnection of deployables as well as some considerations on satellite direct access.

3.3.5 IP-Based Interconnection Backbones

The interconnection of the multiple and diverse components (e.g. radio sites, data centres hosting the mobile core networks and service delivery platforms, PPDR deployables, emergency control centres and PSAPs, interconnection of regional/national PPDR networks, etc.) advocates for the use of IP-based interconnection solutions. IP backbones consisting of fibre, copper, microwave, satellite and other links deployed in a redundant topology are central components. The interconnect infrastructure can be fully or partly owned by government agencies as well as rely on the use of interconnection services provisioned by private carriers. Interconnection frameworks such as the IP Packet Exchange (IPX) promoted by GSMA are gaining strong consensus among the commercial industry and are certainly a potential solution to be considered for the interconnection of a number of regional/national PPDR networks in order to allow roaming/migration and interoperable communications services and associated applications within a secure framework. A further description of IP-based interconnection technologies and frameworks is addressed under Chapter 5.

3.3.6 Applications and User Equipment

At the end of the day, the multimedia- and data-enriched applications enabled by mobile broadband connectivity (see Chapter 2) are the ‘visible’ tools to PPDR responders. The introduction of smartphones and other sort of devices with high computational capabilities and the adoption of common standards are expected to pave the way towards a rich ecosystem of interoperable PPDR applications. As of today, the customization of applications and services within the PMR industry business model is mostly based on vendors’ proprietary interfaces. Therefore, many of the applications are limited, use expensive hardware and might lock-in the user to a single manufacturer. In contrast, in the commercial domain, the proliferation of software-enabled devices together with interchangeable peripherals has given consumers and enterprise customers the ability to personalize how they receive media information, communicate with others and configure their homes, workplaces and automobiles. The majority of devices and applications is interoperable because of open standard technologies such as Bluetooth, USB, Wi-Fi and published software development kits (SDKs). These standards have expanded the market to thousands of developers and greatly facilitated the proliferation of specialized products and information.

A central effort towards the standardization of a comprehensive application architecture for the delivery of critical communications services on top of IP-based connectivity is being carried out by ETSI [51], in close cooperation with the 3GPP. ETSI is specifying a reference model of a critical communications system (CCS), defining the functional elements along with the interfaces and reference points among them. A central element of the CCS architecture is the critical communications application (CCA), which can be understood as the service delivery platform providing the communications services (e.g. mission-critical push-to-talk services) to critical communications users. The CCA includes capabilities on the terminal side and on the infrastructure side. Further details on the ETSI CCS reference model are provided in Chapter 5.

Enabling the development of applications by as many stakeholders as possible, in a secure and reliable manner, promises to empower the public safety communications marketplace in the same way the mobile broadband application ecosystem has empowered consumers today. In this regard, initiatives such as the Application Community (AppComm) [52] promoted by APCO International can be instrumental to favour the development of an application ecosystem for the PPDR community. AppComm provides a collection of applications related to public safety and emergency response for use by the general public and first responders. AppComm is also a forum where public safety professionals, the general public and app developers can discuss and rate apps, identify unmet needs and submit ideas for apps they would like to see built. With this initiative, APCO is determined to play a major leadership role in supporting the development of a diverse, practitioner-driven public safety app ecosystem fostered through the collaborative efforts of public safety professionals and app developers. To further foster this sort of PPDR application ecosystems, open standard-based solutions for terminals' client applications downloading and installation are also a must (e.g. the likes of the popular applications' stores in the commercial domain), together with other post-manufacturing configuration of terminals through mobile device management (MDM) software solutions (e.g. Open Mobile Alliance Device Management (OMA DM) standards, widely adopted in commercial networks). Another central element of the PPDR application ecosystem is the operating system used in the users' devices. One compelling candidate is the Android platform, which is being enhanced rapidly (e.g. support of SELinux kernel security module, Samsung's Knox security software) and adopted by 'stringent' users in terms of security such as the Federal Bureau of Investigation (FBI) in the United States [53].

Another area that deserves further consideration is the standardization of functional frameworks and interfaces for dispatch centre control systems in the context of emerging broadband wireless technologies. In this context, the requirements for the functionality and interfaces for command and control consoles connected to an LTE network have been developed by the NPSTC [54]. These 'console' systems are primarily located in emergency control centres (ECC) and public safety answering points (PSAPs), though they may also be located in other facilities (e.g. hospital emergency departments) as well as be used at the scene of a major incident as either a wired or wireless console device. The best practices and requirements provided in the document are intended to describe the features and functionality for console-based dispatch operations that involve broadband services. They are intended to capture the operational requirements of dispatch and console operator functionality with the objective of fully leveraging the features and functionality of the LTE network.

Last but not least, the commercial availability of devices for public safety use is of great importance. User-friendly rugged devices must be available and able to handle rough environments. It is also important to have the proper types of devices for a particular mission. Different levels of ruggedization and security, exceeding those of current consumer UEs, are anticipated as well as PPDR-specific complements (e.g. wearables such as smart glasses and smart helmets) and functionalities (e.g. hands-free with voice recognition, emergency button functionality that provides services similar to the emergency buttons in PMR radios). In this respect, a key challenge is the integration of many diverse components into the appropriate gear in a way that it can best serve the needs of PPDR practitioners [55]. Additionally, the devices must have support for the required frequencies, which can vary. Device applications must be easy to use and support the typical requirements of a public safety mission, such as dynamically changing priority depending on the situation and possessing the ability to gracefully adapt to lower bandwidth in order to make sure that services are available during hard radio conditions. Besides interoperability and certification, one major challenge for terminal manufacturers is to provide dual-mode terminals that support broadband LTE and narrowband PPDR networks.

3.3.7 Spectrum

The need for spectrum suitable for the support of emerging broadband applications for PPDR has been recognized for many years. The public safety community is well aware of these needs. Numerous studies have already substantiated these requirements in different countries and regions across the world [56–59]. Exclusive or primary allocations[5] of spectrum for broadband PPDR have also been enforced in some countries (the United States, Australia, Canada, etc.). The typical amount of spectrum being allocated is 10+10MHz or 5+5MHz in the 700-MHz or 800-MHz bands. In Europe, PPDR agencies and industry have also identified a need in the range of 10+10 MHz [60], and spectrum regulatory authorities have started the process of finding a proper spectrum allocation [61], though changes in the current spectrum regulatory framework for PPDR are not expected before 2016. This amount of dedicated spectrum is estimated to be enough to satisfy PPDR needs for mission-critical communications in most operational scenarios. However, it is also recognized that no amount of spectrum used by a conventional cellular network is likely to satisfy a localized, short-notice spike in demand that might result from a major incident such as a terrorist attack in a central business district or major urban centre. Furthermore, it would be highly economically inefficient to try, and dimension spectrum provisions around what might be a once-in-a-generation event. For these reason, other ways to increase capacity in a more effective manner are also essential. In this regard, making additional spectrum available in higher frequency bands (e.g. 4.9 or 5GHz) as

[5] A primary allocation mean than other secondary users might be permitted insofar as they do not interfere with the primary PPDR use.

well as adopting dynamic spectrum sharing solutions (e.g. opportunistic access to TV spectrum, licenced secondary access models with pre-emption capabilities, etc.) can bring additional capacity to better cope with a surge of PPDR traffic demand and enable extremely high data rates (including multiple video streams) in localized hot spots (e.g. around an incident site). In addition to spectrum for dedicated LTE networks and transportable systems, there could also be additional spectrum requirements to cater for broadband transmissions in D2D operation mode, air–ground–air (AGA) links and microwave links needed for backhauling of PPDR systems. Further details on regulatory and technical aspects related to the use of dedicated and dynamically shared spectrum for PPDR use are addressed in Chapter 6.

3.4 Current Initiatives

The need to provide reliable broadband communications for emergency services is currently recognized by many government agencies around the world, and some of them have already taken different actions towards the materialization of the future PPDR mobile broadband communications systems.

A pioneering role is being played by some of the nations represented by the partner associations of the APCO Global Alliance (Australia, Canada, New Zealand, the United Kingdom and the United States) [2]. In particular, a key milestone was set out by the United States when a first swath of 5 + 5-MHz spectrum was allocated to mobile broadband PPDR back in 2007 and a first attempt was made to create a nationwide public safety-grade network though the auctioning of the spectrum block (i.e. the D-Block) that is contiguous to the PPDR allocation. Later in 2012, a new legislation was enacted in the United States that established a single, nationwide governance body (the FirstNet), assigned the D-Block also to PPDR (resulting in the total 10 + 10-MHz block currently available), and provided up to $7B funding for the development of a nationwide, interoperable public safety broadband network. In close cooperation with the United States, Canada has already allocated 20 MHz in the 700-MHz band to match the US allocation. Australia has also reserved 10 MHz in the 800-MHz band for possible allocation to public safety agencies.

In Europe, the main efforts towards a European harmonized solution for broadband PPDR communications are currently localized at the regulatory level and primarily targeted to identify and evaluate suitable spectrum bands (both below and above 1 GHz) for European-wide harmonization of spectrum. This regulatory effort is being conducted within the Electronic Communications Committee (ECC) of the European Conference of Postal and Telecommunications Administrations (CEPT), which has also developed its view of the future European broadband PPDR systems and established a transition roadmap towards broadband PPDR communications in Europe. In parallel to spectrum harmonization activities, some European countries have already initiated some actions towards the delivery of mobile broadband PPDR services over commercial networks, such as the PPDR communications service provider in Belgium (Astrid), who has launched an MVNO service for data-centric applications. In addition to Belgium, other European countries such as Finland and France have also announced plans that consider the deployment of an MVNO model in the initial stages to leverage commercial networks' capacity and enable a progressive deployment of dedicated networks. At the same time, the UK Home Office (HO) has initiated a procurement process to replace the current voice services currently provided by a TETRA network with a new national voice and broadband mobile communications service. Also remarkable is an action established within the EU Framework Programme for Research and Innovation HORIZON 2020 under the challenge of 'secure societies' [74] and which is intended to develop the core set of specifications, roadmap for research and tender documents to be used as a basis for national procurements for interoperable next-generation PPDR broadband communications systems across Europe. The expected impact of this action is to create an EU interoperable broadband radiocommunications system for public safety and security deployed by 2025.

Initiatives are also ongoing in some Middle East countries. For example, Qatar has already established a fully functional dedicated PPDR LTE network. In the UAE, the regulatory authority has already designated spectrum for broadband PPDR in the 700-MHz band.

At industry level, in order to achieve cohesion and foster the adoption of a common ecosystem, industry organizations such as the TCCA have established tentative roadmaps for LTE as a technology evolution for TETRA and other existing mission-critical systems. The TCCA's envisioned roadmap is based on the predicted time frame for the availability of suitable standards as well as on the analysis of different delivery options that the PPDR sector may adopt (e.g. use of dedicated or public networks). TCCA plays an active role in the coordination of the different PPDR end-user organizations that participate in standardization activities (NIST, NPSTC and APCO from the United States; UK HO; German Ministry of Interior; etc.). Their joint efforts are instrumental to drive in a consistent manner the requirement specification phases of the work that is being addressed in 3GPP and other standard-setting bodies (ETSI, OMA).

A further insight into some of the aforementioned initiatives is given in the following.

3.4.1 Deployment of a Nationwide Dedicated LTE Broadband Network in the United States

In February 2012, the US Congress enacted the Public Law 112-96 'The Middle Class Tax Relief and Job Creation Act of 2012' to create a nationwide interoperable public safety broadband network. The act includes the following:

- The public safety broadband network will be based on a single national architecture based upon the LTE technology.
- The governing framework for the deployment and operation of this high-speed network dedicated to public safety is the new FirstNet, an independent authority within the National Telecommunications and Information Administration (NTIA) under the US Department of Commerce.
- FirstNet will hold the spectrum licence for the network and is charged with taking 'all actions necessary' to build, deploy and operate the network, in consultation with federal, state, tribal and local public safety entities and other key stakeholders.
- The act allocates the 700-MHz D-Block Band 14 (758–763 and 788–793 MHz) to FirstNet for the construction of a single wireless nationwide public safety broadband network.
- Non-public safety entities will be allowed to lease the spectrum on a secondary basis.

FirstNet [62] is tasked with cost-effectively creating a nationwide network and providing wireless services to public safety agencies across the country. FirstNet has the Public Safety

Advisory Committee (PSAC) to assist it. The PSAC has access to NPSTC, APCO and a host of other organizations and local resources. FirstNet is also working with the Public Safety Communications Research (PSCR) programme [63] and standards organizations on network requirements and on defining how standards can support building future networks as public safety grade.

The US Congress allocated $7B in funding to FirstNet for the deployment of this network as well as $135M for a new State and Local Implementation Grant Program (SLIGP) administered by NTIA to support state, regional, tribal and local jurisdictions' efforts to plan and work with FirstNet to ensure the network meets their wireless public safety communications needs. To contain costs, FirstNet is committed to leverage on existing telecommunications infrastructure and assets. This includes exploring public–private partnerships that can help to support and accelerate the creation of this new advanced wireless network. In addition, FirstNet has stated that it will explore ways to make the PPDR spectrum available to other users in times when there is excess capacity while preserving priority access to first responders. The legislation that established FirstNet stipulated that FirstNet would be self-sustaining and that any fees collected by FirstNet shall not exceed the amount necessary to recoup expenses. FirstNet is working to establish a pricing model that should attract users and ensure the network is self-sustaining. Remarkably, there is no requirement for the public safety community to subscribe to the FirstNet network. Through the assessment of fees, FirstNet must generate sufficient funds to enable the organization to operate, maintain and improve the network each year. Besides the public safety community, other federal agencies (e.g. US Department of Homeland Security (DHS)) consider the forthcoming national broadband network as a way to expand its own mission capabilities.

In a first stage, FirstNet is committed to provide mission-critical, high-speed data services though the LTE network to supplement the voice capabilities of today's LMR networks. In time, FirstNet plans to offer Voice over LTE (VoLTE) for daily public safety telephone communication, as long as this technology matures.

FirstNet has already signed four Spectrum Manager Lease Agreements (SMLAs) with Broadband Technology Opportunities Program (BTOP) awardees. The BTOP administered by NTIA provided funding for seven public safety projects in 2010 to deploy mobile broadband. These funds were partially suspended 2 years later, after the Congress enacted the law creating FirstNet. The suspension was needed to ensure that any further activities would be consistent with the mandates of the new law. FirstNet reviewed the proposed BTOP projects and determined that there was value in continuing to support them. As a result, FirstNet reached SMLAs with the Los Angeles Regional Interoperable Communications System (LA-RICS) Authority; Adams County, Colorado (ADCOM 911); the state of New Jersey; and the state of New Mexico. In this context, several public safety LTE systems can go operational along 2015 such as the over 200-site network being built by the LA-RICS. In addition, FirstNet has also approved a similar SMLA with the state of Texas for the Harris County LTE public safety network, which is funded through a federal port security grant, not a BTOP award. Harris County was the first county to go live with a private LTE system for public safety in 2012. Before the SMLA, a special temporary authority (STA) from the FCC was in place to operate the Harris County network.

In September 2014, FirstNet has issued a request for information (RFI) with a draft statement of objectives (SOO) to seek input from interested parties regarding specific topics, which are intended to help FirstNet develop a comprehensive network acquisition strategy. The RFI includes technical questions related to the building, deployment, operation and maintenance of the nationwide network; ways to accelerate speed to the market; priority and pre-emption implementation; opt-out RAN integration and reliability and restoration as well as life cycle. Some of the key goals of the RFI are to minimize public safety user fees; deliver advanced, resilient wireless services; and maximize the value of excess network capacity to keep costs low for public safety. The key outcomes of this market research phase should help to develop the final state of the request for proposals (RFP), which is expected to be released during 2015. In addition, once the RFI process is concluded, FirstNet plans to begin the opt-in, opt-out process for the states, which will have to decide whether to opt in and pay to access the FirstNet network or opt out and either build their own public safety LTE network by using FirstNet's 700MHz, Band 14 spectrum and linking to the FirstNet core or simply go ahead without a dedicated public safety broadband network at 700MHz.

FirstNet is a major step within the US DHS's vision [64] of the evolution of public safety communications as it transitions from today's technology to the desired long-term state of convergence of mission-critical voice and data. Figure 3.4 depicts the conceptual framework for building wireless broadband communications while maintaining LMR networks to support mission-critical voice. According to the picture, LMR networks, commercial broadband networks and a nationwide public safety wireless broadband network are at present evolving in parallel. As communications further evolve, public safety will continue to use the reliable mission-critical voice communications offered by traditional LMR systems. At the same time, agencies will begin to implement emerging wireless broadband services and applications. During the transition period, public safety will begin building out a dedicated public safety

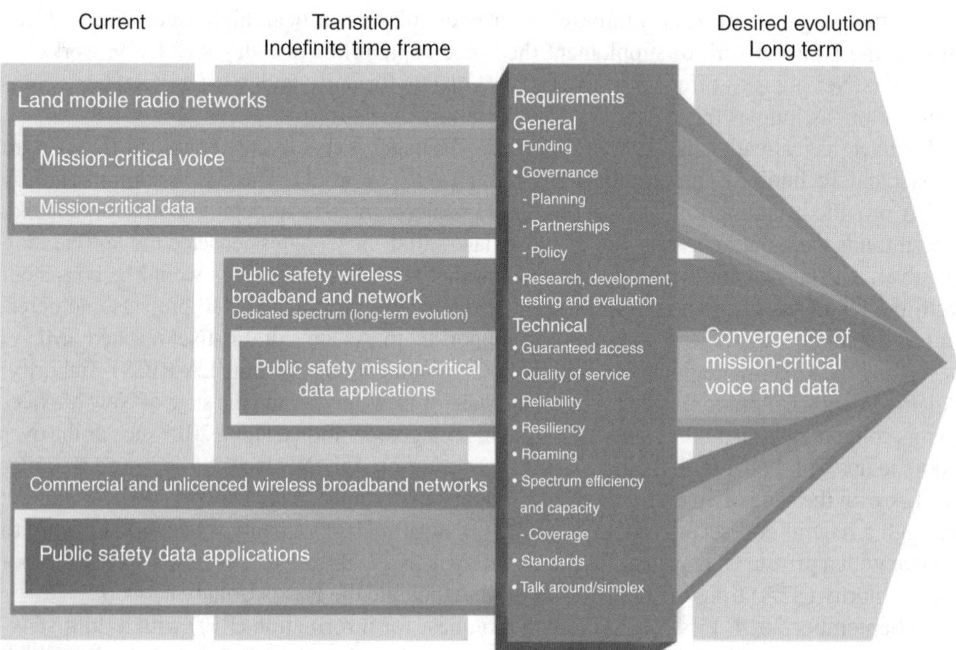

Figure 3.4 Public safety communications evolution by describing the long-term transition towards a desired converged future. Reproduced from Ref. [64].

wireless broadband network, and public safety organizations will begin to transition from commercial broadband services to the public safety dedicated network. If and when the technical and non-technical requirements (listed in the vertical box) can be met and are proven to achieve mission-critical voice capability, it is desired that over time, agencies will migrate entirely to this 'converged network'. However, convergence will be a long-term and gradual transition as agencies integrate new technologies, rather than replace existing systems. The pace of convergence will vary from agency to agency and will be influenced by operational requirements, existing systems and funding levels. During this migration period, solutions for connecting traditional LMR with broadband systems will be necessary. Even when the nationwide public safety network is capable of meeting public safety requirements, some agencies may need to operate separate LMR systems until the public safety wireless broadband network is fully deployed in their regions. Therefore, additional investments will continue to be necessary for both LMR and a dedicated public safety wireless broadband network simultaneously.

3.4.2 CEPT ECC Activities for a European-Wide Harmonization of Broadband PPDR

The Electronic Communications Committee (ECC) is one of three business committees of the CEPT, an organization where expert policy makers and regulators from 48 countries across the whole of Europe collaborate to create a stronger and more dynamic market in the electronic communications and postal sectors. The primary objective of the ECC is to harmonize the efficient use of the radio spectrum, satellite orbits and numbering resources across Europe. It takes an active role at the international level, preparing common European proposals to represent European interests in the ITU and other international organizations.

Within ECC, the Frequency Management Project Team 49 (FM PT 49) [61] is working on radio spectrum issues concerning PPDR applications and scenarios, in particular concerning high-speed broadband communications capabilities requested by PPDR organizations. The primary challenge is to identify and evaluate suitable bands (both below and above 1 GHz) for European-wide harmonization of spectrum by taking into account cross-border communications issues and PPDR application requirements and with a focus on medium- and long-term (before year 2025) spectrum realization. FM PT 49 work is being addressed in cooperation (through liaisons) with ETSI and other organizations (e.g. Law Enforcement Working Party (LEWP) of the European Council, Public Safety Communications (PSC) Europe).

FM PT 49 delivered CEPT Report 199 [57] in May 2013, which focuses on the definition of the applications and network-related requirements of broadband PPDR networks, the specification of typical PPDR operational scenarios, the usage of BB PPDR applications and the assessment of the spectrum needs for a WAN.

CEPT Report 199 also elaborates on the concept of future European broadband PPDR systems. According to the proposed concept, future European BB PPDR systems to cope with mission-critical as well as in non-mission-critical situations will consist of the following two central elements:

1. **BB PPDR WAN**. BB PPDR WAN should provide a coverage level meeting the national requirements and support high mobility PPDR users. Initially, it is expected that BB PPDR WAN systems will operate together with narrowband TETRA and TETRAPOL networks, and those networks will continue to provide voice and narrowband services for at least the

coming decade. In the future, the broadband technology will be capable of supporting the PPDR voice services as well as the data applications.

2. **BB PPDR temporary additional capacity**. BB PPDR temporary additional capacity (also known as 'hot spot' or LAN) should provide additional local coverage at the scene of the incident through the deployment of the necessary communications facilities in addition to those available through the WAN. This additional capacity should be provided through, for example, ad hoc networks or additional temporary BSs of the WAN and to support low mobility PPDR users.

CEPT Report 199 provides spectrum requirements of BB PPDR WAN. Instead, due to the fact that there are no commonly agreed requirements on temporary additional capacity, it does not address the assessment of spectrum requirements for BB PPDR temporary additional capacity by ad hoc networks using different frequencies from the ones used in the WAN.

CEPT Report 199 explicitly recognizes that countries may have widely varying BB PPDR WAN needs. To accommodate these different needs, the report claims that the operating bands of future equipment should be wide enough to cover both the minimum spectrum requirement calculated for BB PPDR WAN which would facilitate cross-border operations and additional individual national needs (e.g. for DR). In order to find a solution to the problem of achieving harmonization while maintaining countries' sovereign right to choose the most suitable solution for broadband PPDR according to national needs, the concept of 'flexible harmonization' has been introduced. This concept includes three major elements:

1. Common technical standard (i.e. LTE).
2. National flexibility to decide how much spectrum should be designated for PPDR within a harmonized tuning range
3. The harmonization should enable national choice of the most suitable service provision model (either dedicated, commercial or hybrid).

Based on the above concept, in order to establish a pan-European family of cross-border functioning BB PPDR networks, it is not required to designate identical bands for this purpose but rather to choose the suitable bands within the harmonized frequency range(s) and to adopt a common technology. This will allow a border-crossing broadband PPDR terminal to find its corresponding BB PPDR network in the visited country.

Assuming the 'flexible harmonization' concept as a basis for the evolution of today's PPDR communications towards a broadband future, a transition roadmap reflecting the current vision of the future evolution's milestones mapped onto the timeline up to and beyond year 2025 has been developed within FM PT 49 [65]. The roadmap, shown in Table 3.7, may assist CEPT administrations in their national planning for the provision of broadband PPDR services.

3.4.3 Hybrid Approaches Taking Off in Belgium and Some Other European Countries

In Belgium, ASTRID [66] is the operator of the national radiocommunications, paging and dispatching network designed for emergency and security services. ASTRID is a government-owned corporation founded in 1998. The ASTRID radio network is based on TETRA technology.

Table 3.7 Transition roadmap towards BB PPDR communications in Europe.

Year	Expected milestones
2014	• Trials by European PPDR organizations based on commercial networks started, first MVNO implementation by a PPDR organization • 3GPP Release 12 approved: ProSe (direct mode communications), GCSE (dynamic groups of mobile users), driven by European and US PPDR stakeholders • CEPT technical work (part A of CEPT Report in response to the EC Mandate on 700 MHz, band plan and block edge mask) for the 700 MHz completed (July 2014), with the exception of the technical studies of the optional 2×5 MHz of a dedicated PPDR spectrum • Trials of dedicated PPDR LTE networks in the 400 MHz
2015	• CEPT technical work for the 400 MHz completed • ECC Report B on 'harmonized conditions and frequency bands for BB PPDR' approved, towards a new ECC decision or revision of ECC/DEC/(08)05 • ETSI approves the suite of first standards for critical communications • OMA 'PTT over cellular' standardization • First contract with a commercial operator for nationwide broadband PPDR service is expected • WRC-15: co-primary mobile allocation in 694–790 MHz
2016	• CEPT work on the refinement of the technical conditions for 700-MHz band to possibly increase the international harmonization (e.g. in line with Resolutions of WRC-15) completed • EU and ECC decisions on the harmonization of the 694–790-MHz band approved • ECC regulatory framework for PPDR is revised by adding spectrum within the 700-MHz band as a new harmonized broadband PPDR frequency range and 'upgrading' parts of the 400-MHz band to the broadband PPDR frequency range based on the LTE technology (either through revision of ECC/DEC/(08)05 or by developing a new ECC decision) • 3GPP Release 13 approved: MCPTT (mission-critical push-to-talk over LTE may be expected), isolated E-UTRAN operation (resilience) • LTE equipment compliant with 3GPP Rel.12 commercially available
2017	• Review of the first results of initial implementations based on commercial networks • Combined LTE-700/TETRA infrastructure solutions commercially available • Authorizations to mobile broadband of the 694–790-MHz range issued in a number of European countries; some countries may opt for dedicated solutions for PPDR • 3GPP is expected to create a new band class for LTE in the 400-MHz band
2018–2020	• Roll-out of first commercial LTE networks in the 700-MHz range • LTE equipment compliant with Rel.13 with enhanced PPDR functionality is commercially available • Combined LTE/TETRA terminal equipment commercially available • First implementations of hybrid solutions based on commercial LTE networks in 700-MHz range • Possible first implementations of dedicated PPDR LTE networks in the 400-MHz range • 3GPP Release 14 approved: possible enhancements of PPDR functionalities in Rel.12/13 • PPDR operational procedures are gradually adjusted to include broadband communications • First trial of cross-border interoperability and roaming between PPDR broadband LTE networks according to the flexible harmonization concept in the 700-MHz range

(*continued*)

Table 3.7 (*continued*)

Year	Expected milestones
2020–2025	• LTE equipment compliant with Rel.14 with full PPDR functionality is commercially available • Mission-critical broadband communications are introduced as trials • Voice (non-mission critical) and data integration within LTE networks
Beyond 2025	• Mission-critical voice and data are provided via LTE networks based on either commercial, hybrid or dedicated solutions, subject to national decisions • TETRA/TETRAPOL networks are gradually phased out • DMO trials are conducted

Reproduced from Ref. [65].

The network is used by the Belgian emergency and security services, alongside with public service organizations and companies that provide assistance (e.g. hospitals, ambulances) or may have to deal with public safety-related problems as part of their operations (e.g. public transport firms, water and energy distribution companies, money transportation companies, security firms).

In April 2014, ASTRID launched a broadband data service called Blue Light Mobile [67] that allows its subscribers to use the commercial 3G networks for data-centric applications. To that end, ASTRID takes on the role of MVNO and manages its own SIM cards. These SIM cards give ASTRID's subscribers the status of international roamers on Belgium's three commercial cellular networks (Proximus, Mobistar and Base) and eleven networks in four neighbouring countries (the Netherlands, Germany, Luxembourg and France). While ASTRID's SIM cards have a 'preferred' network, they will automatically switch to another network whenever the coverage is lost. A VPN client programme is provided to the users to create a secure connection (a kind of 'tunnel') that guarantees the confidentiality and integrity of the data transfer between the mobile terminal and ASTRID's data centres. In the case of using the service as TETRA backup function, the terminals needs to be compatible with 3G/4G and TETRA.

Blue Light Mobile is seen as a temporary solution to the problem of supplying PPDR users with mobile broadband. However, 'temporary' could mean 5–10 years [9]. While no other similar services are in operation in other countries, the MVNO model is also considered in other European countries such as Finland and France.

In Finland, VIRVE, the Finnish TETRA operator, has already established a roadmap towards the implementation of a government-controlled hybrid of dedicated and commercial LTE networks to eventually offer critical voice and broadband data by 2030 [68]. VIRVE's current TETRA network provides critical voice and messaging services to all PPDR agencies ranging from social services to defence forces. With regard to the spectrum needed in the new network, some amount of dedicated spectrum in the 700-MHz band will be designated for public safety needs. This assignment is expected to be harmonized with the other European Union countries. With regard to the involvement of commercial networks, the view is that the use of dedicated network(s) in incident-rich areas where the population is located (e.g. urban areas as well as alongside the main highways) and relying on the commercial networks in scarcely populated areas is regarded as the most economical approach. In this regard, ensuring that commercial networks will meet fundamental authority requirements such as capability to guarantee authorities' priority access at all times in addition to the increased network availability and reliability is likely to be pursued by adding specific requirements into the commercial frequency licencing terms. On this basis, a reasonable time window for the transition from TETRA to broadband in Finland could begin with the availability of critical voice services over LTE early next decade and could end when the current TETRA network reaches its end of life, somewhere in the first half of the 2030s. Building out the nationwide TETRA coverage took several years, and it was even longer until all the separate analogue systems were shut down. Thus, a long period of parallel networks with narrowband TETRA services and LTE broadband should be seen as an asset rather than a burden. In this context, five steps are envisioned [68]:

Step 1. To set up a data MVNO to address the increased everyday data requirements. This will be accomplished by extending the subscriber and service provisioning system to support provisioning users on a broadband network.

Step 2. To control subscribers in an owned LTE core. In this second step, the critical voice and messages will run in the narrowband network, and high-speed non-critical (but secure) data will run in the commercial broadband network.

Step 3. To expand the owned LTE core to an owned dedicated broadband radio access in chosen locations, providing critical-grade data services.

Step 4. To connect the TETRA and the LTE network once the critical voice over LTE standardization is ready and the TETRA supplier supports group call over LTE functionality in the TETRA side. In this way, the large development investments in TETRA group communications functionalities, such as prioritization, could be used. Then, the same voice services would be available both in narrowband and broadband networks. Nevertheless, while in the dedicated networks this would be on a critical service level, in the commercial operators' networks, it would just be up to the levels that they could provide.

Step 5. To dismantle the TETRA radio access once broadband service availability and reliability meet public safety's requirements. In some (most of all rural) areas, this might take place first when the narrowband network spare parts stock runs out.

During these five steps, the narrowband TETRA network will transform to a TETRA critical voice service server, the operator will gain knowledge and understanding about how to operate a broadband network, and the users will have access to high-speed data service that enables them to benefit from data applications and to develop information-centric ways of working.

In France, the Ministry of Interior has also revealed a hybrid strategy towards the deployment of mobile broadband PPDR based on [69]:

- A dedicated network for PPDR critical communications
- Commercial network for non-critical and broadband transmissions managed by a MVNO

Starting from the current situation with two national TETRAPOL networks that serve different PPDR agencies together with several TETRA networks used by other critical infrastructure operators (e.g. airports, railways), France is seizing the opportunity to move towards a unique dedicated broadband PPDR network for voice and data communications. The reasons argued to justify the option of a dedicated solution instead of a commercial contract with a MNO are the following [69]:

- A commercial network is unable to maintain the availability of a dedicated network.
- The service can be guaranteed even in case of crisis with a dedicated network.
- Higher security levels can be ensured on a dedicated network.
- The legal obligation is limited for MNOs.
- The coverage of commercial networks is not global.
- A commercial SLA cannot replace state responsibility.

On this basis, France is also planning to designate dedicated frequencies for PPDR. In particular, France is considering some amount of spectrum in the 700-MHz range to take advantage of the economies of scale of the LTE commercial ecosystem expected to boost in this band, along with an additional amount of spectrum in the 400-MHz range to reuse part of the existing infrastructure. Together with the envisioned dedicated network for PPDR critical communications, commercial networks would also be used for non-critical and broadband PPDR communications services based on an MVNO model.

3.4.4 LTE Emergency Services Network in the United Kingdom

In the United Kingdom, the HO has already initiated the process to replace the TETRA system that provides mission-critical communications for public safety agencies and other government organizations in Great Britain (England, Scotland and Wales) since 2005. This TETRA system is a private network with dedicated spectrum owned and operated by Airwave. This system covers 99% of the land mass and 98% of the population in United Kingdom. It serves all three emergency services (3ES, i.e. police, firefighters and ambulances) and other national users that pay subscriptions fees. According to officials from the UK HO [70], the performance of the TETRA system is 'very good' but 'extremely expensive' for users, particularly when compared to the plummeting per-minute costs of commercial wireless airtime. The cost of the Airwave service, together with the fact that contracts associated with the Airwave system are scheduled to expire from 2016 to 2020, has motivated the UK HO to seek for a replacement for critical voice, as well broadband data services in a cost-effective manner.

The emergency services mobile communications programme (ESMCP) [71] is the cross-government, multi-agency programme that will deliver the communications system of the future to the emergency services and other public safety users (known as sharers). This system will be named the Emergency Services Network (ESN) and will be expected to provide integrated critical voice and broadband data services to all 3ES. These services require a mobile communications network capable of providing the full coverage, resilience, security and public safety functionality required by the 3ES.

The ESN will replace these services delivered under the current service contract(s). A number of these service contracts operate across the 3ES and other users. The new service contracts are expected to be awarded during 2015 to facilitate the commencement of service delivery from late 2016 as existing service contracts with Airwave begin to expire.

The ESMCP aims to maximize the sharing of commercial infrastructure with the emergency services. The contractual structure defined by the UK HO consists of a set of four contracts between the central government and the commercial suppliers:

Lot 1, ESN delivery partner (DP) – transition support, cross-lot integration and user support: a delivery partner (DP) to provide programme management services for cross-lot ESN integration, programme management services for transition, training support services, test assurance for cross-lot integration and vehicle installation design and assurance

Lot 2, ESN user services (US) – a technical service integrator to provide end-to-end systems integration for the ESN: provide public safety communications services (including the development and operation of public safety applications) and provide the necessary telecommunications infrastructure, user device management, customer support and service management

Lot 3, ESN mobile services (MS) – a resilient mobile network: a network operator to provide an enhanced mobile communications service with highly available full coverage in the defined lot 3 area (in GB), highly available extended coverage over the lot 4 telecommunications network and technical interfaces to lots 2 and 4

Lot 4, ESN extension services (ES) – coverage beyond the lot 3 network: a neutral host to provide a highly available telecommunications network in the defined lot 4 areas to enable the lot 3 supplier to extend their coverage

Such contracts include clauses on every aspect from government step-in upon failure, SLAs on minimum availability, contract transfer limits, limits on force majeure claims and so on. Remarkably, among the suppliers competing for the network (lot 3) are Airwave Solutions Ltd, EE Ltd, Telefonica UK Ltd, UK Broadband Networks Ltd and Vodafone Ltd. This reflects that MNOs in the United Kingdom seem to see a business model within the mission-critical market, especially if the government pays to harden the networks to meet PPDR standards [9]. Under this contract, it is assumed that the government could buy capacity at wholesale rates for its PPDR services at much lower prices than individual subscribers. Also, the bidding framework requires tenders from multiple MNOs, introducing competition into the price offers. According to UK HO officials [72], the transition to the ESN should begin in 2016 to enable completion by 2020. The contract for the service is estimated to be worth up to £1.2 billion [73].

With regard to spectrum, UK government's policy is to divest itself of spectrum and for users (including government) to pay market rates [74]. Accordingly, the preferred direction is to minimize the requirement for dedicated spectrum and to ensure that any spectrum used is in harmonized bands to allow use of commercial off-the-shelf (COTS) equipment. Dedicated spectrum is only envisioned if justified either as the only mean of providing the required operational capability or as a mean of achieving a better overall commercial outcome. In particular, dedicated spectrum may be needed for direct mode-type applications, air-to-ground support or a private network should the preferred option fail. In this way, the ECS in the United Kingdom is likely to use 800-MHz band frequencies assigned through auctioning to the commercial MNOs.

3.4.5 TCCA

The TETRA MoU Association Ltd, now known as the TCCA, was established in December 1994 to create a forum that could act on behalf of all interested parties, representing users, manufacturers, application providers, integrators, operators, test houses and telecom agencies.

Nowadays, the TCCA represents more than 160 organizations from all continents of the world. All the governments in Europe are members of TCCA. Half of the Board of TCCA are national governments' representatives. Many PPDR operators and other critical users in Europe take TCCA's views into consideration to define their future roadmaps.

A Critical Communications Broadband Group (CCBG) [75] was established within TCCA to provide support information and guidelines for critical communications users, operators and other interested parties who are considering the implementation of mission-critical mobile broadband services. Remarkably, the TCCA's CCBG is working with public safety, transportation, utilities and other key stakeholder groups worldwide to [76]:

- Drive the standardization of common, global mobile broadband technology solutions for critical communications users
- Lobby for appropriate (and as far as possible harmonized) spectrum for deployment of critical communications broadband networks

In this context, the CCBG is working towards the definition of a robust LTE migration roadmap for PPDR and other critical communications network solutions, initially for data services. Sustained on the fact that standardization, together with consequent conformance and interoperability testing, has been a fundamental aspect in the worldwide success of TETRA, TCCA strongly supports the development of common, global standards, based on LTE, for the future of critical communications worldwide. According to TCCA, the potential market is much larger than just for PPDR alone. TCCA supports the idea that nations will ultimately have one or more private critical LTE networks, operating in dedicated spectrum and complemented with public MNO services. A number of white papers and reports (see [77–79]) have been issued by TCCA's CCBG on issues related to delivery options for mission-critical broadband and discussion on practical standardization and roadmap considerations.

The roadmap reproduced in Figure 3.5 has been established by TCCA to show the phasing from the existing mission-critical voice networks to mission-critical broadband [80].

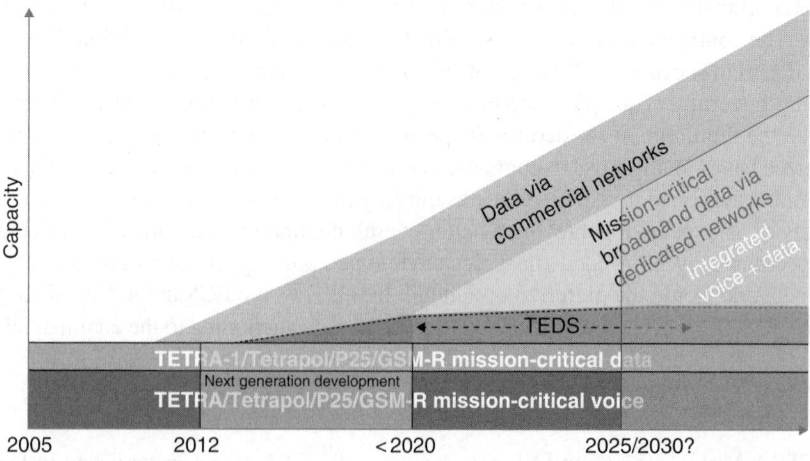

Figure 3.5 Roadmap and timescales based on practical implementation of a harmonized European solution. Reproduced from Ref. [80].

The roadmap points out that existing TETRA/TETRAPOL/P25/GSM-R networks are needed until 2025–2030 for mission-critical voice. These technologies are also able to deliver limited mission-critical data functionality. Enhancements such as TETRA Enhanced Data Services (TEDS) are nowadays available and offer greater (wideband) data capacity, so that some countries may introduce TEDS to some extent. Therefore, it is envisioned that most countries will continue operating their PMR networks for at least another 10–15 years and, in parallel, will start (or have already started) to deliver broadband data applications mostly via commercial networks, which can already provide data broadband functionalities today. In this way, the period until 2020 would be used primarily for preparing a mission-critical broadband solution (e.g. harmonized frequency band, technology readiness). When this work is ready around 2020, dedicated networks can be realized. Afterwards, the roadmap gives an indication around 2025–2030 for the broadband networks also to deliver mission-critical voice. In any case, this period is seen as uncertain since the migration of voice services from legacy PMR networks to the broadband network is not only dependent on technology maturity (full TETRA functionality is expected to take longer to replicate) but also on the fact that broadband coverage has achieved similar or better coverage than existing narrowband networks. By then, commercial networks will still be used (e.g. for non-mission critical) in a hybrid model.

References

[1] ETSI TR 103 064 V1.1.1, 'Business and Cost considerations of Software Defined Radio (SDR) and Cognitive Radio (CR) in the Public Safety domain', April 2011.
[2] R. Ferrús, R. Pisz, O. Sallent and G. Baldini, 'Public Safety Mobile Broadband: A Techno-Economic Perspective', Vehicular Technology Magazine, IEEE, vol. 8, no. 2, pp.28, 36, June 2013.
[3] APCO Global Alliance, 'Updated Policy Statement – 4th Generation (4G) Broadband Technologies for Emergency Services', October 2013. Available online at http://apcoalliance.org/4g.html (accessed 27 March 2015).
[4] NPSTC, 700 MHz Public Safety Broadband Task Force Report and Recommendations, September 2009.
[5] TETRA and Critical Communications Association (TCCA) Board, 'Statement to 3GPP and other interested parties regarding adoption of LTE', October 2012. Available online at http://www.tandcca.com/Library/Documents/LTEBoardstatement.pdf (accessed 27 March 2015).
[6] Balazs Bertenyi, 'LTE Standards for Public Safety – 3GPP view', Critical Communications World, 21–24 May 2013.
[7] Public Safety Homeland Security Bureau, Federal Communications Commission, 'The Public Safety Broadband Wireless Network: 21st Century Communications for First Responders', March 2010.
[8] TETRA + Critical Communications Association (TCCA), 'Mobile Broadband in a Mission Critical Environment', January 2012.
[9] Simon Forge, Robert Horvitz and Colin Blackman, 'Study on use of commercial mobile networks and equipment for "mission-critical" high-speed broadband communications in specific sectors', Final Report, December 2014.
[10] Ryan Hallahan and Jon M. Peha, 'Quantifying the Costs of a Nationwide Public Safety Wireless Network', Telecommunications Policy, Elsevier, vol. 34, no. 4, pp. 200–220, 2010.
[11] Federal Communications Commission, 'A broadband network cost model: a basis for public funding essential to bringing nationwide interoperable communications to America's first responders', OBI Technical Paper no. 2, May 2010.
[12] Francesco Pasquali, 'The TETRA business case', TETRA Association Board Member, March 2007.
[13] GSMA White Paper, 'Mobile infrastructure sharing'. Available online at http://www.gsma.com/publicpolicy/wp-content/uploads/2012/09/Mobile-Infrastructure-sharing.pdf.
[14] Alcatel Lucent, 'A How-to Guide – FirstNet Edition', 2012. Available online at http://enterprise.alcatel-lucent.com/private/images/public/si/pdf_publicSafety_howto.pdf (accessed 27 March 2015).

[15] F. Craig Farrill, 'FirstNet Nationwide Network (FNN) Proposal', First Responders Network Authority, Presentation to the Board, 25 September 2012. Available online at http://www.ntia.doc.gov/files/ntia/publications/firstnet_fnn_presentation_09-25-2012_final.pdf.

[16] R. Ferrús and O. Sallent, 'Extending the LTE/LTE-A Business Case: Mission- and Business-Critical Mobile Broadband Communications', Vehicular Technology Magazine, IEEE, vol. 9, no. 3, pp. 47, 55, September 2014.

[17] Radio Spectrum Committee (RSC), European Commission, 'Public consultation document on use of spectrum for more efficient energy production and distribution', May 2012.

[18] Ryan Herrallahan and Jon M. Peha, 'The business case of a network that serves both public safety and commercial subscribers', Telecommunications Policy, Elsevier, vol. 35, no. 3, pp. 250–268, April 2011.

[19] Christian Mouraux, 'ASTRID High Speed Mobile Data MVNO', PMR Summit, Barcelona, 18 September 2012.

[20] R. Baldini, R. Ferrús, O. Sallent, P. Hirst, S. Delmas and R. Pisz, 'The evolution of Public Safety Communications in Europe: the results from the FP7 HELP project', ETSI Reconfigurable Radio Systems Workshop, Sophia Antipolis, France, 12 December 2012.

[21] R. Ferrús, O. Sallent, G. Baldini and L. Goratti, 'LTE: The Technology Driver for Future Public Safety Communications', Communications Magazine, IEEE, vol. 51, no. 10, pp. 154, 161, October 2013.

[22] 3GPP TS 22.153 V11.1.0, 'Multimedia priority service (Release 11)', June 2011.

[23] PROSIMOS project public deliverables. Available online at http://www.prosimos.eu (accessed 27 March 2015).

[24] Roberto Gimenez, Inmaculada Luengo, Anna Mereu, Diego Gimenez, Rosa Ana Casar, Judith Pertejo, Salvador Díaz, Jose F. Monserrat, Vicente Osa, Javier Herrera, Maria Amor Ortega and Iñigo Arizaga, 'Simulator for PROSIMOS (PRiority communications for critical SItuations on MObile networkS) Service', Towards a Service-Based Internet. ServiceWave 2010 Workshops, Lecture Notes in Computer Science, 2011.

[25] Ryan Hallahan and Jon M. Peha, 'Compensating Commercial Carriers for Public Safety Use: Pricing Options and the Financial Benefits and Risks', 39th Telecommunications Policy Research Conference, September 2011.

[26] Ryan Hallahan and Jon M. Peha, 'Policies for Public Safety Use of Commercial Wireless Networks', 38th Telecommunications Policy Research Conference, October 2010.

[27] Xu Chen, Dongning Guo and J. Grosspietsch, 'The public safety broadband network: a novel architecture with mobile base stations', Proceedings of the 2013 IEEE International Conference on Communications (ICC), pp. 3328, 3332, 9–13 June 2013. http://ieeexplore.ieee.org/stamp/stamp.jsp?tp=&arnumber=6655060&isnumber=6654691 (accessed 27 March 2015).

[28] SNS Research Report, 'HetNet Bible (Small Cells, Carrier WiFi, DAS & C-RAN): 2014–2020 – Opportunities, Challenges, Strategies, & Forecasts'. Available online at http://www.snstelecom.com/hetnet (accessed 27 March 2015).

[29] E. Lang, S. Redana and B. Raaf, 'Business impact of relay deployment for coverage extension in 3GPP LTE-advanced', ICC Workshops 2009, IEEE International Conference on Communications (ICC), pp. 1, 5, 14–18 June 2009. Available online at http://ieeexplore.ieee.org/stamp/stamp.jsp?tp=&arnumber=5208000&isnumber=5207960 (accessed 27 March 2015).

[30] ICT Regulation Toolkit, Information for Development Program (infoDev) and International Telecommunication Union (ITU). Available online at http://www.ictregulationtoolkit.org/en/index.html (accessed 27 March 2015).

[31] ITU Telecommunication Development Sector, 'Exploring the value and economic valuation of spectrum', Broadband Series, April 2012.

[32] Alexander Grous, 'Socioeconomic Value of Mission Critical Mobile Applications for Public Safety in the UK: 2×10 MHz in 700 MHz', Centre for Economic Performance, London School of Economics and Political Science, November 2013. Available online at http://www.tandcca.com/Library/Documents/Broadband/LSE%20PPDR%20UK.pdf (accessed 27 March 2015).

[33] Alexander Grous, 'Socioeconomic Value of Mission Critical Mobile Applications for Public Safety in the UE: 2×10 MHz in 700 MHz in 10 European Countries', Centre for Economic Performance, London School of Economics and Political Science, December 2013. Available online at http://www.tandcca.com/Library/Documents/Broadband/LSE%20PPDR%20EU.PDF (accessed 27 March 2015).

[34] WIK-Consult, 'PS Spectrum harmonisation in Germany, Europe and Globally', Study for the German Federal Ministry of Economics and Technology, December 2010.

[35] Coleago Consulting, 'German Spectrum Auctions Results', August 2010. Available online at http://coleago.wordpress.com/2010/08/04/german-spectrum-auctions-results/ (accessed 27 March 2015).

[36] Bob Lovett, 'Emergency services mobile communications programme', BAPCO Update, November 2013.

[37] J.A. Hoffmeyer, 'Regulatory and Standardization Aspects of DSA Technologies – Global Requirements and Perspectives', in Proceedings of the First IEEE International Symposium, New Frontiers in Dynamic Spectrum Access Networks (DySPAN 2005), Baltimore, pp. 700–705, 2005.
[38] R. Ferrús, O. Sallent, G. Baldini and L. Goratti, 'Public Safety Communications: Enhancement Through Cognitive Radio and Spectrum Sharing Principles', Vehicular Technology Magazine, IEEE, vol. 7, no. 2, pp. 54–61, June 2012.
[39] Federal Communications Commission (FCC), 'Third memorandum opinion and order – unlicensed operation in the TV broadcast bands', Document FCC 12-36, April 2012.
[40] Ofcom, 'Implementing geolocation: summary of consultation responses and next steps', September 2011.
[41] ECC Report 169, Description of practices relative to trading of spectrum rights of use, Paris, May 2011.
[42] COM(2012) 478 Final, 'Promoting the shared use of radio spectrum resources in the internal market', European Commission, September 2012.
[43] CEPT ECC FM Project Team 53 on 'Reconfigurable Radio Systems (RRS) and Licensed Shared Access (LSA)'. Available online at http://www.cept.org/ecc/groups/ecc/wg-fm/fm-53/page/terms-of-reference (accessed 27 March 2015).
[44] New Work Item (NWI) in ETSI TC RRS, 'Mobile broadband services in the 2300 MHz – 2400 MHz frequency band under Licensed Shared Access regime', May 2012.
[45] Dawinderpal Sahota, 'Spectrum sharing could threaten operator investment says GSMA', Telecoms.com, 11 February 2014. Available online at http://www.telecoms.com/221681/spectrum-sharing-could-threaten-operator-investment-says-gsma/?utm_source=rss&utm_medium=rss&utm_campaign=spectrum-sharing-could-threaten-operator-investment-says-gsma (accessed 27 March 2015).
[46] Marguerite Reardon, 'Hurricane Sandy disrupts wireless and Internet services', CNET, October 2012. Available online at http://www.cnet.com/news/hurricane-sandy-disrupts-wireless-and-internet-services/ (accessed 27 March 2015).
[47] David Kahn, 'Will First Responders Be Able to Communicate When the Next Hurricane Sandy Hits?', HSToday.us, 12 March 2013. Available online at http://www.hstoday.us/blogs/best-practices/blog/will-first-responders-be-able-to-communicate-when-the-next-hurricane-sandy-hits/c5d9ae9cec463da2ed8b3de53dd67009.html (accessed 27 March 2015).
[48] TCCA, 'Mobile Broadband for Critical Communications Users: a review of options for delivering Mission Critical solutions', December 2013.
[49] Ericsson White Paper, 'Public safety mobile broadband', February 2014.
[50] NPSTC, position paper on 'Why can't public safety just use cell phones and smart phones for their mission critical voice communications?', April 2013. Available online at http://psc.apcointl.org/wp-content/uploads/Why-cant-PS-just-use-cell-phones-NPSTC-041513.pdf (accessed 27 March 2015).
[51] ETSI TR 103 269-1 V1.1.1, 'TETRA and Critical Communications Evolution (TCCE); Critical Communications Architecture; Part 1: Critical Communications Architecture Reference Model', July 2014.
[52] APCO International's online Application Community. Available online at http://appcomm.org/ (accessed 27 March 2015).
[53] Federal Manager's Daily Report, 'FBI Issues Solicitation for 26,500 Samsung KNOX Licenses', Federal Manager's Daily Report, 23 June 2014. Available online at http://www.fedweek.com/federal-managers-daily-report/fbi-issues-solicitation-for-26500-samsung-knox-licenses/ (accessed 27 March 2015).
[54] National Public Safety Telecommunications Council (NPSTC), 'Public Safety Broadband Console Requirements', 30 September 2014.
[55] Donny Jackson, 'Integration of fire gear, communications needed, but it will take time', Urgent Communications, 16 July 2013.
[56] J. Scott Marcus, 'The need for PPDR Broadband Spectrum in the bands below 1 GHz for the TETRA + Critical Communication Association', October 2013.
[57] CEPT ECC Report 199, 'User requirements and spectrum needs for future European broadband PPDR systems (Wide Area Networks)', May 2013.
[58] John Ure, 'Public Protection and Disaster Relief (PPDR) Services and Broadband in Asia and the Pacific: A Study of Value and Opportunity Cost in the Assignment of Radio Spectrum', June 2013.
[59] US NPSTC, 'Public Safety Communications Assessment 2012–2022: Technology, Operations, & Spectrum Roadmap', Final Report, 5 June 2012.
[60] ETSI TR 102 628, 'Additional spectrum requirements for future Public Safety and Security (PSS) wireless communication systems in the UHF frequency', August 2010.

[61] CEPT ECC FM49 on 'Radio Spectrum for Public Protection and Disaster Relief (PPDR)', Working documents. Available online at http://www.cept.org/ecc/groups/ecc/wg-fm/fm-49/page/terms-of-reference (accessed 27 March 2015).
[62] FirstNet's. Available online at www.firstnet.gov (accessed 27 March 2015).
[63] Public Safety Communications Research (PSCR) Program, US Department of Commerce – Boulder Laboratories. Available online at http://www.pscr.gov/ (accessed 27 March 2015).
[64] US Department of Homeland Security, 'Public Safety Communications Evolution', November 2011.
[65] Draft ECC Report 218, 'Harmonised conditions and spectrum bands for the implementation of future European broadband PPDR systems', April 2014.
[66] ASTRID. Available online at http://www.astrid.be/ (accessed 27 March 2015).
[67] Blue Light Mobile Service. Available online at http://bluelightmobile.be/en/home (accessed 27 March 2015).
[68] Jarmo Vinkvist, Tero Pesonen and Matti Peltola, 'Finland's 5 steps to Critical Broadband', RadioResource International Magazine (RRImag.com), Quarter 4, 2014.
[69] Vincent Lemonnier, 'LTE for critical communications', ETSI Summit on Critical Communications, 20 November 2014.
[70] Donny Jackson, 'UK seeks to replace TETRA with LTE as early as 2016', Urgent Communications, 6 June 2013.
[71] UK Home Office, Promotional material on 'Emergency services mobile communications programme'. Available online at https://www.gov.uk/government/publications/the-emergency-services-mobile-communications-programme (accessed 27 March 2015).
[72] Gordon Shipley, 'UK Emergency Services Mobile Communications Programme (ESMCP)', PSCR Public Safety Broadband Stakeholder Conference, June 2013.
[73] Ian Weinfass, 'Airwave replacement on track for 2016', Police Oracle, 11 November 2014. Available online at http://www.policeoracle.com/news/Airwave-replacement-on-track-for-2016_86158.html (accessed 27 March 2015).
[74] HORIZON 2020 work programme 2014–2015, 'Secure societies – protecting freedom and security of Europe and its citizens', European Commission Decision C (2014)4995 of 22 July 2014.
[75] TCCA Broadband Group Page. Available online at http://www.tandcca.com/assoc/page/18100 (accessed 27 March 2015).
[76] Tony Gray (Chairman, TCCA Critical Communications Broadband Group (CCBG)), 'Assessing and Delivering the Fundamentals for Critical Communications Broadband Networks', LTE World Summit, 2014. Available online at http://www.gsacom.com/downloads/pdf/PSCCSzone_Tony_Gray_TCCA_June_2014.php4 (accessed 27 March 2015).
[77] TETRA and Critical Communications Association (TCCA), 'Mobile Broadband for Critical Communications Users: a review of options for delivering Mission Critical solutions', December 2013. Available online at http://www.tandcca.com/Library/Documents/Broadband/MCMBB%20Delivery%20Options%20v1.0.pdf (accessed 27 March 2015).
[78] TETRA and Critical Communications Association (TCCA), 'The Strategic Case for Mission Critical Broadband', December 2013. Available online at http://www.tandcca.com/Library/Documents/Broadband/MCMBB%20Strategic%20Case%20v1_0.pdf (accessed 27 March 2015).
[79] TETRA and Critical Communications Association (TCCA), 'Mission Critical Mobile Broadband: practical standardisation and roadmap considerations', February 2013. Available online at http://www.tandcca.com/Library/Documents/CCBGMissionCriticalMobileBroadbandwhitepaper2013.pdf (accessed 27 March 2015).
[80] Hans Borgonjen, 'European PPDR Broadband situation', CEPT/ECC CPG-PTA meeting, Mainz, January 2013.

4

LTE Technology for PPDR Communications

4.1 Standardization Roadmap towards Mission-Critical LTE

LTE [1] is part of the GSM evolutionary path established by the 3rd Generation Partnership Project (3GPP) for mobile broadband, following EDGE, UMTS, HSPA and HSPA Evolution (HSPA+). The LTE standardization work within 3GPP started in 2004. The first specification (LTE Release 8) was frozen in December 2008 and was the basis for the first wave of LTE equipment that entered in the market in late 2009. From Release 8, enhancements are being introduced in subsequent releases. The production of a new 3GPP standards release usually takes between 18 and 24 months, and the introduction of new major capabilities into the standards typically spans several releases. At the time of writing, the latest frozen release is Release 11, with the core network protocols stable in December 2012 and radio access network (RAN) protocols stable in March 2013. Work is in progress towards LTE Release 12 (functional freeze date[1] set for March 2015) and LTE Release 13 (functional freeze date set for March 2016). From Release 10 onwards, the technology is referred to as LTE-Advanced since it fulfils the requirements set by ITU for IMT Advanced (i.e. 4G) systems. Some new capabilities brought by LTE Release 10 such as carrier aggregation (CA) have been already commercially deployed along 2014.

The standards work developed at 3GPP is evolving today's mobile broadband technologies by aiming to address the immense challenge facing mobile network operators in order to cope with the continued increase and high growth projections in mobile data traffic in the coming years [2]. In particular, mobile video is expected to generate much of the mobile traffic growth,

[1]These are the indicative functional freeze dates provided by 3GPP (www.3gpp.org). According to 3GPP, after 'freezing', a release can have no further additional functions added. However, detailed protocol specifications may not yet be complete at the functional freeze date and require additional time. Moreover, considerable number of refinements and corrections can be expected for at least 2 years following the freeze dates.

Mobile Broadband Communications for Public Safety: The Road Ahead Through LTE Technology, First Edition.
Ramon Ferrús and Oriol Sallent.

increasing also the need for much higher bit rates compared to other mobile content types. In this context, the main requirements set out for LTE technology were about achieving higher spectral efficiency and higher peak data rates, as well as flexibility in frequency and bandwidth for operators to be able to deploy LTE across their diverse spectrum holdings. LTE Release 8 includes capabilities enabling downlink/uplink peak data rates of up to 300/75 Mb/s with 20 MHz bandwidth, operation in both TDD and FDD modes and supports scalable bandwidth from 1.4 up to 20 MHz. LTE-Advanced further improves the supported bit rates and spectral efficiency in a cost-efficient way, reaching peak data rates of up to 3/1.5 Gb/s over an aggregated 100 MHz bandwidth.

LTE has been designed to provide a high-rate, very-low-latency IP connectivity solution with a clear focus on the commercial business and consumer markets. The IP connectivity service delivered over LTE access can be utilized by almost any application relying on IP communications, enabling a large number of services to be provided over LTE networks. Indeed, the LTE ecosystem has matured rapidly, becoming the prevalent standard for mobile broadband communications worldwide.

However, while the aforementioned features make LTE a suitable technology for deploying a rich number of mobile broadband applications for PPDR, including video delivery, important features are still lacking to turn the LTE standard into a full mission-critical technology [3, 4] that could, in the longer term, be well positioned to replace current narrowband PMR technologies. To that end, there are several work and study items currently in progress within 3GPP aimed at extending the LTE specifications to add support for the functionalities expected by PPDR first responders as well as other mission-critical or business-critical users (e.g. utilities, transportation). These new features include:

- **Group communications system enablers**, together with push-to-talk (PTT) voice application and its evolution towards multimedia (voice, data, video, etc.) group communications. Enablers for group communications are covered under Release 12 specifications, and support for a mission-critical push-to-talk (MCPTT) application is included in Release 13.
- **Proximity-based services**, which enable device-to-device communications without the need to have coverage from a network infrastructure (off-network operation). ProSe support was planned for Release 12 specifications, but some features are being addressed under Release 13. Notice that device-to-device communications, together with group communications and PTT, are among the key requirements for 'mission-critical voice' discussed in Chapter 1.
- **Isolated network operation**, enabling any base station to act alone in routing calls and messages between such parts of the network that stay operational after, for example, a disaster had partially caused some network equipment to fail. These features are planned for Release 13.
- **High-power terminals**, enabling the use of high transmit power in specific bands to increase the coverage range. A higher power transmit class has already been specified under Release 11.
- **Prioritization and quality-of-service (QoS) control features**. LTE technology already provides a suite of standard capabilities for prioritization and QoS control since Release 8. Additional enhancements to support priority services have been addressed in subsequent releases up to Release 11.

In addition to the above features, the capabilities of LTE technology enabling the active sharing of RAN equipment by multiple operators, known as RAN sharing, are also relevant for the PPDR sector as long as they can be instrumental for the realization of PPDR service delivery models based on network sharing practices with commercial operators, as discussed in Chapter 6. In this regard, the technological enhancements being introduced in LTE RAN sharing are expected to bring further flexibility in sharing network resources so that they could permit the deployment of networks (or parts of them) that are dynamically shared between critical and non-critical users. New RAN sharing capabilities are referred to as **RAN sharing enhancements** and are expected to form part of Release 13.

Figure 4.1 illustrates the abovementioned list of features already supported or being introduced in LTE specifications. After a brief introduction to the LTE technology in Section 4.2, these items are further described in Sections 4.3–4.8.

Organizations such as the National Public Safety Telecommunications Council (NPSTC), TETRA and Critical Communications Association (TCCA) and the ETSI Technical Committee on TETRA and Critical Communications Evolution (ETSI TC TCCE) closely cooperate with the 3GPP to provide requirements and technical inputs to guide the specification of the necessary LTE enhancements [5]. Complementing the 3GPP work, there are other standard development organizations (SDOs) that work on application layer standards (i.e. over the top

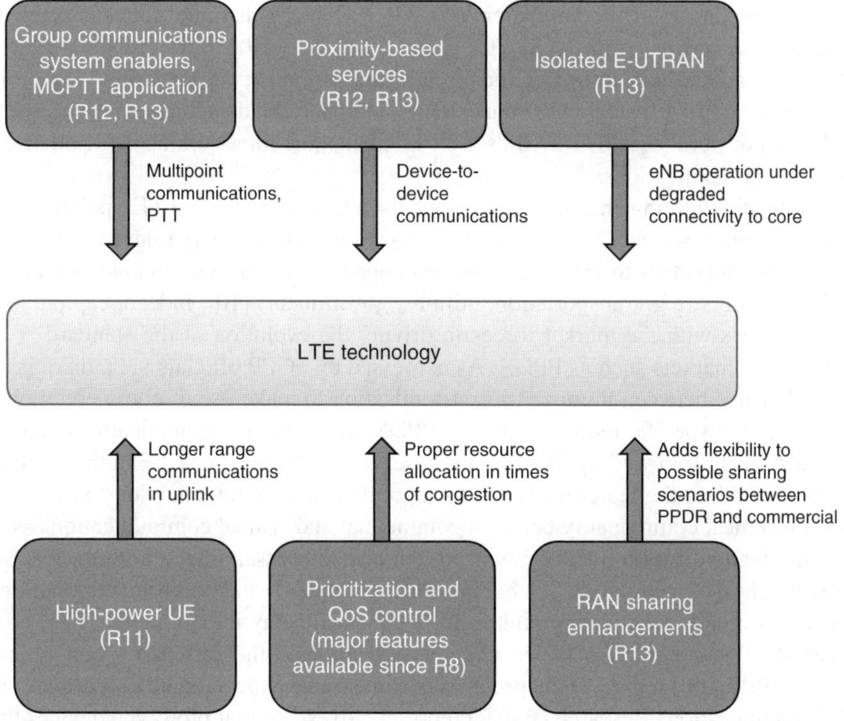

Figure 4.1 Features supported or being introduced in LTE specifications especially relevant for PPDR and critical communications.

of network layer access technology standards such as LTE) with a focus for specific extensions for PPDR communications. One of these SDOs is ETSI TC TCCE, which has defined a critical communications architecture reference model [6] and, on this basis, develops the architecture for a generic mission-critical service equivalent to the existing narrowband technologies, which could be used over a broadband IP bearer, with specific focus for LTE [7]. Another SDO to playing an important role in the specification of application layer components that can be incorporated in mission-critical communications solutions is the Open Mobile Alliance (OMA). The OMA is a global organization that delivers open specifications for creating interoperable services that work across all geographical boundaries, on any bearer network. OMA has recently conducted the specification of a Push-to-Communicate for Public Safety (PCPS) application, consolidating a previous solution known as Push-to-Talk over Cellular (PoC) adopted in the commercial domain. OMA is currently working with 3GPP to find the best method for the effective transfer of this specification to 3GPP so that this can be leveraged and further continued within 3GPP to meet the requirements of the PPDR community [8]. In this regard, a new working group (WG SA6) has been set up within the 3GPP in late 2014 to specifically undertake the standardization work for applications in the mission-critical communications space [9], in what represents a major consensus effort from the industry towards the establishment of a globally recognized body for the standardization critical communications applications. In particular, WG SA6 is responsible for the definition, evolution and maintenance of the technical specifications for application layer functional elements and interfaces supporting critical communications, MCPTT being its initial focus of work. Close collaboration is expected from ETSI TCCE and OMA as well as other SDOs dealing with PPDR-related standardization (e.g. Telecommunications Industry Association (TIA) for P25) to make use of expertise available in other groups to support the development of specifications under the responsibility of WG SA6 as well as to support interworking with other critical communications applications.

It is worth highlighting that the improvements introduced in the LTE specifications, in particular proximity services and group communications, are not only relevant for the PPDR sector but also important to raise new business opportunities in the commercial and other professional sectors (e.g. transportation, utilities, government) [10]. Indeed, capitalizing and seeking synergies with the market forces in driving the evolution of the standard is a vital aspect for niche markets such as PPDR. As expressed by 3GPP officials [11], there is a need to strike a balance between more or less customization to make use of commercial products while meeting the specific requirements for PPDR and critical communications. Therefore, the approach being followed by 3GPP is to preserve the strengths of LTE in the commercial domain while adding the features needed to support critical communications and seeking to maximize technical commonality between commercial and critical communications aspects.

Once the standardization work is finalized, there is a necessary delay until the availability in the marketplace of technology that both meets the standards and is sufficiently mature to be relied upon for mission-critical communications. Typical delay in commercial LTE products and other 3GPP standards is between one and 2 years from the date that specifications are frozen by 3GPP. In this regard, some preliminary time frame expectations, according to Public Safety Communications Research (PSCR) representatives, are that prototype devices that are designed to deliver mission-critical voice over LTE will be in their labs for evaluation in 2015 [12]. In the TCCA's opinion [3], the earliest that LTE technology suitable for PPDR and other critical data communications use could be available for purchase is 2018.

4.2 LTE Fundamentals

LTE is a radio access technology (RAT) designed to provide a high-bit-rate, very-low-latency IP connectivity service to IP-based packet data networks (PDNs) such as the public Internet or private networks either managed by the network operator or by a third party. LTE technology turns a mobile network into a versatile communications platform over which many applications and services built around the IP-based communications model, including real-time video and multimedia services, can be readily deployed by means of software applications running on terminals (e.g. LTE smartphones) and on servers and/or other computing devices reachable from the accessed PDNs. The LTE IP connectivity service has been designed to be able to provide differentiated treatment to IP traffic flows that might have different QoS requirements in terms of required bit rates as well as acceptable packet delays and packet loss rates. Therefore, it is said that LTE provides a QoS-enabled IP connectivity service.

An LTE network, referred to as Evolved Packet System (EPS) in the 3GPP specifications, consists of two main parts [13]: a RAN based on orthogonal frequency-division multiple access (OFDMA) technology, named Evolved UMTS Radio Access Network (E-UTRAN), and an enhanced packet-switched core network, named Evolved Packet Core (EPC). The E-UTRAN is mainly in charge of radio transmission functions, while session and mobility management functions are handled by the EPC.

The E-UTRAN consists of base stations named evolved Node Bs (eNBs) that provide the radio interface towards the UE and are directly connected to the EPC. Optionally, eNBs can be directly interconnected with each other in their vicinity for, for example, handover optimization purposes. Unlike UMTS and GSM systems in which the radio protocol stack functions were split between the base station and a central radio controller, the LTE eNB runs the full radio protocol stack. This characteristic is used to affirm that LTE follows a 'flat' architecture, in contrast with the hierarchical architecture of controllers and base stations used in the predecessor technologies. As part of the radio protocol stack embedded in the eNB, the Radio Resource Control (RRC) protocol is used between the UE and the eNB to control the operation of the radio interface (e.g. delivery of system information messages to UEs, activation/deactivation of bearer services, mobility control, etc.).

The EPC comprises a network entity (NE) named Mobility Management Entity (MME) to handle control functions (e.g. authentication of users and location management). Additionally, the EPC comprises two entities through which the user data traffic is transferred: a Serving Gateway (S-GW), which anchors user traffic from/to E-UTRAN into the EPC, and a PDN Gateway (P-GW), which provides the IP connectivity to the external IP networks. There is a set of protocols between the UE and the MME, referred to as non-access-stratum (NAS) protocols, to cope with session and mobility management. The operation of the EPC is assisted by the Home Subscriber Server (HSS), a central database that contains, among others, user subscription-related information.

The LTE IP connectivity service is realized through the establishment of the so-called EPS bearer services. The EPS bearer service represents the level of granularity for QoS control in E-UTRAN/EPC and provides a logical transmission path with well-defined QoS properties between the UE and the P-GW. IP-based service control platforms such as 3GPP IP Multimedia Subsystem (IMS) and related application servers can be used on top of this QoS-aware LTE connectivity service to support a diverse range of services (e.g. telephony,

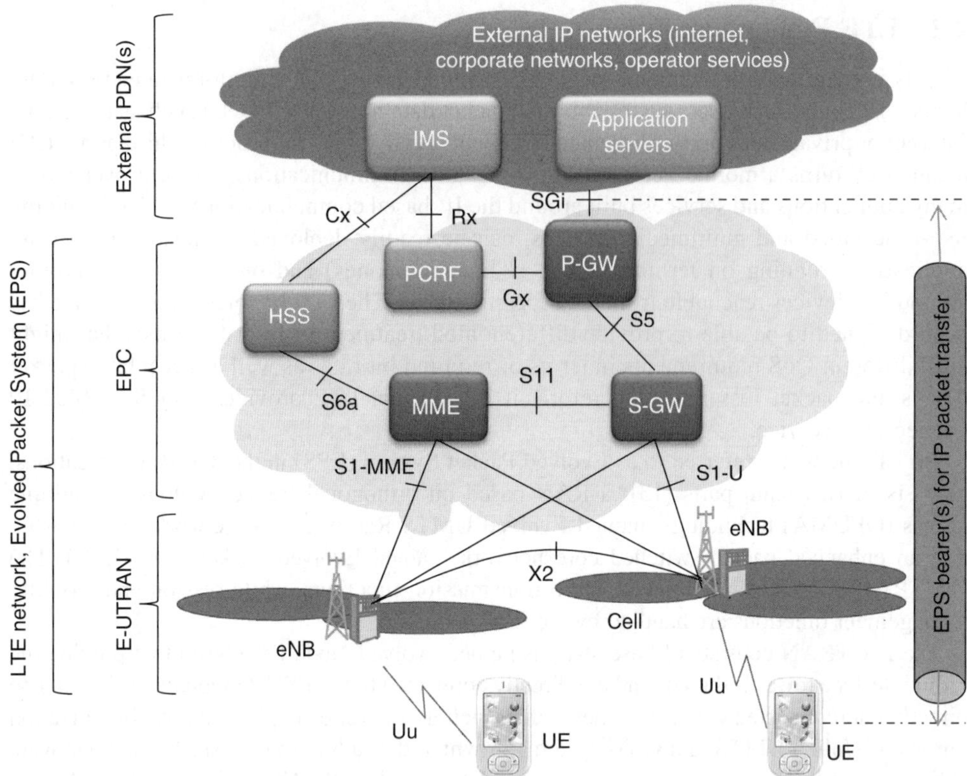

Figure 4.2 Basic architecture of an LTE network.

videoconferencing, video streaming, messaging, etc.). The IMS, also referred to as IP Multimedia Core Network (IM CN) Subsystem, enables mobile network operators to offer their subscribers multimedia services based on and built upon Internet applications, services and protocols. The IMS is mostly based on IETF 'Internet standards' to achieve access independence (e.g. IMS-based services can be delivered over any IP-based connectivity access network, with LTE being one of these possible access technologies) and to maintain a smooth interoperation with other terminals across the Internet. 3GPP has also specified a Policy and Charging Control (PCC) system, which provides operators with advanced tools for service-aware QoS and charging control. The PCC architecture, by means of a Policy and Charging Rules Function (PCRF) NE, enables control of the EPS bearer services (e.g. QoS settings) for both IMS and non-IMS services. Figure 4.2 illustrates the main LTE network components and interfaces among them as well as the concept of EPS bearer service between the UE and the external IP networks.

The following subsections are intended to provide an overview of the main aspects that are helpful to have a basic background and so facilitate the reading of the rest of this chapter covering the specific LTE features relevant for PPDR. For a more in-depth description of the LTE technology, the reader is referred to Refs [13–17].

4.2.1 Radio Interface

The LTE radio interface, identified as Uu in Figure 4.2, is based on the orthogonal frequency-division multiplexing (OFDM) transmission technique. OFDM transmits data on a large number of parallel, narrowband subcarriers. The use of relatively narrowband subcarriers (subcarrier spacing in LTE is 15 kHz), in combination with a cyclic prefix, makes OFDM transmission inherently robust to time dispersion on the radio channel without a requirement to resort to advanced and potentially complex receiver-side channel equalization.

For the downlink, OFDM transmission simplifies the receiver baseband processing with reduced terminal cost and power consumption as consequences. This is especially important considering the wide transmission bandwidths of LTE, and even more so in combination with advanced multi-antenna transmission, such as spatial multiplexing enabled by multiple-input/multiple-output (MIMO) antenna configurations. At a given time, the set of subcarriers being transmitted can carry information addressed to different users. This multiplexing technique based on OFDM is termed as OFDMA.

For the uplink, where the available transmission power is significantly lower than for the downlink, the situation is somewhat different. Rather than the amount of processing power at the receiver, one of the most important factors in the uplink design is to enable highly power-efficient transmission. This improves coverage and reduces terminal cost and power consumption at the transmitter. For this reason, single-carrier transmission based on discrete Fourier transform (DFT)-precoded OFDM is used for the LTE uplink. DFT-precoded OFDM has a smaller peak-to-average power ratio than regular OFDM, thus enabling less complex and/or higher-power terminals. As in the downlink, user multiplexing is also supported in the uplink across the frequency domain. The transmission/multiplexing technique used in the uplink is commonly referred to as single-carrier frequency-division multiple access (SC-FDMA).

The LTE transmitted signal can be viewed as a two-dimensional entity, with a subcarrier axis (frequency dimension) and a symbol axis (time dimension), as illustrated in Figure 4.3.

In the time dimension, the transmitted signal is organized into frames of 10 ms duration. Each frame is split into 10 subframes of 1 ms duration each. The subframe is the time granularity at which users(s) can be scheduled to transmit/receive in LTE, which is known as the Transmission Time Interval (TTI). A further division of the subframe results in the slot definition, which lasts for 0.5 ms. Over this time structure, each slot has room for the transmission of seven or six OFDM symbols, depending on whether a normal or extended cyclic prefix is used (the cyclic prefix is a temporal extension of the OFDM symbol that helps to combat multipath propagation in the radio channel). The normal cyclic prefix is of 4.7 μs, suitable for most deployments, while the extended cyclic prefix of 16.7 μs can be more appropriate for highly dispersive environments. Therefore, the duration of the OFDM symbol is 71.4 or 83.4 μs, which equals to the inverse of the subcarrier spacing (1/(15 kHz) = 66.7 μs) plus the corresponding cyclic prefix.

In the frequency dimension, the total number of occupied subcarriers depends on the LTE channelization being used. In this regard, LTE defines the following set of transmission bandwidths: 1.4, 3, 5, 10, 15 and 20 MHz that results in 72, 180, 300, 600, 900 and 1200 occupied subcarriers (i.e. subcarriers used for data and reference signals, not counting subcarriers left for, e.g. guard band). Within a time slot, subcarriers are grouped in blocks of 12, forming what is known as the Resource Block (RB). Indeed, RBs are the central unit for resource allocation in LTE (i.e. the resource scheduler in LTE works with a granularity of 1 RB, so that resources are allocated to users as multiples of RBs).

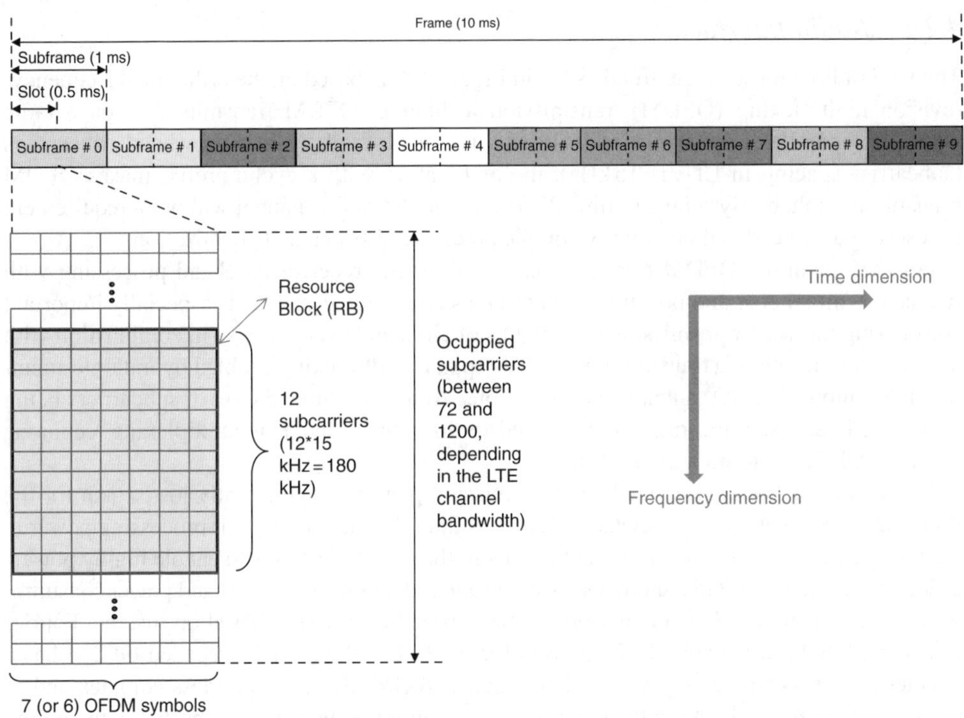

Figure 4.3 Time and frequency dimensions of the LTE radio signal.

This fine-grained resource resolution in both the time and frequency domains allows LTE to exploit channel-dependent scheduling in both dimensions to benefit, rather than suppress, rapid channel-quality variations due to fading, thereby achieving more efficient utilization of the available radio resources. LTE transmissions also exploit link adaptation techniques that make use of turbo coding schemes with variable coding rates and different modulations such as quadrature phase-shift keying (QPSK), 16-QAM or 64-QAM. Indeed, a radio resource scheduler is a central function in the operation of the LTE radio interface, and largely, it determines the overall system performance, especially in a highly loaded network. Of note is that in LTE both the downlink and uplink transmissions are controlled by the scheduler implemented at the eNB and are supported over shared channels, as opposed to the use of dedicated channels (i.e. unlike previous GSM and UMTS system, LTE does not add support for dedicated channels).

The shorter TTI supported by LTE compared to 3G technologies (HSPA uses a TTI of 2 ms) directly results in reduced user plane latency as well as shorter round-trip time for control signalling in the radio interface that enables the use of fast hybrid Automatic Repeat reQuest (ARQ) processes and the fast delivery of channel-quality feedback from terminals to eNBs.

A summary of the discussed parameters of the LTE radio interface is provided in Table 4.1, together with an estimation of the raw peak data rate achievable under the different bandwidth options.

Table 4.1 LTE radio transmission overview information.

Access scheme	DL	OFDMA
	UL	SC-FDMA (DFT-spread OFDM)
Bandwidth		1.4, 3, 5, 10, 15, 20 MHz
Subcarrier spacing		15 kHz
Occupied subcarriers		72,180,300,600,900,1200
ODFM symbol duration		71.4 or 83.4 μs
Subframe duration (i.e. scheduler TTI)		1 ms
Modulation		QPSK, 16-QAM, 64-QAM
(Raw) peak bit rate (for each channel bandwidth and assuming 64-QAM modulation and no spatial multiplexing)		6, 15, 25, 50, 75, 100 Mb/s

A central feature to highlight about LTE radio interface is its high degree of spectrum flexibility in terms of:

- **Duplex arrangement**. LTE can be deployed in both *paired* and *unpaired* spectrum. To that end, LTE supports both frequency- and time-division-based duplex (FDD and TDD) arrangements. Unlike previous technologies such as UMTS that also defined support for both duplexing modes, LTE enables both modes within a single RAT with minimum technological deviations, facilitating the manufacturing of equipment capable of supporting both FDD and TDD.
- **Operating bands**. LTE is able to operate in a wide range of frequency bands, from as low as 450 MHz band up to 3.8 GHz. The E-UTRA operating bands specified up to Release 12 are shown in Table 4.2 [18]. From a radio access functionality perspective, this has limited impact, and LTE physical layer specifications do not assume any specific band. What may differ, in terms of specification, are mainly more specific radio-frequency (RF) requirements such as the allowed maximum transmit power, requirements/limits on out-of-band emission, etc., due to potentially different external constraints coming from spectrum regulations.
- **Channel arrangement**. Various channel bandwidths are available in the LTE technology allowing for spectrum flexibility (1.4, 3, 5, 10, 15 and 20 MHz). Indeed, the LTE physical layer specifications are bandwidth agnostic and do not make any particular assumption on the supported transmission bandwidths beyond a minimum value. Hence, LTE specifications allow for transmission bandwidth ranging from around 1 MHz up to beyond 20 MHz in steps of 180 kHz (the size of the RB). Therefore, though RF requirements have been only specified so far for the previously mentioned set of transmission bandwidths, additional transmission bandwidths could be easily supported by updating only the RF specifications [13].
- **Carrier Aggregation (CA)**. CA is a feature that was first introduced for LTE-Advanced in Release 10 where multiple Release 8 component carriers were allowed to be aggregated together intra-band and offer a means to increase both the peak data rate and throughput. In subsequent releases, inter-band CA was also specified, where the component carriers can be in different frequency bands. The number of CA schemes is growing from 3 in Release 10 to over 20 in Release 11, a clear indication of the great interest in developing this capability. Within Release 12, 3GPP is working on procedures for allowing UEs to aggregate both TDD and FDD spectrum jointly [19].

Table 4.2 E-UTRA operating bands [18].

E-UTRA operating band	Uplink (UL) operating band BS receive UE transmit	Downlink (DL) operating band BS transmit UE receive	Duplex mode
	$F_{UL\_low} - F_{UL\_high}$ (MHz)	$F_{DL\_low} - F_{DL\_high}$ (MHz)	
1	1920–1980	2110–2170	FDD
2	1850–1910	1930–1990	FDD
3	1710–1785	1805–1880	FDD
4	1710–1755	2110–2155	FDD
5	824–849	869–894	FDD
6	830–840	875–885	FDD
7	2500–2570	2620–2690	FDD
8	880–915	925–960	FDD
9	1749.9–1784.9	1844.9–1879.9	FDD
10	1710–1770	2110–2170	FDD
11	1427.9–1447.9	1475.9–1495.9	FDD
12	699–716	729–746	FDD
13	777–787	746–756	FDD
14	788–798	758–768	FDD
15	Reserved	Reserved	FDD
16	Reserved	Reserved	FDD
17	704–716	734–746	FDD
18	815–830	860–875	FDD
19	830–845	875–890	FDD
20	832–862	791–821	FDD
21	1447.9–1462.9	1495.9–1510.9	FDD
22	3410–3490	3510–3590	FDD
23	2000–2020	2180–2200	FDD
24	1626.5–1660.5	1525–1559	FDD
25	1850–1915	1930–1995	FDD
26	814–849	859–894	FDD
27	807–824	852–869	FDD
28	703–748	758–803	FDD
29	N/A	717–728	FDD
30	2305–2315	2350–2360	FDD
31	452.5–457.5	462.5–467.5	FDD
…			
33	1900–1920	1900–1920	TDD
34	2010–2025	2010–2025	TDD
35	1850–1910	1850–1910	TDD
36	1930–1990	1930–1990	TDD
37	1910–1930	1910–1930	TDD
38	2570–2620	2570–2620	TDD
39	1880–1920	1880–1920	TDD
40	2300–2400	2300–2400	TDD
41	2496–2690	2496–2690	TDD
42	3400–3600	3400–3600	TDD
43	3600–3800	3600–3800	TDD
44	703–803	703–803	TDD

Another relevant feature of the LTE radio interface is the support of relaying capabilities. E-UTRAN supports relaying by having a relay node (RN) wirelessly connected to an eNB serving the RN, referred to as donor eNB (DeNB), via a modified version of the E-UTRA radio interface, denoted as Un interface [20]. One of the main benefits of relaying is to provide extended LTE coverage in targeted areas at low cost, as discussed in Chapter 3.

On the one hand, the RN supports the eNB functionality – meaning that it terminates the radio protocols of the E-UTRA radio interface as well as the protocols that connect a conventional eNB to the EPC or other eNBs (e.g. S1 and X2 interfaces). On the other hand, the RN also supports a subset of the UE functionality in order to wirelessly connect to the DeNB as a 'special' UE (e.g. RRC and NAS protocols). It is worth noting that inter-cell handover of the RN is not supported in the current LTE specifications, so that RNs are mainly intended to be fixed or nomadic but not mobile. A simplified illustration of the architecture supporting RNs is shown in Figure 4.4.

An RN can operate in *inband* or *outband* relaying modes. In *inband* relaying, the relay backhaul link (Un interface) and the relay access link (Uu interface) share the same carrier frequency, whereas different carrier frequencies are employed for the backhaul and access links in *outband* relaying. In addition, the RN could have its own physical cell identity (cell ID) and so appear as a distinct conventional cell to the terminals. This would be the case of a non-transparent relay, also known as Layer 3 relay. On the contrary, the RN may not have a cell ID and consequently not being 'seen' by the served UEs. This case is known as transparent relay or Layer 2 relay. On this basis, RNs are classified into Type 1, Type 1a and Type 1b, all of them being Layer 3 relays, and Type 2, which is a Layer 2 relay. Type 1 is an *outband* relay, Type 1a is an *inband* relay able to operate in full duplex (this turns to be the most complex node since isolation is required between access and backhaul transmissions), and Type 1b is an *inband* relay that uses time-division multiplexing (TDM) to support the access and backhaul link on the same frequency carrier.

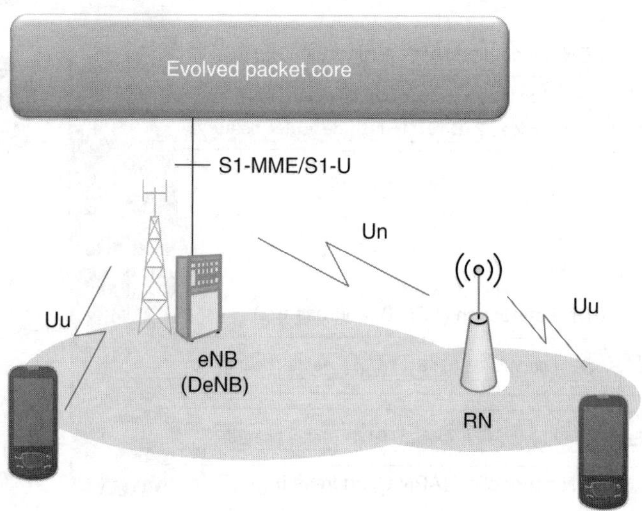

Figure 4.4 Illustrative view of the E-UTRAN architecture supporting RNs.

4.2.2 Service Model: PDN Connection and EPS Bearer Service

The QoS-enabled IP connectivity service provided by an LTE network between a given UE and a given external PDN is denoted as 'PDN connection'. The LTE network operator may provide access to different types of PDNs, through which different types of services might be reached. One type of PDN could be, for example, the public Internet. Another type of PDN could be, for example, a private IP network belonging to the mobile network operator to offer, for example, multimedia telephony services via IMS service delivery platforms. An LTE UE may access a single PDN at a time, or it may have multiple PDN connections open simultaneously.

Each PDN connection has its own IP address (one IPv4 and/or one IPv6 address) so that all IP packets belonging to the same PDN share a common address (or a pair of addresses, if both IPv4 and IPv6 addresses are used simultaneously over the same PDN connection). A PDN is identified by a parameter named Access Point Name (APN), which is a character string that is used when selecting the PDN for which to set up the IP connectivity. One PDN connection is always established when the terminal attaches to the EPS. During the network attach procedure, the terminal may indicate the APN for the LTE network to select the PDN that the user wants to access (i.e. APN of the PDN). In case the terminal does not provide the APN, a default value from user's subscription profile would be used. The APN is also used by the network to choose the P-GW providing access to that PDN (there can be several P-GW providing access to a given PDN, as well as a given P-GW that provides access to several PDNs). The concept of a PDN connection and associated parameters (IP address(es), APN) is illustrated in Figure 4.5, showing a particular case where a terminal is configured with three simultaneous PDN connections.

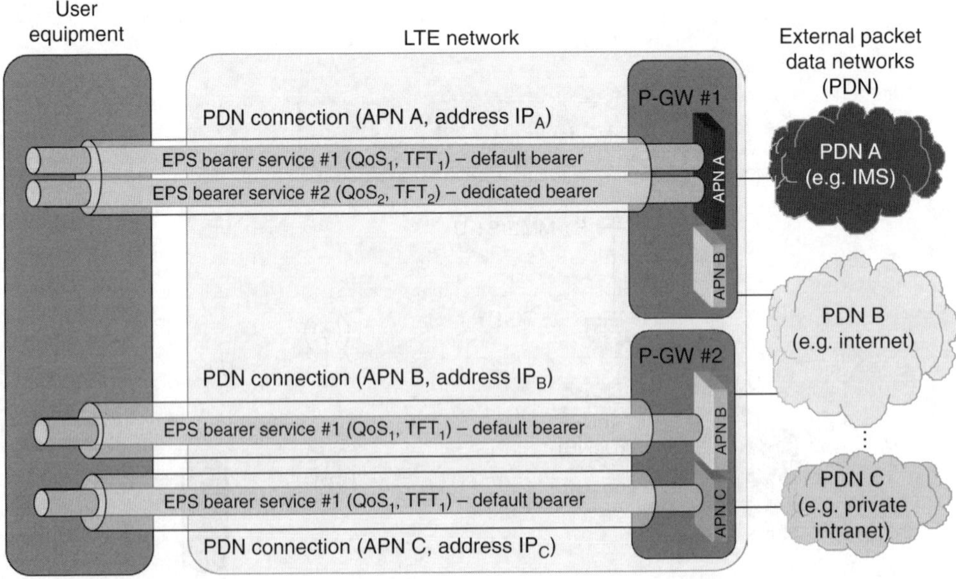

Figure 4.5 LTE service model: PDN connections and EPS bearer services.

Different QoS treatments can be enforced to different IP packet flows within the same PDN connection: this is implemented through the concept of EPS bearer service. 3GPP specifications define a bearer service as a type of telecommunications service that provides the capability of transmission of signals between two network points. On this basis, an EPS bearer service, or EPS bearer for short, provides a logical transport channel between the UE and the PDN for the transmission of IP packets. Each EPS bearer is associated with a set of QoS parameters that describes the properties of the transport channel, for example, bit rates, delay and packet loss rate. All conformant traffic sent over the same EPS bearer will receive the same packet forwarding treatment (e.g. scheduling policy, queue management policy, rate shaping policy, radio protocol stack configuration, etc.). In order to provide different QoS treatments to two distinct IP packet flows, they need to be sent over separate EPS bearers. Therefore, the EPS bearer is the level of granularity for bearer level QoS control in the network. A description of the QoS parameters used to specify a given treatment is provided below in this section.

The mapping of IP traffic onto the different bearers is done based on packet filters called Traffic Flow Template (TFT). Therefore, together with a set of QoS parameters, each EPS bearer is associated with packet filter information that allows the UE (in uplink) and the P-GW (in downlink) to identify which packets belong to a certain IP packet flow aggregate. This packet filter information is typically a 5-tuple defining the source and destination IP addresses, source and destination port as well as protocol identifier (e.g. UDP or TCP), while it is also possible to use additional parameters related to an IP flow (e.g. type of service/traffic class).

One EPS bearer service is always established when the UE connects to a PDN, and that remains established throughout the lifetime of the PDN connection to provide the UE with the so-called 'always-on' IP connectivity to the PDN. That bearer is referred to as the default bearer. Any additional EPS bearer service that is established to the same PDN is referred to as a dedicated bearer. Therefore, each PDN connection has, at least, one default EPS bearer established and a number of additional dedicated EPS bearers if traffic differentiation is needed. The association of the EPS bearer service with some specific parameters (QoS, TFT) is illustrated in Figure 4.5, showing the particular case where PDN connection with *APN A* is using two EPS bearers, one default and one dedicated, to provide two different QoS treatments between the UE and *PDN A*, while the other two PDN connections in the example rely only on the default EPS bearer, thus providing the same treatment to all the packet flow aggregate transported over these connections.

As explained earlier, an EPS bearer has an associated QoS profile that determines its expected behaviour. QoS parameters defined in LTE are illustrated in Figure 4.6. EPS bearers can be classified as Guaranteed Bit Rate (GBR) bearers or non-GBR bearers. In the former, GBR resources sufficient to deliver a given GBR value are allocated and reserved (e.g. by an admission control function in the RAN) at bearer establishment/modification. In the latter, such reservation is not enforced. GBR bearers are characterized through four parameters: QoS Class Identifier (QCI), Allocation and Retention Priority (ARP), GBR and Maximum Bit Rate (MBR). In the case of non-GBR bearers, only QCI and APR parameters are associated with a bearer, and two additional parameters, UE Aggregate Maximum Bit Rate (UE-AMBR) and Access Point Name Aggregate Maximum Bit Rate (APN-AMBR), are used to characterize the aggregate behaviour of all non-GBR bearers that a UE may have established. Further details on these parameters are given in the following subsections.

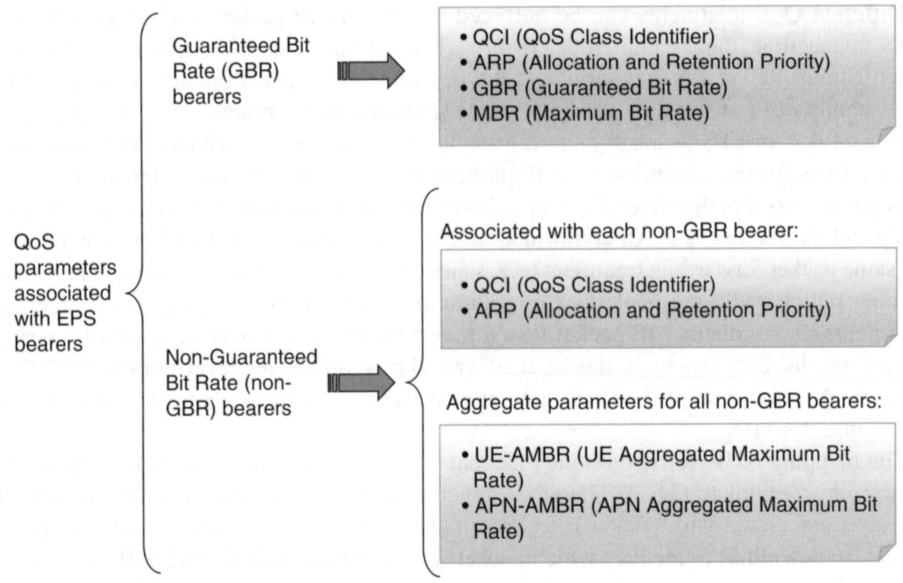

Figure 4.6 QoS parameters in LTE.

4.2.2.1 QCI

The QCI parameter is a scalar value that is used as a reference to establish eNB-specific parameters that control the bearer level packet forwarding treatment (e.g. scheduling weights, queue management thresholds, configuration of the retransmission modes, etc.). The specific configuration of the packet forwarding process in eNB is managed by the network operator and not signalled on any control interface. Rather, the goal of standardizing a QCI is to ensure that applications/services mapped to a given QCI receive the same treatment in multivendor network deployments and in case of roaming. The current standardized list consists of the nine QCI values specified in 3GPP TS 23.203 [21] and reproduced in Table 4.3, which are defined in terms of the following performance characteristics: (i) resource type (GBR or non-GBR), (ii) priority, (iii) packet delay budget (PDB) and (iv) packet error loss rate (PELR).

The PDB defines an upper bound for the time that a packet may be delayed between the UE and the P-GW. The fulfilment of a given PDB determines the configuration of the scheduling and link layer functions (e.g. the setting of scheduling priority weights). PDB is primarily associated with the delay introduced by the radio interface, though part of the PDB also accounts for the delay in the network side (delay estimates of roughly 10 ms on the network side for most cases and up to roughly 50 ms for a roaming scenario between Europe and the US West Coast are pointed out in 3GPP TS 23.203 [21]). In any case, the PDB values given in Table 4.3 are quite conservative. Actual packet delays, in particular for GBR traffic, are typically much lower than the PDB specified for a QCI as long as the UE has sufficient radio channel quality.

While scheduling between different packet aggregates is to be primarily based on the PDB (e.g. the scheduler serves the traffic so that the corresponding PDB is satisfied), the priority parameter (a value of 1 representing the highest priority) is used to give preference to

Table 4.3 Standardized QCI characteristics [21].

QCI	Resource type	Priority	Packet delay budget (PDB) (ms)	Packet error loss rate (PELR)	Example services
1	GBR	2	100	10^{-2}	Conversational voice
2		4	150	10^{-3}	Conversational video (live streaming)
3		3	50	10^{-3}	Real-time gaming
4		5	300	10^{-6}	Non-conversational video (buffered streaming)
5	Non-GBR	1	100	10^{-6}	IMS signalling
6		6	300	10^{-6}	Video (buffered streaming) TCP based (e.g. www, e-mail, chat, ftp, p2p file sharing, progressive video, etc.)
7		7	100	10^{-3}	Voice, video (live streaming) Interactive gaming
8		8	300	10^{-6}	Video (buffered streaming) TCP based (e.g. www, e-mail, chat, ftp, p2p file sharing, progressive video, etc.)
9		9			

higher-priority traffic when the target set by the PDB can no longer be met for one or more traffic aggregates.

On the other hand, the PELR parameter defines an upper bound for the rate of non-congestion-related packet losses. A PELR value applies completely to the radio interface between a UE and eNB since other losses are considered negligible. The purpose of the PELR is to allow for appropriate link layer protocol configurations (e.g. configuration of the retransmission modes).

As it can be observed in Table 4.3, there are four QCIs for GBR bearers and five for non-GBR bearers. According to 3GPP TS 23.203 [21], QCI 9 is to be typically used for the default bearer in the case of 'default'/'non-privileged subscribers'. Note that the AMBR parameter (that limits the maximum achievable bit rate) can be used in this case as the 'tool' to provide subscriber differentiation between subscribers with the same QCI on the default bearer (e.g. different maximum bit rates enforced per user). With similar characteristics as QCI 9 except for the priority value, QCI 8 could then be used for the default bearer of a kind of 'premium subscribers'. In turn, there is QCI 6 that only differs in the priority value with regard to QCI 8 and 9. If the network supports Multimedia Priority Service (MPS),[2] then QCI 6 could be used for the prioritization of non-real-time data (most typically TCP-based services/applications) of the MPS subscribers. Thus, in the case of congestion, MPS subscribers using default EPS bearers with QCI 6 would be favoured in front of 'premium subscribers' and 'default

[2] Described later in Section 4.5.4.

subscribers'. The rest of QCIs (1, 2, 3, 4, 5 and 7) can be used for operator-controlled services, that is, services that require a dedicated EPS bearer with specific traffic forwarding behaviour. For example, to transfer the IP packets containing the voice frames in Voice over LTE (VoLTE) services, dedicated EPS bearers with QCI 1 could be used.

4.2.2.2 ARP

The ARP parameter is used to decide whether a bearer establishment or modification request can be accepted or needs to be rejected in case of resource limitations. The ARP parameter is especially relevant for the admission control decision-making of GBR EPS bearer services. It can also be used to decide which existing bearers to pre-empt during resource limitations. The ARP encodes information about:

- Priority level (scalar with 15 levels, 1 being the highest level of priority)
- Pre-emption capability (flag with 'yes' or 'no')
- Pre-emption vulnerability (flag 'yes' or 'no')

Once successfully established, the APR value associated with the EPS bearer does not have any impact on the bearer level packet forwarding treatment (e.g. scheduling and rate control). Such packet forwarding treatment is solely determined by the other EPS bearer QoS parameters (QCI and bit rate parameters).

3GPP establishes that the ARP priority levels 1–8 should only be assigned to resources for services that are authorized to receive prioritized treatment within an operator domain (i.e. that are authorized by the serving network (SN), regardless of whether this is the home or visited network for some subscribers). The ARP priority levels 9–15 may be assigned to resources that are authorized by the home network and thus applicable when a UE is roaming. This ensures that future releases may use ARP priority levels 1–8 to indicate, for example, emergency and other priority services within an operator domain in a backward compatible manner. This does not prevent the use of ARP priority levels 1–8 in roaming situation in case appropriate roaming agreements exist that ensure a compatible use of these priority levels.

It is worth mentioning that the ARP may be used to free up capacity in exceptional situations, for example, a disaster situation [22]. In such a case, an LTE eNB may drop bearers with a lower ARP priority level to free up capacity if the pre-emption vulnerability information allows this.

4.2.2.3 Rate Limiting Parameters

For non-GBR bearers, rate limiting can be enforced through the Aggregate Maximum Bit Rate (AMBR) parameters. In particular, the APN-AMBR can be used to enforce a maximum aggregate bit rate across all of the UE bearers for one APN (i.e. all non-GBR bandwidths flowing through the UE and particular external IP network). Another rate limiting control for non-GBR bearers is the per UE-AMBR. This rate liming control is enforced across all of non-GBR bearers that are associated with a UE, independent of the bearer's termination point (i.e. all non-GBR bandwidths flowing through the UE and any external IP network).

The LTE network will allow rates up to the minimum of the APN-AMBR and UE-AMBR values for a UE, and if exceeded, data rates can be throttled. Once the UE's aggregate bit rate falls below the maximum value, the system will no longer throttle data.

For GBR bearers, the offered bit rate is controlled through two parameters: the GBR and an MBR associated with each bearer. The GBR value is the minimum bandwidth provided by the network should the bearer be admitted. The admission process has to allocate enough bandwidth to assure the delivery of data up to the value of the GBR. This bandwidth is available to the UE independent of the network congestion levels. On the other hand, the MBR is the absolute maximum amount of bandwidth that the GBR bearer can utilize once it has been admitted. The MBR allows for additional bandwidth utilization above the GBR value assuming there are resources available in the network. Once the MBR bandwidth is exceeded, the network can throttle the excessive bandwidth usage. The GBR and MBR limits essentially create a minimum and maximum amount of bandwidth that can be used for a given GBR bearer.

4.2.3 PCC Subsystem

3GPP specifies the functionality for PCC with the following main functions [21]:

- Policy control, which comprises QoS control (i.e. selection of the QoS parameters for the EPS bearers) and gating control (i.e. the process of blocking or allowing packets, belonging to a service data flow/detected application's traffic, to pass through the network)
- Flow-based charging for network usage, including charging control and online credit control, for service data flows and application traffic

The PCC subsystem enables a centralized control to ensure that services and applications running on top of the LTE network are provided with the appropriate transport, that is, service-aware QoS configuration of the EPS bearers, while providing also the means to control charging on a per-service basis. Indeed, the PCC subsystem is an access-agnostic framework that could be applicable to a number of accesses other than LTE (i.e. GPRS, UMTS but also non-3GPP access technologies).

The reference network architecture for the PCC subsystem for LTE access is shown in Figure 4.7. The central element of this architecture is the PCRF. This element encompasses both policy control decision and flow-based charging control functionalities. Decisions made by the PCRF are in the form of PCC rules that are sent to the LTE network over the Gx interface for enforcement in the Policy and Charging Enforcement Function (PCEF) located within the P-GW. Criteria such as the QoS subscription information may be used together with policy rules such as service-based, subscription-based or predefined PCRF internal policies to derive the authorized QoS to be enforced for a service data flow. The Gx interface is also used for the PCEF to provide the PCRF with user- and access-specific information. The usage of the monitoring control capability is also supported to enable dynamic policy decisions based on the total network usage in real time. Another important capability is the application detection and control feature, which comprises the request to detect the specified application traffic, to report to the PCRF on the start or stop of application traffic and to apply the specified enforcement and charging actions.

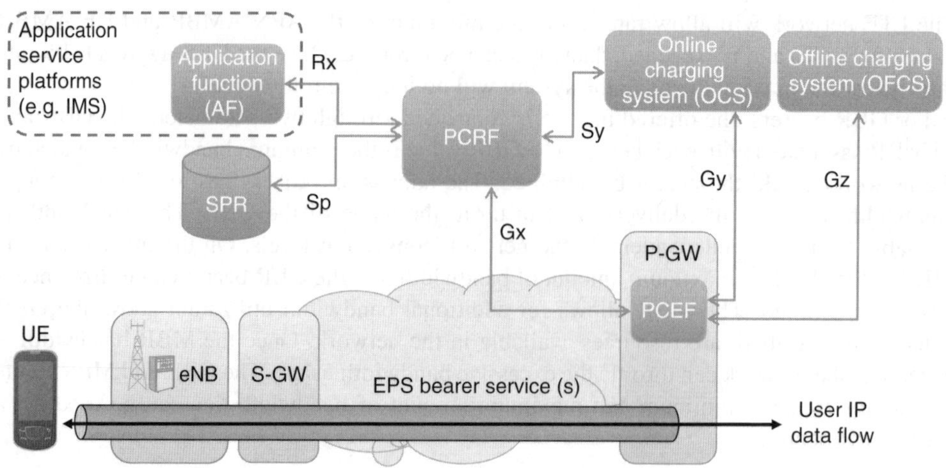

Figure 4.7 PCC architecture.

The PCRF terminates an interface named Rx over which external application servers (e.g. residing on service delivery platforms such as IMS) can send service-related information, including resource requirements for the associated IP flow(s). In this regard, the term application function (AF) is the generic term used to refer to the functional entity that interacts (or intervenes) with applications or services that require dynamic PCC. Typically, the application level signalling for the service passes through or is terminated in the AF (i.e. within the IMS platform, there is a functional entity that takes on the role of the AF for the interaction between IMS services and the PCC subsystem). The AF can also subscribe to certain events that occur at the traffic plane level (e.g. IP session termination, access technology type change, etc.). Decisions made by the PCRF also rely on subscriber-related information, obtained through the interface Sp from a Subscriber Profile Repository (SPR).

On the side of the charging-related functionality, the Online Charging System (OCS) is a credit management system for pre-paid charging, while the Offline Charging System (OFCS) is used for post-paid charging. The PCEF performs measurements of user data plane traffic (e.g. traffic volume and/or time duration) and interacts with the OCS over the Gy interface to check out credit and report credit status. For online charging, there is also the Sy reference point between PCRF and OCS that enables the transfer of policy counter status information related to subscriber spending (e.g. notifications of spending limit reports from OCS to PCRF). For offline charging, the PCEF reports charging events to the OFCS over the Gz reference point. This information is used by the OFCS to generate charging data records (CDRs) to be transferred to the billing system.

The use of the PCC functions for real-time service control is enabling innovative types of service to be offered. Examples are: fair usage, allowing operators to limit (throttle) the bandwidth available to the heaviest users on their unlimited data plans; RAN congestion, facilitating premium customers to receive guaranteed bandwidth during periods of RAN congestion; bill-shock prevention, allowing operators to warn users that they have exceeded their usage allocation or that their charges have exceeded a specified amount; 'freemium', through which operators can apply a 'zero-rated' tariff to traffic for specific applications/ websites, such as Facebook, Twitter and the like; time-based usage, enabling offers of discounted or free data usage (data passes) based on time of day (peak or off-peak plans) that can be monthly, weekly, daily or hourly; and tiered services, allowing operators to provide a range of appropriately priced packages with limits based on access speed, download cap, time of day and so on [23].

In the context of PPDR communications, the PCC subsystem enables the deployment of a powerful framework for dynamic policies that can be adapted to specific incident needs, as discussed further in the Section 4.5.

4.2.4 Security

The increasingly rapid evolution and growth in the complexity of new systems and networks, coupled with the sophistication of changing threats and the presence of intrinsic vulnerabilities, present demanding challenges for maintaining the security of communications systems and networks. This is particularly relevant in the context of mobile networks such as LTE, since the wireless interface that terminals use to access the network makes these systems more exposed to different attacks (e.g. eavesdropping of wireless transmissions, or even manipulation, by third parties; denial-of-service attacks, preventing legitimate users from getting access to the system).

Inherited and evolved from the previous GSM and UMTS systems, LTE specifications adopt a security architecture organized in five functional areas [24], as illustrated in Figure 4.8:

Network access security (I) provides users with secure access to services and protects against attacks on the radio access interfaces.

Network domain security (II) enables nodes to securely exchange signalling data and user data and protects against attacks on internal network interfaces.

User domain security (III) provides secure access to terminals. This is typically implemented by using a PIN code for the user to get access to the network services. Of note is that LTE does not specify a new type of Subscriber Identity Module (SIM) card but leverages the one specified by UMTS and known as UMTS SIM.

Application domain security (IV) enables applications in the user and service/application provider domains to securely exchange messages. Application level security traverses on top of the user plane transport provided by EPS and as such is transparent to the LTE network.

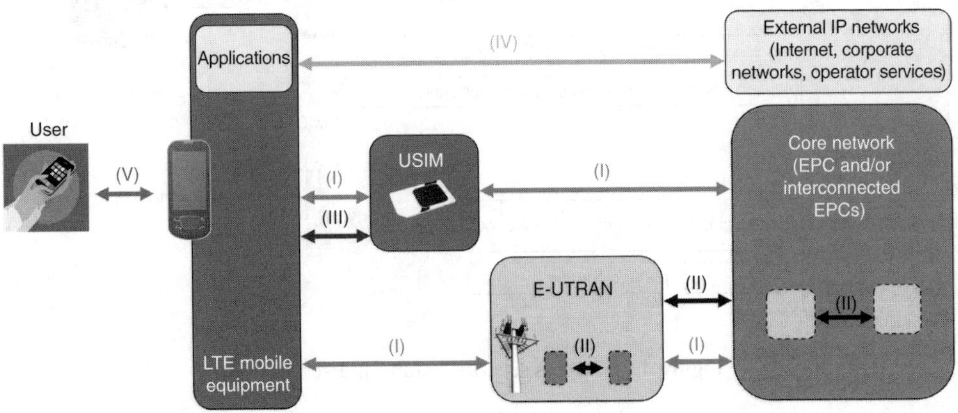

Figure 4.8 Overview of the security functional areas defined for LTE systems.

Visibility and configurability of security (V) allow the user to learn whether a security feature is in operation or not and whether the use and provision of services should depend on the security feature (e.g. use of a symbol on the terminal display that let the user know whether encryption is being applied or not).

Each of these areas represents a group of security features needed to face certain threats and accomplish certain security objectives. The security features in (I) and (II) are the subject of this overview since they are the ones directly related to the LTE network itself. More details on the other security domains can be found in 3GPP TS 33.401 [24] and TS 33.102 [25].

4.2.4.1 Network Access Security

Access security in E-UTRAN consists of the following components:

- Mutual authentication between UE and network
- Key derivation to establish the keys for ciphering and integrity protection
- Ciphering, integrity and replay protection of NAS signalling (i.e. session management and mobility management signalling) between UE and MME
- Ciphering, integrity and replay protection of RRC signalling between UE and eNB
- Ciphering of the user plane between the UE and the eNB
- Use of temporary identities in order to avoid sending the permanent user identity (i.e. International Mobile Subscriber Identity (IMSI)) over the radio link

Figure 4.9 illustrates these components (except the use of temporary identities).

Mutual authentication covers both user authentication (i.e. the property that the SN corroborates the user identity of the user) and network authentication (i.e. the property that the user corroborates that he/she is connected to an SN that is authorized by the user's home

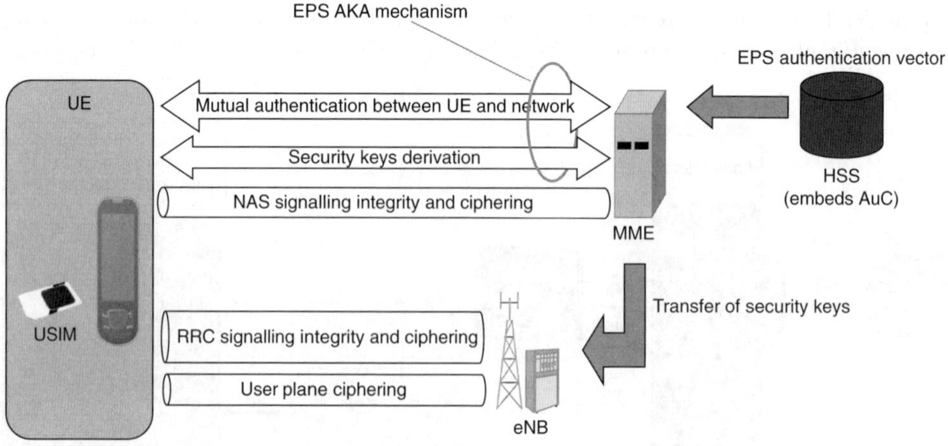

Figure 4.9 Access security features in E-UTRAN.

network to provide services to him/her). Mutual authentication in E-UTRAN is based on the fact that both the USIM card and the network (the authentication centre (AuC) function embedded within the HSS) have access to the same secret key K. This is a permanent key that never leaves the USIM or the HSS/AuC and, as such, is not directly used to protect any traffic. Instead, other keys are generated from the key K in the terminal and in the network during the authentication procedure, which are the ones used for the ciphering and integrity protection services. The mechanism for authentication as well as for key generation in E-UTRAN is named EPS Authentication and Key Agreement (EPS AKA). EPS AKA is performed when the user attaches to the EPS via E-UTRAN access. The procedure takes place between the UE and one MME entity in the SN, which receive an EPS authentication vector (AV) for that particular user (IMSI) from the home network's HSS/AuC. The AV consists of the parameters (generated with the key K as input, but not explicitly contained within the AV) needed for the mutual authentication and for the derivation of the ciphering and integrity keys to be used within the MME (for the NAS signalling) and within the serving eNB (for user plane and RRC signalling). For those readers familiar with UMTS, the EPS AKA resembles the procedure used in UMTS and referred to as UMTS AKA. However, there a few differences to highlight:

- The keys generated within the home network's HSS/AuC in EPS AKA are tied to a given SN by including a serving network identity (SN ID). This ensures that a key derived for one SN cannot be (mis)used in a different SN.
- Larger key sizes are possible. E-UTRAN supports not only 128-bit keys, but it is prepared to also use 256-bit keys.
- Additional protection is added against compromised base stations. The keys used in the air interface are updated/refreshed each time the UE changes its point of attachment or when transitioning from idle to active states.

As far as signalling/control plane protection is concerned, NAS signalling between the UE and MME is protected end to end with integrity and confidentiality features. Moreover, integrity and confidentiality protection is also provided for the radio network signalling (RRC signalling, which is also used to encapsulate NAS signalling) over the radio path between the UE and the eNB. Regarding the user plane (i.e. transfer of user IP packets), confidentiality protection is provided between the UE and the eNB, though, unlike the signalling/control plane, integrity protection is not supported.

The LTE standard allows the usage of different cryptographic algorithms for ciphering and integrity protection. Two sets of security algorithms were initially developed for LTE: one set based on Advanced Encryption Standard (AES) and the other on SNOW 3G. The principle being adopted is that the two should be as different from each other as possible to prevent similar attacks being able to compromise them both. The ETSI Security Algorithms Group of Experts (SAGE) is responsible for specifying the algorithms. The key length is 128 bits, with the possibility to introduce 256-bit keys in the future if necessary. In 2011, a third algorithm, ZUC, was approved for its use in LTE. The encryption algorithm 128-EEA3 and the integrity algorithm 128-EIA3, based on ZUC, were finalized in 2012. The UE and the network need to agree on which algorithm to use for a particular connection.

Finally, the identity protection capabilities of LTE also deserve some attention. As in UMTS networks, LTE provides:

- User identity confidentiality: the property that the permanent user identity (IMSI) of a user to whom services are delivered cannot be eavesdropped on the radio access link
- User location confidentiality: the property that the presence or the arrival of a user in a certain area cannot be determined by eavesdropping on the radio access link
- User untraceability: the property that an intruder cannot deduce whether different services are delivered to the same user by eavesdropping on the radio access link

To achieve these identity protection objectives, the user is normally identified by a temporary identity (e.g. Globally Unique Temporary Identifier (GUTI)) by which he/she is known by the visited SN. To avoid user traceability, which may lead to the compromise of user identity confidentiality, the user should not be identified for a long period by means of the same temporary identity. In addition, it is required that any signalling or user data that might reveal the user's identity is ciphered on the radio access link. Only when the user cannot be identified by means of a temporary identity, a user identification mechanism should be invoked by the SN to retrieve the IMSI from the connected terminal/user. This mechanism is initiated by the MME that requests the user to send its permanent identity. The user's response contains the IMSI in clear text. This represents a breach in the provision of user identity confidentiality.

More details on LTE access security can be found in 3GPP TS 33.401 [24], together with the security provisions for the interworking between E-UTRAN and UTRAN and GERAN. Besides, security features in the case of getting access to EPS from non-3GPP RATs are covered in 3GPP TS 33.402 [26].

4.2.4.2 Network Domain Security

In 2G systems (e.g. GSM), no solution was specified on how to protect traffic in the core network. This was not perceived as a problem since the specific protocols and interfaces used for circuit-switched (CS) traffic were typically only accessible to large telecom operators. With the introduction of packet data services (e.g. GPRS, EPS) in mobile communications systems as well as IP transport in general, the signalling of 3G/4G networks runs over networks and protocols that are more open and accessible. Therefore, 3GPP has developed specifications on how IP traffic is to be secured within the core network as well as between a core network and some other (core) networks to be interconnected. These specifications are known as network domain security for IP-based control planes (NDS/IP) [27].

A central concept introduced in the NDS/IP specification is the notion of *security domain*. A security domain is a network or part of it that is managed by a single administrative authority. Within a security domain, the same level of security and usage of security services can be expected. Typically, a network operated by a single network operator or a single transit operator will constitute one security domain, although an operator may at will organize its network into separate subnetworks from a security standpoint. The border between the security domains is protected by Security Gateways (SEGs). The SEGs are responsible for enforcing the security policy of a security domain towards other SEGs in the destination security domain. The network operator may have more than one SEG in its network in order to avoid a single point of failure or for performance reasons. An SEG may be defined for interaction towards all reachable security domain destinations, or it may be defined for only a subset of the reachable destinations. All NDS/IP traffic shall pass through an SEG before entering or leaving the security domain. The basic idea to the NDS/IP architecture is to provide hop-by-hop security. Therefore, chained-tunnel and hub-and-spoke connection models are used to enforce hop-by-hop-based security protection within and between security domains. The use of hop-by-hop security makes it easy to operate separate security policies internally in a security domain and towards other external security domains.

The traffic between SEGs is protected at network layer, using the IPsec protocols defined by IETF as specified in RFC-4301 [28]. IPsec offers a set of security services, which is determined by the negotiated IPsec security associations (SAs). An SA can be defined as a simplex 'connection' that affords security services to the traffic carried by it. In this way, the IPsec SA defines which security protocol to be used, the mode and the endpoints of the SA. Security protocols that form part of the IPsec framework are Authentication Header (AH) and Encapsulating Security Payload (ESP). These protocols can operate in either *tunnel and transport* modes (the tunnel mode encapsulates the full IP packet to be protected into another IP packet, while the transport mode does not). For NDS/IP networks, the IPsec protocol between SEGs shall always be ESP in tunnel mode. Within a security domain, the use of transport mode is also allowed. For NDS/IP networks, it is further mandated that integrity protection/message authentication together with anti-replay protection shall always be used. Therefore, the security services provided by NDS/IP are:

- Data integrity
- Data origin authentication
- Anti-replay protection
- Confidentiality (optional)
- Limited protection against traffic flow analysis when confidentiality is applied

To set up the IPsec SAs between SEGs, Internet Key Exchange 1 (IKEv1) or Internet Key Exchange 2 (IKEv2) is used for key management and distribution. The main purpose of IKEv1 and IKEv2 is to negotiate, establish and maintain the SAs between parties that are to establish secure connections (from Release 11 onwards, support of IKEv2 is required; thus, it is likely that support of IKEv1 in SEGs will no longer be mandatory in future 3GPP releases). Thus, the concept of an SA is central to IPsec protocols and IKEv1/IKEv2. Security services are afforded to an SA by the use of AH, or ESP protocols, but not both. If both AH and ESP protection are applied to a traffic stream, then two SAs must be created and coordinated to effect protection through iterated application of the security protocols. To secure typical, bidirectional communication between two IPsec-enabled systems, a pair of SAs (one in each direction) is required. IKE explicitly creates SA pairs in recognition of this common usage requirement.

The profiling of the protocols ESP, IKEv1 and IKEv2 for the NDS/IP application is specified in 3GPP TS 33.210 [27]. These profiles provide the minimum set of features that must be supported and are required for interworking purposes.

An example scenario employing the NDS/IP architecture is shown in Figure 4.10. In this case, the traffic transfer between security domains A and B is secured by an ESP SA in tunnel mode between SEG A and SEG B. The IKE connection is used to establish the needed IPsec

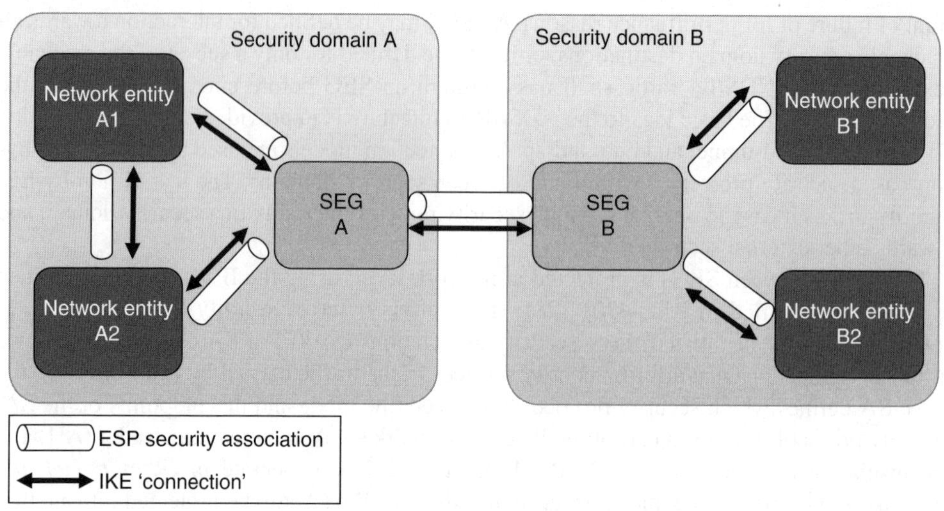

Figure 4.10 3GPP NDS/IP architecture for IP network layer security.

SAs. SEGs will normally maintain at least one IPsec tunnel available at all times to a particular peer SEG. Inside each domain, the NEs may be able to establish and maintain ESP SAs as needed towards an SEG or other NEs. All NDS/IP traffic from an NE in one security domain towards an NE in a different security domain will be routed via the corresponding SEG and will be afforded hop-by-hop security protection towards the final destination. Although NDS/IP was initially intended mainly for the protection of control plane signalling only, it is possible to use similar mechanisms to protect the user plane traffic. As an extension to the NDS/IP framework, 3GPP TS 33.310 [29] defines an inter-operator public key infrastructure (PKI) based on the use of certificates that can be used to support the establishment of the IPsec connections in NDS/IP.

The policy control granularity afforded by NDS/IP is determined by the degree of control with respect to the ESP SA between NEs and SEGs. The normal mode of operation is that only one ESP SA is used between any two NEs and SEGs, and therefore, the security policy will be identical to all secured traffic passing between NEs. This is consistent with the overall NDS/IP concept of security domains, which should have the same security policy in force for all traffic within the security domain. However, operators may decide to establish more than one ESP SA between two communicating security domains should finer-grained security granularity be required. The actual inter-security domain policy is determined by roaming agreements when the security domains belong to different operators and the SEGs are responsible for enforcing security policies for the interworking between networks. In addition to NDS/IP security services, the security may include filtering policies and firewall functionality not specified within 3GPP NDS/IP. Indeed, simple filtering may be needed before the SEG functionality. The filtering policy must allow key protocols such as Domain Name Service (DNS) and Network Time Protocol (NTP) to pass, as well as IKEv1/IKEv2 and IPsec ESP in tunnel mode protocols. Unsolicited traffic shall be rejected. In addition, SEGs shall be physically secured and offer capabilities for secure storage of long-term keys used for IKE authentication.

4.2.5 Roaming Support

Roaming is defined as the use of mobile services from another operator, which is not the home operator (i.e. the operator with which the user holds its subscription). Roaming is a key capability supported in current mobile communications networks, especially exploited to provide service to users when abroad.

LTE supports two roaming architectures [22, 30]:

1. **Home-routed roaming**. In this architecture, the roamer's traffic is routed back to the home network to enable the use of home resources. Therefore, the IP connectivity service is provided by the gateway functionality (i.e. P-GW) located in the home network.
2. **Local breakout roaming**. In this architecture, the IP connectivity service is provided by the P-GW located in the visited network. Local breakout architecture allows for AF serving the roaming user to be both on the visited operator network and the home network.

Both architectures are illustrated in Figure 4.11. The three relevant interfaces for roaming support in LTE networks are the following:

1. **Interface S6a**. This interface is used to exchange the data related to the location of the mobile station and to the management of the subscriber between the MME in the visited network and the HSS in the home network. The main service provided to the mobile

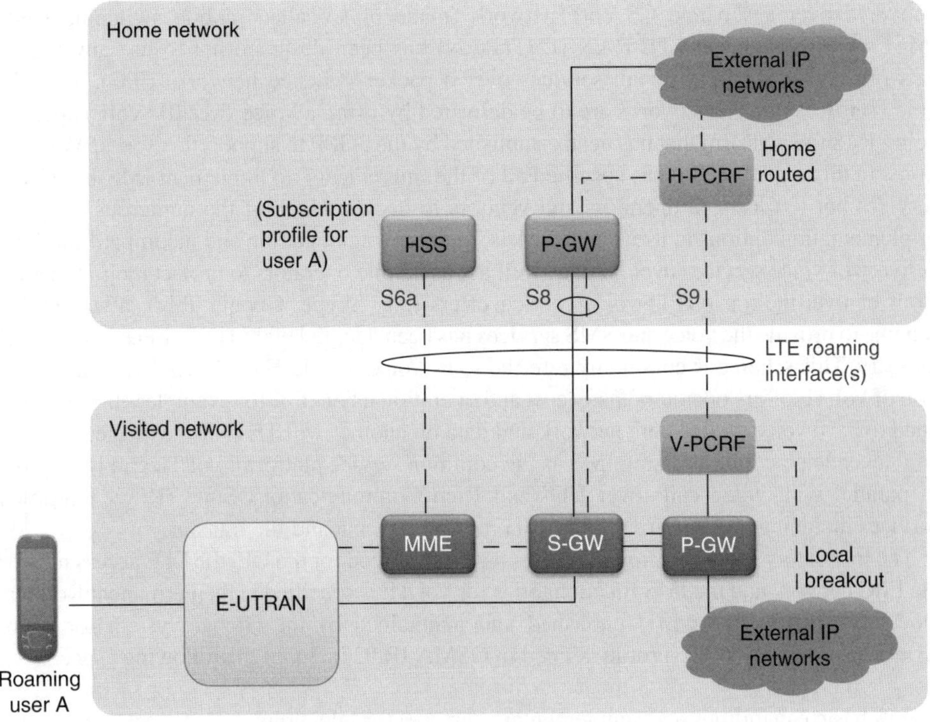

Figure 4.11 Roaming architectures supported in LTE networks.

subscriber is the capability to transfer packet data within the whole service area. The MME informs the HSS of the location of a mobile station managed by the latter. The HSS sends to the MME all the data needed to support the service to the mobile subscriber. Exchanges of data may occur when the mobile subscriber requires a particular service, when he/she wants to change some data attached to his/her subscription or when some parameters of the subscription are modified by administrative means. This interface is needed for both home-routed and local breakout roaming architectures.

2. **Interface S8**. This interface provides the user and control plane between the S-GW located in the visited network and the P-GW located in the home network. S8 is the inter-network variant of the S5 interface depicted in Figure 4.2. This interface is only needed for home-routed architecture.
3. **Interface S9**. This interface is needed to deploy the PCC functionalities described in Section 4.2.3 in roaming scenarios. It provides transfer of QoS PCC information between the Home PCRF (H-PCRF) and the Visited PCRF (V-PCRF) in order to support the local breakout function.

4.2.6 Voice Services over LTE

The fact that LTE is an all-IP technology makes that the voice service has to be delivered in a different way as it is done in the previous CS-based technologies. Indeed, GSM and UMTS RAN support the establishment of voice services though dedicated connections that are delivered over a separate CS core network (made of so-called mobile switching centres (MSCs)). However, the LTE RAN (E-UTRAN) has been designed not to use any CS core network and instead deliver all services over a packet-switched network (EPC). Therefore, voice services in LTE networks are to be delivered by using a Voice over IP (VoIP) approach, being IMS the platform that has been established by the 3GPP to support this sort of VoIP solutions. In this regard, 3GPP has specified all of the 'ingredients' to implement IMS-based voice services but left it up to operators and vendors to decide which of the numerous alternative implementation options to use. On this basis, to avoid fragmentation and incompatibility in the delivery of voice services over LTE, as well as for mobile operators to protect their revenues in front of over-the-top (OTT) voice service offers (e.g. Skype, Google Talk), a standardized scheme to provide the voice and SMS services has been adopted within the wireless commercial industry. This scheme is commonly referred to as Voice over LTE (VoLTE). The implementation of VoLTE offers operators many cost and operational benefits, for example, eliminating the need to have voice on one core network and data on another. VoLTE is also expected to unlock new revenue potential: utilizing IMS as the common service platform, VoLTE can be deployed in parallel with video calls over LTE and Rich Communications Suite (RCS) multimedia services including video share, multimedia messaging, chat and file transfer.

The mandatory set of functionalities that has been agreed for the UE, the LTE access network, the EPC network and the IMS functionalities for VoLTE is specified in the Permanent Reference Document (PRD) IR.92 [31] published and maintained by the Global System for Mobile Association (GSMA). The profile defined in GSMA IR.92 includes the following aspects:

- IMS basic capabilities and supplementary services for telephony
- Real-time media negotiation, transport and codecs

- LTE radio and EPC capabilities
- Functionality that is relevant across the protocol stack and subsystems

An illustrative view of the UE and network protocol stacks forming the scope of the VoLTE solution is depicted in Figure 4.12. In the upper layers, the VoLTE implementation relies on the use of a set of Internet-based protocols, including Session Initiation Protocol (SIP) and Extensible Markup Language (XML) Configuration Access Protocol (XCAP). SIP is the core protocol for session management (e.g. service registration, activation, etc.). XCAP is used for the configuration of supplementary services (e.g. line identification, forwarding, barring). Support for the Adaptive Multi-Rate (AMR) codec specified by 3GPP is mandated, which is also used in GSM and UMTS. This provides advantages in terms of interoperability with legacy systems (i.e. no transcoders are needed). Additionally, the AMR Wideband (AMR-WB) codec has to be used for high-definition (HD) voice service offers. Voice frames are sent encapsulated in the Real-time Transport Protocol (RTP). UE and the network must support both IPv4 and IPv6 for all protocols that are used for the VoLTE application (e.g. SIP, SDP, RTP, RTCP and XCAP/HTTP). In the network layers, VoLTE leverages the QoS capabilities defined for EPS bearer services, which specify different QoS classes. Optimization features available in LTE to make voice operation more efficient include semi-persistent scheduling (SPS) and TTI bundling. SPS reduces control channel overhead for applications that require a persistent and periodic radio resource allocation. Meanwhile, TTI bundling (transmitting in a set of contiguous TTI) can improve uplink coverage. Optimizations also include support for vocoder rate adaptation, a mechanism with which operators can control the codec rate based on network load, thus dynamically trading off voice quality against capacity. The support of Robust Header Compression (ROHC) is also mandated to reduce the overhead associated with the IP and transport layer headers. The VoLTE solution also mandates the support of Single Radio Voice Call Continuity (SRVCC) capability, which provides seamless handover of voice calls from VoLTE to 2G/3G access. Complementing PRD IR.92, GSMA has also delivered PRD IR.94 [32] that defines a minimum mandatory set of features on top of GSMA PRD IR.92 to implement conversational video services.

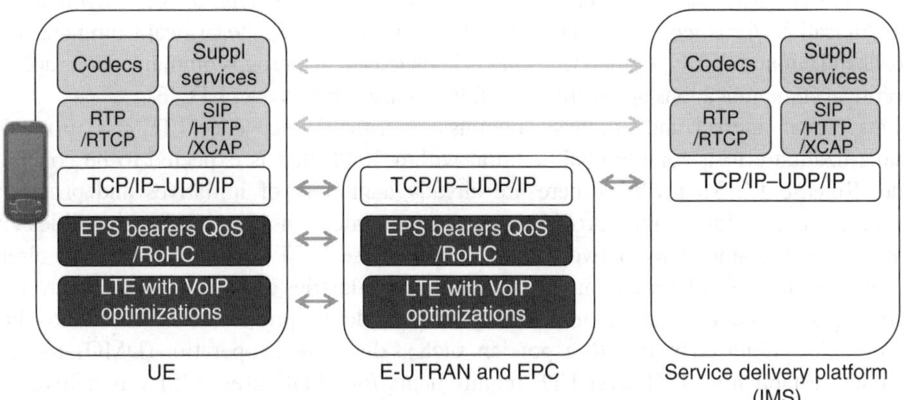

Figure 4.12 UE and network protocol stacks in VoLTE.

A detailed description of the VoLTE solution can be found in Ref. [33], together with some illustrative figures about VoLTE performance in terms of radio coverage, radio capacity and end-to-end latency. With regard to radio coverage, [33] claims that similar voice coverage as in UMTS systems can be achieved by using TTI bundling and retransmission techniques. In terms of capacity, considering the use of header compression techniques and 12.2 kb/s voice codecs, between 40 and 50 simultaneous users can be served in 1 MHz of spectrum well above the efficiencies of 10 users/MHz achieved in systems such as GSM with similar voice codecs. With regard to latency performance, mouth-to-ear delay budgets in the range of 160 ms are achievable, which are lower than those delivered in today's CS voice calls.

VoLTE services have been already launched along 2014 across the North America and Asia-Pacific regions, with new deployments expected to accelerate in 2015 and beyond. Moreover, consumer-grade HD voice over LTE is already a reality, significantly increasing voice intelligibility, which is anticipated to become a very attractive feature to mission-critical voice, particularly when compared to currently achieved PMR voice quality standards.

4.3 Group Communications and PTT

Group communications with PTT features are central capabilities in mission-critical voice services, as discussed in Chapter 1. These capabilities are well supported in current PMR technologies but thus far have not received much attention in the commercial domain. While it is considered that the initial primary focus of embracing LTE for PPDR is to deliver broadband data applications, the progressive maturity of the VoLTE technology together with the introduction of these mission-critical voice features in LTE paves the way for a mid- to long-term migration scenario in which both mission-critical voice and data services will eventually be supported over a common communications platform.

Indeed, there is global interest in LTE as a voice solution for the PPDR community. Some relevant PPDR associations have also made public their positions regarding the support of mission-critical voice over LTE networks. In particular, in the policy statement issued by APCO Global Alliance to endorse LTE as the global standard for emergency communications broadband networks, it is stated that *'Advanced 4G systems' cellular services will provide a comprehensive and secure IP-based mobile broadband network solution for wireless data, video and* voice communications *for emergency services worldwide'* [34]. As well, the Critical Communications Broadband Group (CCBG) of the TCCA has acknowledged that group communications and PTT are critical communications applications for future voice services over LTE [3].

This global interest and emphasis on mission-critical voice over LTE has resulted in standardization efforts for a global solution within 3GPP that is expected to be concluded under Release 13. Meanwhile, there are already a number of initiatives and proprietary products on the market that provide group communications and PTT capabilities over commercial IP connectivity networks (LTE, 3G). These 'PTT over LTE' solutions already available or under development may well serve some industries and disciplines. However, it is becoming a fragmented market, and current solutions do not provide some of the most basic functionalities required by PPDR responders such as direct mode operation (DMO). Indeed, a document establishing PTT over LTE requirements for PPDR users [35] was delivered by NPSTC in July 2013 for consideration by FirstNet as it embarks on its mission to deploy the nation's first nationwide public safety broadband network (NPSBN) in the United States as well as by the 3GPP within its specification work.

An overview of the existing initiatives and solutions for PTT over LTE is addressed in the next section. After that, the 3GPP standardization work on group calls and PTT is explained. In particular, the so-called Group Communications System Enablers (GCSE) features being added to the LTE standard as well as the support of a standardized MCPTT application over LTE are covered. The section is concluded with an overview of the current activities at the OMA related to the further enhancement of the OMA's PoC enabler to meet public safety user needs.

4.3.1 Existing Initiatives and Solutions for PTT over LTE

Nowadays, commercial PTT services are already available and the offerings are likely to continue growing. Three main approaches are followed in the current offers [36]:

1. A commercial carrier offers the service as a core feature on its network.
2. A company offers software or a smartphone application that provides PTT on a legacy network or a network that doesn't offer it as a core service.
3. Vendor solutions for the dedicated deployment and operation of a specific user base.

As to the first approach, PTT services have been available for a number of years mainly in the US market, beginning with Nextel Communications launching its Direct Connect service back in 1993. Nowadays, several of the leading commercial carriers in the United States offer PTT voice services, also referred to as PoC. Indeed, PTT services from AT&T and Verizon Wireless are also delivered over their LTE networks. However, each network is using a different PTT technology, which is not cross network compatible.

On the second approach, there are numerous smartphone applications that emulate PTT on a device using an LTE network. Several companies have developed enterprise-level software solutions with such functionality. Two examples are WAVE and Voxer Walkie-Talkie PTT applications, which offer a PTT subscription service for mobile workforce communications.

Finally, as to the third approach, manufacturers in the PMR industry domain are already developing technology with PTT over LTE together with PTT bridges for the interworking with the legacy PMR networks. Indeed, companies such as Alcatel-Lucent, Cassidian, Harris, Motorola, Thales and others have been demonstrating PTT over LTE and cross-network LTE since 2012 (e.g. BeOn by Harris, Broadband Push to Talk by Motorola, etc.). However, once again, each vendor is using a different technology for its PTT solution. These solutions are not compatible with each other so that, until a common interface is not decided upon, the use of PTT over LTE remains proprietary to each vendor.

4.3.2 3GPP Standardization Work

The following Work Items (WIs) have been established within 3GPP to develop specifications for group communications and PTT application over LTE:

- 'Group Communication System Enablers for LTE (GCSE_LTE)' [37], initiated in Release 12
- 'Service Requirements Maintenance for Group Communication System Enablers for LTE (SRM_GCSE_LTE)' [38], initiated in Release 13
- 'Mission Critical Push to Talk over LTE (MCPTT)' [39], initiated in Release 13

Table 4.4 lists the main 3GPP technical specifications and reports related to the above WIs.

Table 4.4 3GPP documents covering group communications system enablers and MCPTT over LTE.

Work Item (WI)	Related technical specifications/reports	Comments
GCSE_LTE and SRM_GCSE_LTE	TS 22.468 – 'Group Communication System Enablers for LTE'	Normative requirements document (Stage 1) (see Note 1 below)
	TR 23.768 – 'Study on Architecture Enhancements to Support Group Communication System Enablers for LTE'	Informative technical report containing candidate architectural proposals for GCSE_LTE
	TS 23.468 – 'Group Communication System Enablers for LTE'	Normative specification work of the functional architecture (Stage 2)
MCPTT	TS 22.179 – 'Mission Critical PTT over LTE'	Normative requirements document (Stage 1)
	TS 23.179 – 'System Architecture Enhancements for Mission Critical PTT over LTE'	Normative specification work of the functional architecture (Stage 2)
	TR 23.779 – 'Study on architectural enhancements to support Mission Critical Push To Talk over LTE (MCPTT) Services'	Technical report to support the specification of Stage 2 architecture definition

Note 1: Standards development in 3GPP progresses through three stages:

- 'Stage 1' refers to the service description from a service user's point of view.
- 'Stage 2' is a logical analysis, devising an abstract architecture of functional elements and the information flows among them across reference points between functional entities.
- 'Stage 3' is the concrete implementation of the functionality and of the protocols appearing at physical interfaces between physical elements onto which the functional elements have been mapped.

In addition, 3GPP often performs feasibility studies, the results of which are made available in technical reports (normally 3GPP internal TRs, numbered xx.7xx or xx.8xx).

WI GCSE_LTE has established the requirements and developed the associated extended features within the 3GPP system group communications services (GCS) using E-UTRAN access. These extensions are denoted as GCSE. Of note is that GCSE do not cover the specification of a particular GCS but only the 'support' within the LTE network to deploy any GCS. From the perspective of the GCSE, a GCS is basically conceived as a fast and efficient mechanism to distribute the same content to multiple users in a controlled manner. A GCS is expected to be able to deliver different types of media. Examples of media are conversational-type communication (voice, video), streaming (video), data (messaging) or a combination of them. The requirements established for the development of the GCSE features are specified in 3GPP TS 22.468 [40]. Sources of input requirements for GCSE include organizations such as the US NPSTC, TCCA, APCO Global Alliance, the International Union of Railways (UIC) and the ETSI Special Committee EMTEL and Project MESA. The GSCE requirements have been developed ensuring some flexibility to accommodate the likely different operational requirements for various types of critical communications user groups. Based on the requirements established in TS 22.468 [40] and on the analysis of different technical solutions for the GCSE reported in 3GPP TR 23.768, the architecture enhancements for 3GPP systems to support GCS are detailed in TS 23.468 [41]. Of note is that some of the functions requested by Stage 1 in TS 22.468 were not handled in Release 12 and these are addressed under Release 13 [38].

While GCSE are being introduced without any specific application in mind, WI MCPTT is complementing this work with further features that are needed to support an MCPTT service over LTE (i.e. the MCPTT service is a particular realization of a GCS). Therefore, the MCPTT service is intended to leverage the GSCE features to provide PTT voice communications with comparable performance to the PTT functionality currently available in PMR/LMR. The ultimate goal is to come up with a single, widely adopted global standard for a MCPTT application, avoiding as much as possible the current fragmentation.

The GCSE and MCPTT extensions are designed to complement the capabilities and features associated with the support of proximity-based services (ProSe), explained later in Section 4.4. Therefore, applications like MCPTT are intended to work also when terminals are out of network coverage (off-network scenarios), even though not all functions could be available in this case. A further insight into the features within the GCSE and MCPTT application is provided in the following subsections.

4.3.3 GCSE

The specification of the GCSE in 3GPP has been developed to meet the following main requirements [40]:

- Interoperability. Interfaces to the provided enablers shall be open so that group communications between users from different agencies in different territories using application layer clients provided by different manufacturers can interoperate. Furthermore, the network shall provide a third-party interface for group communication and a mechanism for a group member that is not connected via a 3GPP network to communicate in a group communication that it is a member of.
- Performance. The system should provide a mechanism to support a group communications end-to-end set-up time less than or equal to 300 ms. It is assumed that this value is for an uncontended network where there are no presence checking and no acknowledgements requested from receiver group member(s). The end-to-end set-up time is defined as the time elapsed since a group member initiates a group communications request on a UE until this group member can start sending voice or data information. The time elapsed since a UE requests to join an ongoing group communication until it receives the group communication should also be less than or equal to 300 ms. The end-to-end delay for media transport for group communications should be less than or equal to 150 ms.
- Priority and pre-emption. The system shall provide a mechanism to support a number of priority levels for group communication. The network operator shall be able to configure each group communications priority level with the ability to pre-empt lower-priority group communications and non-group communications traffic.
- Flexibility to accommodate the different operational requirements. The service should allow flexible modes of operation as the users and the environment they are operating in evolve. GCS is expected to support, voice, video or, more general, data communication. Also, GCSE should allow users to communicate to several groups at the same time in parallel (e.g. voice to one group, different streams of video or data to several other groups).
- Resource efficiency and scalability. The system shall provide a mechanism to efficiently distribute data for group communication. The number of receiver group members in any

area may be very large (as an illustrative scenario, 3GPP TS 22.468 states that, based on real-life scenarios, at least 36 simultaneous voice group communications involving a total of at least 2000 participating users in an area with up to 500 users being able to participate in the same group could be expected). The system shall support multiple distinct group communications in parallel to any one UE. The mechanisms defined shall allow future extension of the number of distinct group communications supported in parallel.

- Interaction with proximity-based services (ProSe). If EPC and E-UTRAN support ProSe, the EPC and E-UTRAN shall be able to make use of 'ProSe Group Communication' and the public safety 'ProSe UE-to-Network Relay' for group communication, subject to operator policies and UE capabilities or settings. (More details on the ProSe features are given in Section 4.4.)
- High availability of group communication. The system shall be capable of achieving high levels of availability for group communications utilizing GCSE, for example, by seeking to avoid single points of failure in the GCSE architecture and/or by including recovery procedures from network failures.

A high-level view of the overall architecture of a group communications system on top of the 3GPP EPS is shown in Figure 4.13. The architecture is split into two separate layers [41]:

1. The **application layer**, which holds the core functionalities of the group communications service. This functionality might be distributed between a GCS Application Server (GCS AS) on the network side and a GCS Client Application (GCS CA) running on the terminals. The application layer can be seen as the 'user' of the GCSE features. Application level interactions between the GCS CA and the GCS AS are out of scope of this GCSE specification.

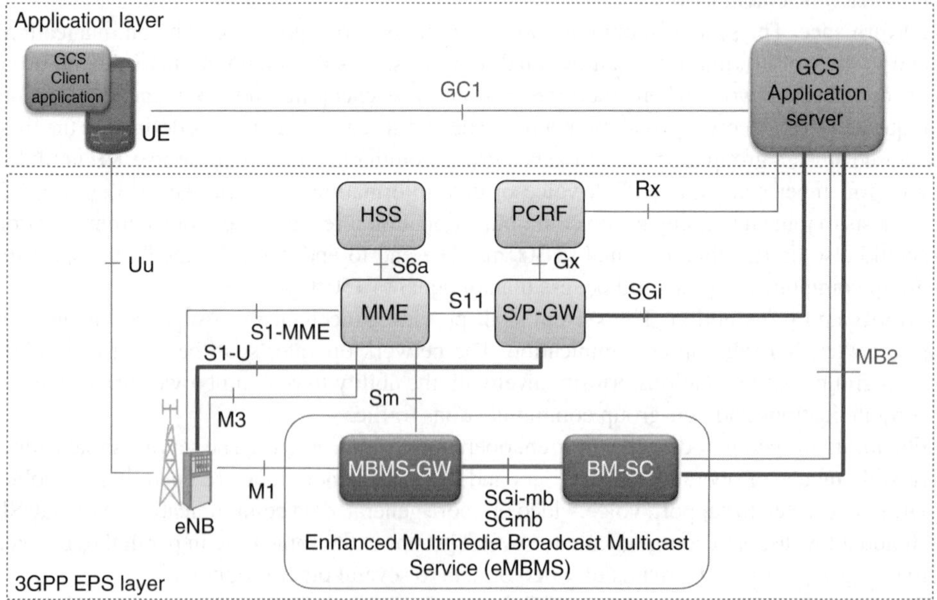

Figure 4.13 High-level architecture view of a group communications system over the 3GPP EPS.

2. The **3GPP EPS layer**, which primarily provides the information delivery service between the application layer entities. This delivery service provided by the 3GPP EPS layer encompasses both unicast and multicast delivery. Therefore, besides the core entities (i.e. MME, S-GW, P-GW, HSS and PCRF) that are central to the provision of (unicast) EPS bearer services, the 3GPP EPS layer also includes the functions specified for Multimedia Broadcast Multicast Service (MBMS) [42]. MBMS is the solution developed within 3GPP to allow the same content to be sent to a large number of subscribers at the same time, resulting in a more efficient use of network resources than each user requesting the same content and having the content unicast to each user. This approach enables the application layer to use a mix of unicast EPS bearer services and MBMS bearer services to support the GCS.

The support of MBMS requires the addition of new capabilities to existing functional entities of the 3GPP architecture together with the introduction of two new functional entities: the Broadcast Multicast Service Centre (BM-SC) and the MBMS Gateway (MBMS-GW). The BM-SC provides functions for MBMS user service provisioning and delivery. The BM-SC serves as an entry point for the content provider, allowing it to authorize and initiate MBMS bearer services within the network and schedule and deliver MBMS transmissions. Furthermore, the MBMS-GW provides the interface for the entities that are actually using the MBMS bearers (through SGi-mb (user plane) and SGmb (control plane) reference points) and supports IP multicast distribution of MBMS user plane data to eNBs (M1 reference point in Figure 4.13). MBMS on LTE is also denoted as evolved MBMS (eMBMS). A tutorial on MBMS over LTE is provided in Ref. [43].

On this basis, two reference points are central in the proposed architecture:

1. **GC1**. It is the reference point between the GCS CA in the UE and the GCSE AS on the network side. Through GC1, application domain signalling (e.g. for group admission and floor control, group management aspects such as group creation, deletion, modification, group membership control, etc.) is exchanged. This application signalling could be based on IMS-style signalling (e.g. SIP signalling supported in VoLTE services). Specification of this interface is left for Release 13 in the context of the MCPTT service.
2. **MB2**. It is the reference point between the GCS AS and the MBMS functions within the 3GPP EPS layer. MB2 offers access to the MBMS bearer service from a GCS AS. MB2 carries control plane signalling (MB2-C) and user plane (MB2-U) between GCS AS and BM-SC. The MB2 reference point provides the ability for the application to request the allocation/deallocation of a set of Temporary Mobile Group Identities (TMGIs)[3] and to request to activate, deactivate and modify an MBMS bearer. As well, the MB2 reference point provides the ability for the BM-SC to notify the application of the status of an MBMS bearer to the GCS AS. The protocol stack and security requirements for MB2-C/U together with some additional parameters needed by the MBMS procedure as defined in 3GPP TS 23.246 [42] have been specified under Release 12.

The architecture assumes that the GCS AS is not associated with any particular network (regardless of the subscription of the UEs that use the GCS). In this regard, 3GPP TS 23.246 [42]

[3] A TMGI is an identifier allocated to the MBMS bearer service. A TMGI can be used to identify one MBMS bearer service inside one LTE network.

covers both roaming and non-roaming scenarios. The group communications system allows for delivering application signalling and data to a group of UEs either (i) over MBMS bearer services or (ii) over EPS bearers or (iii) over both MBMS and EPS bearer services. QoS parameters of both EPS bearers and MBMS bearers can be controlled from the networks. Indeed, QoS attributes related to the GBR EPS bearer service (e.g. QCI, ARP, GBR and MBR) are also applicable to MBMS bearer services. In uplink direction, each UE establishes an EPS bearer service to transfer application signalling and data to the GCS AS. In downlink direction, the GCS AS may transfer application signalling and data via the UE individual EPS bearer services and/or via MBMS bearer service. The MBMS bearer(s) used for MBMS delivery can be pre-established before the group communications session is set up or can be dynamically established after the group communications session is set up. The GCS UEs register with their GCS AS using application signalling for participating in one or multiple GCS groups. When an MBMS bearer service is used, its broadcast service area (i.e. LTE cells involved in the delivery of the information) may be preconfigured for use by the GCS AS. Alternatively, the GCS AS may dynamically decide to use an MBMS bearer service when it determines that the number of UE for a GCS group is sufficiently large within an area (e.g. within a cell or a collection of cells).When MBMS bearer service is used, GCS AS may transfer data from different GCS groups over a single MBMS broadcast bearer. The application signalling and data transferred via MBMS bearer(s) are transparent to BM-SC and the MBMS bearer service. The GCS AS provides the UEs via GCS application signalling with all configuration information that the UE needs to receive application data via MBMS bearer services and to handle that data appropriately. When a GCS UE using MBMS bearer services moves to an area where the MBMS bearers are not available, the UE informs the GCS AS via application signalling that it changes from MBMS broadcast bearer reception to non-reception and the GCS AS activates the downlink application signalling and data transfer via the UE individual EPS bearer(s) as appropriate. To accomplish service continuity in this switching between EPS bearer(s) and MBMS bearer(s), in both ways, a UE may temporarily receive the same GCS application signalling and data in parallel. The GCS UE application discards any received application signalling or data duplicates.

Figure 4.14 shows an example of a scenario where a combination of unicast and MBMS traffic is used by the GCS AS on the DL to different UEs. Here, UE-1 to UE-3 receive DL traffic over unicast, whereas UE-4 to UE-6 receive DL traffic over eMBMS. The decision about whether a particular GCSE group communication (or UE/receiving group member) is using unicast delivery or multicast delivery is determined by the GCS AS. UL traffic, not shown in the illustration, is always done via unicast.

4.3.4 MCPTT over LTE

The MCPTT over LTE service is intended to support an enhanced PTT service, suitable for mission-critical scenarios, between several users (a group call), each user having the ability to gain access to the permission to talk in an arbitrated manner [44]. The MCPTT service builds on the existing 3GPP transport communications mechanisms provided by the EPS architectures, extended with the GCSE and ProSe capabilities, to establish, maintain and terminate the actual communications path(s) among the users. To the extent feasible, it is expected that the end user's experience to be similar regardless if the MCPTT service is used under coverage of an EPC network or based on ProSe without network coverage.

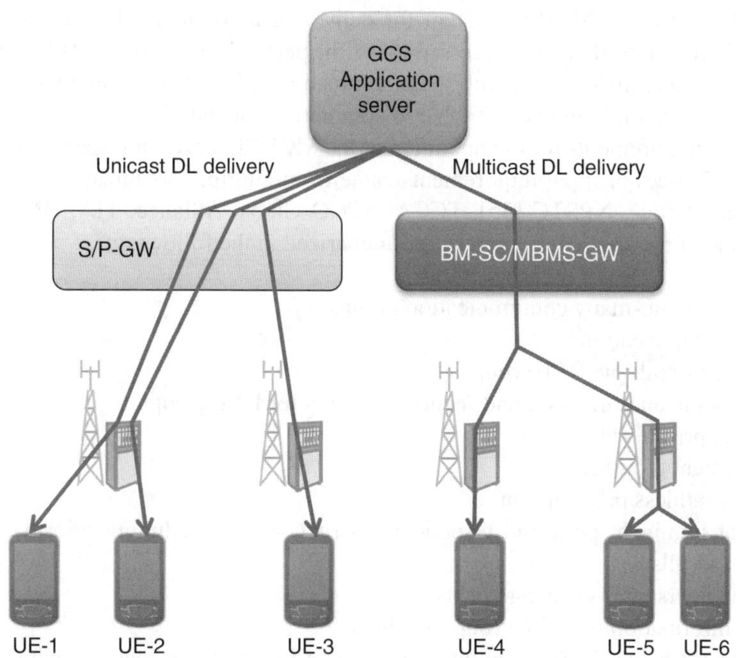

Figure 4.14 Media traffic with unicast and MBMS on DL.

Though the MCPTT service primarily focuses on the use of LTE, there might be users who access the MCPTT service through non-3GPP access technology. Examples are dispatchers and administrators. These special users typically have particular administrative and call management privileges that normal users might not have. In MCPTT, dispatchers can use an MCPTT UE (i.e. LTE access) or a non-3GPP access connection to the MCPTT service based on an interface specifically designed for such a purpose.

The MCPTT service allows users to request the permission to talk (transmit voice/audio) and provides a deterministic mechanism to arbitrate between requests that are in contention (i.e. floor control). When multiple requests occur, the determination of which user's request is accepted and which users' requests are rejected or queued is based upon a number of characteristics (including the respective priorities of the users in contention). MCPTT service provides a means for a user with higher priority (e.g. emergency condition) to override (interrupt) the current talker. MCPTT service also supports a mechanism to limit the time a user talks (hold the floor), thus permitting users of the same or lower priority a chance to gain the floor.

The MCPTT service provides the means for a user to monitor the activity on a number of separate calls and enables the user to switch focus to a chosen call. An MCPTT service user may join an already established MCPTT group call (i.e. late call entry feature). In addition, the MCPTT service provides the user identity of the current speaker(s) and user's location determination features. Moreover, MCPTT users would also be able to communicate with non-MCPTT users using their MCPTT UEs for normal telephony services.

MCPTT is primarily targeting to provide a professional PTT service to, for example, public safety, transport companies, utilities or industrial users with more stringent expectations of performance than the users of a commercial PTT service. In addition, the specification also envisions the use of an MCPTT system for delivering a commercial PTT service to non-professional users. Based on the addressed users, the performance and MCPTT features in use can vary (i.e. more mission-critical specific functionalities such as ambient listening or imminent peril call might not be available to commercial customers).

The service requirements for the operation of the MCPTT service are specified in 3GPP TS 22.179 [44]. Sources of input requirements came from multiple organizations (e.g. FirstNet, the UK Home Office, NPSTC [35], TCCA, APCO Global Alliance, TIA, OMA, ETSI TC TCCE). Some of the key requirements are summarized in the following:

- Support for one-to-many communication groups
- Dynamic group creation
- Monitoring of multiple PTT groups
- Authentication, authorization and security controls for PTT groups
- One-to-one private calls
- Announcement group calls
- Support of ruthless pre-emption
- Support of imminent peril and responder emergency calls including prioritization above normal PTT calls
- Identity and personality management
- Location information for PTT group members
- Support of off-network PTT communications and its operation together with on-network PTT at the same time

The current work in 3GPP centres on the study and evaluation of possible 3GPP technical system solutions for architectural enhancements needed to support MCPTT services based on the above requirements. A general principle being followed in the architectural design is the possibility for network operators to reuse the MCPTT architecture for non-public safety customers that require similar functionality. This could facilitate that these operators may integrate many components of the MCPTT solution with their existing network architecture. This approach requires the functional decomposition of MCPTT into a small number of distinct logical functions (e.g. 'group management functions', 'PTT functions'). As an illustrative example of this work, Figure 4.15 shows some preliminary schematics of the high-level functional entities and main reference points under consideration in Ref. [45] for the implementation of the MCPTT service in both on-network and off-network scenarios. In the on-network operation scenario, an MCPTT server is placed on the network side, on top of the EPS layer. Indeed, the MCPTT server is a specific instantiation of the generic GCS AS discussed in Section 4.3.3. UEs gain access to the MCPTT service by communicating with the MCPTT server via the GC1 interface. This operation is referred to as MCPTT Network Mode Operation (MCPTT NMO). GC1 is based on the SIP for session control (establishment, release, etc.). Additional protocols may be used for centralized floor control or for UE configuration. The on-network operation also considers the case of UE being out of network coverage, but within the transmission range of another UE supporting *ProSe UE-to-Network Relay* capabilities (ProSe features are described in Section 4.4). In this case, PC5 is the functional architecture to support ProSe features between the UEs. On top of this, a GCbis interface is needed between the MCPTT client of the terminal out of coverage and an MCPTT proxy functionality inside the relaying UE. This operation is referred to as Network Mode

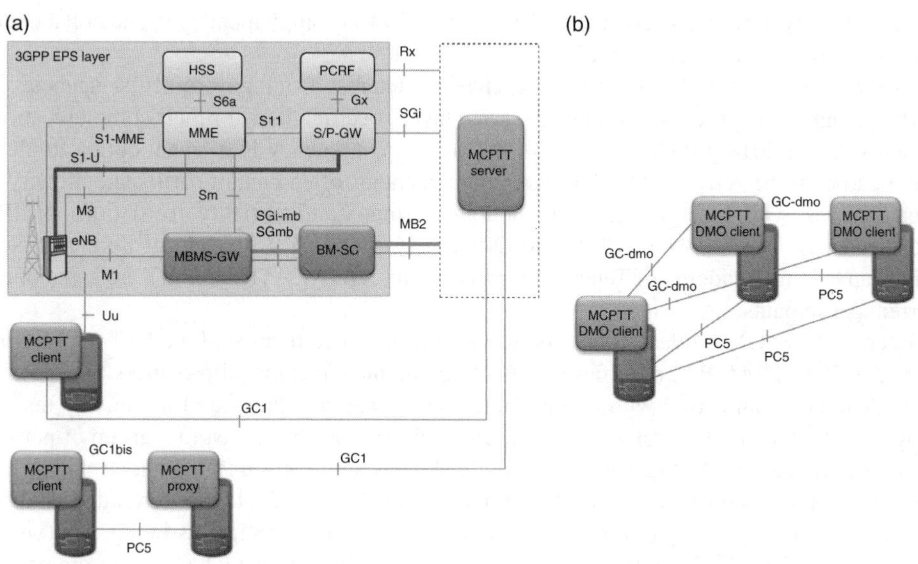

Figure 4.15 High-level functional entities and main reference points for the implementation of the MCPTT service in both (a) on-network scenarios and (b) off-network scenarios.

Operation via Relay (MCPTT NMO-R). In the off-network scenario, an MCPTT DMO client is introduced. This client would run on top of the ProSe one-to-many communications service defined for PC5. It would have functionality for fully decentralized floor control. It may also support functionality for location, presence, group management and status reporting, as identified in the Stage 1 requirements. In this case, GC-dmo is the inter-UE application level interface connecting the MCPTT clients for DMO. At the time of writing, no version of the normative work TS 23.179 is still available.

4.3.5 OMA PCPS

3GPP MCPTT work aims at reusing existing, standardized functionality when possible and justified. Remarkably, the Push-to-Communicate for Public Safety (PCPS) specifications defined by OMA have several components that could provide partial support for the MCPTT standard. As previously noted in Section 4.1, OMA is currently working with 3GPP to find the best method for the effective transfer of this specification to 3GPP so that this can be leveraged and further continued within 3GPP to meet the requirements of the PPDR community and take the specification to its final point in MCPTT.

OMA develops service enablers and Application Programming Interfaces (APIs) in the areas of communications, content delivery, device management (DM) and location, among others. OMA service enablers and APIs are mostly network agnostic, meaning that they are designed to be deployable over any type of network layer. Through the OMA APIs, many fundamental capabilities such as SMS, MMS, location services, presence services, payment and other core network assets in current communications systems can be exposed in a standardized way to the service layer. One of the flagship specifications of OMA is DM, which is widely deployed in commercial handsets. OMA is complementary to and collaborates with standards bodies such as 3GPP.

In recent years, OMA has seen an increasing interest from governmental agencies in participating in the process of building service layer specifications for their communications systems [8]. In 2014, OMA introduced the Government Agency Participant option to allow these parties to be active in OMA (government agencies often cannot participate directly in organizations with foreign legal jurisdictions, strong confidentiality restrictions or IPR requirements). The FirstNet, UK Home Office, UK Met Office, County of Somerset New Jersey and China Academy of Telecommunication Research (CATR) of MIIT are some of the current participants.

Back in 2008, OMA released the first service enabler specifications for a PTT application, named as PoC. OMA PoC specifies a form of communications that allows users to engage in immediate communication with one or more users ('walkie-talkie'-like) in the way that by pressing a button, a talk session with an individual user or a broadcast to a group of participants is initiated. OMA PoC version 2.1 (published as approved in 2011) allows audio (e.g. speech, music), video (without audio), still image, text (formatted and non-formatted) and file sharing with a single recipient (one to one) or among multiple recipients in a group (one to many). The OMA PoC solution has been commercially deployed by a few commercial operators (e.g. AT&T and Bell Canada, both launched the service in 2012). The PoC solution was mainly conceived as a service for consumers and developed to satisfy the functional and performance requirements arisen from the commercial domain, which are far less strict than those imposed by the public safety community [46]. Some details of this solution follow.

The OMA PoC solution builds upon the 3GPP IMS infrastructure and the IP connectivity service provided by the mobile network for the delivery of half-duplex, one-to-one or one-to-many voice services as well as video and data communications [47]. A simplified view of the OMA PoC solution architecture [48] is outlined in Figure 4.16. The central PoC functional entities are:

- **PoC Server**. A network element, which implements the IMS application level network functionality for the PoC service. It operates as an SIP Application Server from the IMS perspective that manages the PoC session set-up and tearing down procedures, talk burst control (i.e. floor control), and enforces policy defined for PoC group sessions.
- **PoC Client**. A functional entity that resides on the UE that supports the PoC service.
- **PoC Box**. A PoC functional entity where PoC session data and PoC session control data can be stored. It can be a network PoC Box or a UE PoC Box.
- **PoC Crisis Event Handling Entity**. A functional entity in the PoC network authorizing PoC users to initiate or join crisis PoC sessions. The PoC Crisis Event Handling Entity enforces the local policy for national security, public safety and private safety applications within a country or a subdivision of a country. The PoC Crisis Event Handling Entity complements the emergency service.

The above PoC functional entities that provide the PoC service use and interact with certain external entities such as a Presence Server that may provide availability information about PoC users to other PoC users or an XML Document Management Servers (XDMS) to manage groups and lists (e.g. contact and access lists). Discovery/Registry, Authentication/Authorization and Security are provided in cooperation with SIP/IP Core. OMA PoC solution

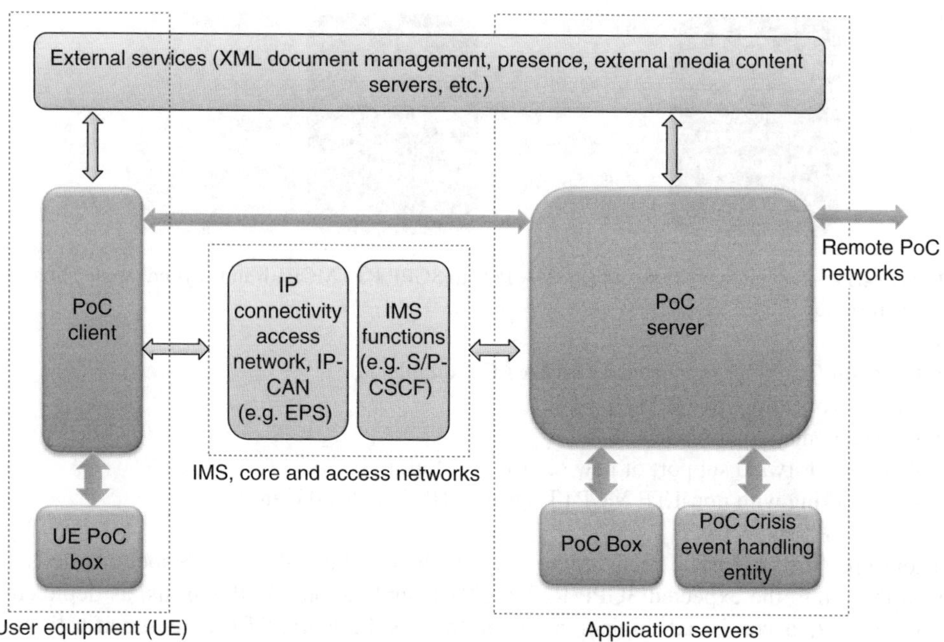

Figure 4.16 PoC service architecture.

makes use of the existing IETF protocol suite. The PoC sessions are managed using SIP and session signalling and bearer transport are performed through RTP/RTCP. On top of RTCP, PoC has defined its own extension – known as Talk Burst Control Protocol (TBCP) – for the purpose of speech channel management. From an IMS and PS Core domain perspective, a PoC session is seen as a classical IMS packet session, set up using SIP signalling through the S-CSCF that the subscriber is currently registered to. In order to support session data and signalling (including SIP and PoC signalling) transport, an EPS bearer over the core and access network is established, from the terminal to P-GW in the case of an EPC core network. OMA PoC also defines an interface to extend PoC services beyond the OMA-defined PoC service and PoC network boundaries, accomplished by interworking with other networks and systems, while not PoC compliant, being able to provide a reasonably comparable capability.

As a continuation of the OMA PoC solution, the PCPS project is consolidating OMA PoC v1.0, v2.0 and v2.1 requirements into a single PCPS v1.0 specification. This work has been done with the collaboration of public safety agencies. In essence, PCPS shares the basic PoC technology platform attributes, updated to operate on LTE networks. PCPS supports most of the MCPTT requirements though enhancements are needed. Of note is that PCPS version 1.0 does not consider support to exploit the new features added within the 3GPP layer with regard to GCSE and ProSe communications. An analysis of the major features missing from PCPS to fulfil the 3GPP MCPTT service requirements has reported the following items [49]:

- Authorization to both UE and application (user); difference in actors
- Group hierarchy and group affiliation
- Personality management

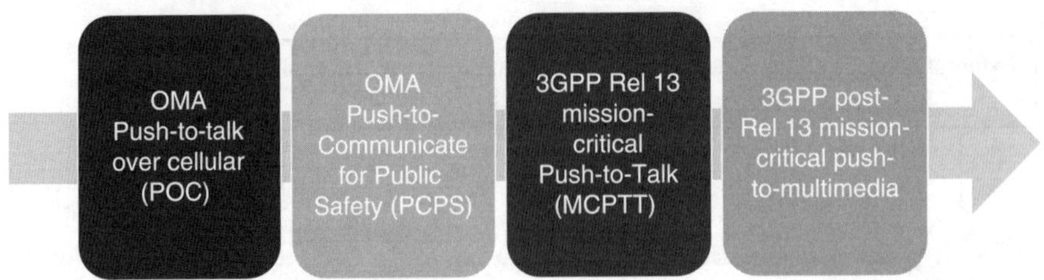

Figure 4.17 Evolutionary view of the OMA PoC to 3GPP R13 MCPTT and beyond application standardization.

- Location (could be supported by OMA LOC enabler)
- Off-network support
- Late call entry performance
- Ability to forward support of new security evolution
- Interworking with non-LTE MCPTT systems (P25, TIA-603, etc.)

Therefore, further work is necessary to advance the baseline OMA PCPS specifications and turn them into the expected 3GPP R13 MCPTT application. On this basis, as depicted in Figure 4.17, a continued augmentation of the resulting MCPTT core specifications is foreseen to progressively develop other applications that might be required for mission-critical communications (e.g. Mission-Critical Push-To-Multimedia) [8]. In addition to the PCPS specifications, other service enablers produced by OMA for location services, presence services, DM and others can be also considered in future versions of critical communications applications.

4.4 Device-to-Device Communications

The capabilities for the devices to communicate directly among them when they are out of network reach (i.e. off-network communications) are central for PPDR users. Nowadays, device-to-device communications, commonly referred to as DMO, is an important means of communicating voice and narrowband data. DMO is now used in current narrowband systems (e.g. TETRA) in several ways [50]:

- When there is no coverage (e.g. in buildings, tunnels, etc.) or when there is a risk to lose terrestrial coverage, which is especially important for the police and fire organizations
- To extend coverage by enabling a low-powered person-worn hand-portable terminal to communicate with a higher-powered vehicle-mounted terminal, which in turn is able to reach the terrestrial infrastructure
- As extra capacity (e.g. in case that the terrestrial network is congested)
- As a fallback when the terrestrial network fails
- For foreign units crossing the border

On this basis, the expectation is that the adoption of a radio interface based on LTE will continue supporting the abovementioned use cases not only for voice communications but also extended to multimedia services.

In addition to the interest from the PPDR community, the delivery of proximity-based applications and services also represent a recent and enormous trend in the commercial domain. The principle of these applications is to discover instances of the applications running in devices that are within proximity of each other and ultimately exchange application-related data. This enables new and enhanced existing apps/services such as social discovery/matching, push/proximate advertising, venue services, geo-fencing, credentialing, proximity-triggered automation, gaming integrating physical world elements and many others [51, 52]. Additionally, connecting devices directly when information is only of local interest (e.g. traffic safety applications, automation, etc.) instead of through the network can be beneficial to both the network operators and the users. In this case, local communications could turn into reduced latencies, higher throughputs, energy savings and improved resource utilization efficiency.

In this context, the 3GPP is seeking to seize the opportunity to become the platform of choice to exploit device-to-device communications and promote a vast array of future and more advanced proximity-based applications. The addition of built-in features for device-to-device communications to a global standard like LTE can readily turn into economies of scale advantages across both PPDR and non-PPDR services. Indeed, device-to-device communications is one of the main areas of focus in LTE Release 12, known in 3GPP terminology as Proximity-based Services (ProSe). ProSe features are aimed at (i) enabling the discovery of mobile devices in physical proximity to each other (i.e. ProSe Discovery) and (ii) enabling optimized communications between them, including the use of a direct communication path between UEs (i.e. ProSe Communication). The term 'LTE Direct' is also often used within the industry to refer to these 3GPP ProSe features.

To some extent, most of the abovementioned benefits intended to be provided by ProSe could also be pursued through the use of other prevalent device-to-device technologies such as Wi-Fi Direct and Bluetooth (Low Energy), complemented by pervasive OTT solutions for tracking the location of the device (e.g. GPS location). Some of the main reasons that justify the development of ProSe in front of these alternative technologies are the following:

- ProSe is conceived as 'network-controlled' device-to-device communications and so is fully integrated within the rest of LTE capabilities.
- ProSe leverages the LTE network infrastructure. The LTE network is intended to assist and supervise ProSe UEs for various functions such as device discovery, radio resource assignment, synchronization and security.
- ProSe is intended to provide a highly power-efficient, privacy-sensitive, spectrally efficient and scalable proximate discovery platform. This allows the discovery to be 'always on' and autonomous, with possible improvement in battery lifetime compared to alternative solutions.
- ProSe relies on the use of LTE/licensed spectrum. Licensed spectrum ensures no interference and very high reliability. Use of controlled – licensed – spectrum is key to provide QoS guarantees, even though the operation in unlicensed bands is not precluded (i.e. ProSe-assisted WLAN Direct Communications).

The next subsection details how standardization work for ProSe is being addressed in 3GPP. After that, the fundamentals of this technology are summarized including requirements and related features for ProSe Discovery and ProSe Communications and describing the main ProSe features specifically introduced to address the PPDR needs.

4.4.1 3GPP Standardization Work

The following WIs have been established within 3GPP to develop specifications for ProSe:

- 'Study on Proximity-based Services (FS_ProSe)' [53], initiated in Release 12
- 'Proximity-based Services (ProSe)' [54], initiated in Release 12
- 'Feasibility Study on LTE Device-to-Device Proximity Services – Radio Aspects (FS_LTE_D2D_Prox)' [55], initiated in Release 12
- 'Enhancements to Proximity-based Services (eProSe)' [56] and 'Enhancements Enhancements to Proximity-based Services – Extensions (eProSe-Ext)' [57], both initiated in Release 13
- 'Study on Security for Proximity-based Services (FS_ProSe_Sec)' [58], initiated in Release 12 and moved to Release 13

Table 4.5 lists the main 3GPP technical specifications and reports related to the above WIs.

Table 4.5 3GPP documents covering ProSe work.

Work Item (WI)	Related technical specifications/reports	Comments
FS_ProSe	TR 22.803 – 'Feasibility study for Proximity Services (ProSe)'	Informative technical report developing use cases for ProSe. Completed under Release 12
ProSe (R12), ePro Se (R13) and eProSe-Ext (R13)	(Modifications to existing TSs) TS 22.115 – 'Service aspects; Charging and billing' TS 22.278 – 'Service requirements for the Evolved Packet System (EPS)' TS 23.401 – 'General Packet Radio Service (GPRS) enhancements for Evolved Universal Terrestrial Radio Access Network (E-UTRAN) Access'	Normative requirements added for ProSe (Stage 1 and Stage 2)
	(New reports and specifications) TR 23.703 – 'Study on Architecture Enhancements to Support Proximity Services (ProSe)' TS 23.303 – 'Architecture Enhancements to Support Proximity Services (ProSe)' TS 33.303 – 'Security for Proximity-based Services (ProSe)'	Informative technical reports and normative specifications (Stage 1 and Stage 2/3) for Prose
FS_LTE_D2D_Prox	TR 36.843 – 'Feasibility Study on LTE Device-to-Device Proximity Services – Radio Aspects'	Feasibility study concerning radio access. Completed under Release 12
FS_ProSe_Sec	TR 33.833 – 'Study on Security Issues to Support Proximity Services'	Informative technical reports

The initial feasibility study on Prose reported in TR 22.803 [59] identified requirements and services that could be provided by the 3GPP system based on UEs being in proximity to each other. These requirements are added to the rest of requirements established for EPS (mainly in 3GPP TS 22.115 [60] and 3GPP TS 22.278 [61]). The feasibility study to support the required architecture enhancements is reported in 3GPP TR 23.703 [62], while the normative Stage 2 specifications (functional architecture) of the ProSe features in EPS are provided in 3GPP TS 23.303 [63]. Security aspects of ProSe are covered in 33 series of 3GPP specifications. As well, specific aspects concerning lawful interception (LI) are addressed under the general LI activities in Release 12. A feasibility study was also addressed for ProSe features on radio aspects. In particular, the feasibility study evaluated LTE device-to-device proximity services in different scenarios (e.g. within/outside network coverage, PPDR-only or general requirements); defined an evaluation methodology and channel models for LTE device-to-device proximity services; identified physical layer options and enhancements to incorporate in LTE the ability for devices within network coverage; considered terminal and spectrum specific aspects, for example, battery impact and requirements deriving from direct device-to-device discovery and communication; evaluated, for non-public safety use cases, the gains obtained by LTE device-to-device direct discovery with respect to existing device-to-device mechanisms (e.g. Wi-Fi Direct, Bluetooth) and existing location techniques for proximal device discovery (e.g. in terms of power consumption and signalling overhead); and investigated the possible impacts on existing operator services (e.g. voice calls) and operator resources. In addition, for the purposes of developing public safety requirements, the study also addressed the additional enhancements and control mechanisms required to realize discovery and communication outside network coverage. Single and multi-operator scenarios including the spectrum sharing case where a carrier is shared by multiple operators were considered both for LTE FDD and LTE TDD operations. The feasibility study is reported in 3GPP TR 36.843 [64].

4.4.2 *ProSe Capabilities*

ProSe is organized around two constituent capabilities [61]:

1. **ProSe Discovery**: a process that identifies that a ProSe-enabled UE is in the proximity of another using the LTE radio interface (with or without the infrastructure network) or using the EPC. The former is denoted as ProSe Direct Discovery. The latter is denoted as ProSe EPC-level Discovery.
2. **ProSe Communication**: a communication between two or more ProSe-enabled UEs in the proximity by means of a ProSe Communication path. This path could be established through the LTE radio interface either directly between the ProSe-enabled UEs or routed via local eNB(s) (i.e. ProSe E-UTRA Communication). The path could also be established over Wi-Fi Direct (i.e. ProSe-assisted WLAN Direct Communication). The case that the path is established directly between the devices is referred to as ProSe Direct Communication.

A ProSe-enabled UE refers to an UE that fulfils ProSe requirements for ProSe Discovery and/or ProSe Communication. Similarly, a ProSe-enabled network is a network that supports any or both capabilities. These capabilities and configuration options are shown in Figure 4.18.

Except for the PPDR case, ProSe Discovery and Communication over the LTE radio interface are intended to be used within network coverage and under continuous operator network control. Furthermore, it is required that the operator includes means to apply

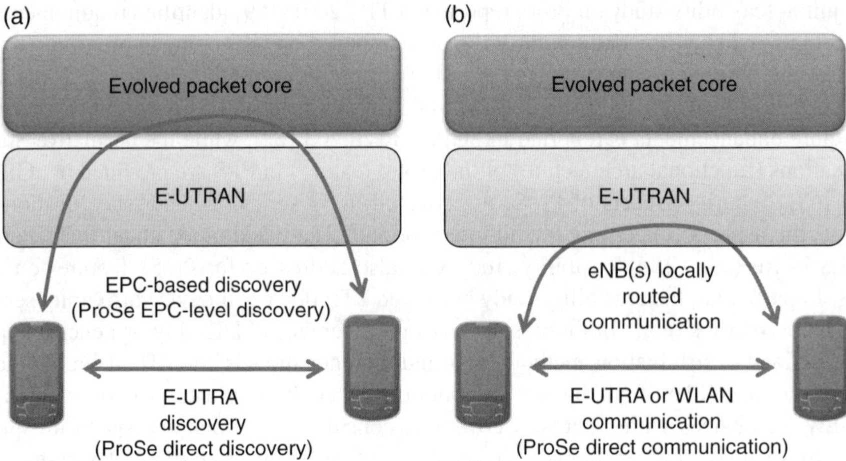

Figure 4.18 Illustration of the ProSe constituent capabilities ((a) Discovery and (b) Communication) and possible configuration options.

regulatory requirements, including LI, as per regional regulation. This represents a full 'operator-centric' approach that is expected to allow mobile network operators to offer and monetize new services that may exploit ProSe capabilities. Indeed, the operator's network and the ProSe-enabled UE provide mechanisms to identify, authenticate and authorize (third-party) application to use ProSe capability features. The operator's network could store information of third-party applications necessary for performing security and charging functions. Indeed, the operator is able to charge for use of ProSe Discovery and/or Communication by an application.

For the PPDR specific usage, ProSe-enabled UEs can establish the communications path directly between two or more UEs, regardless of whether UEs are served by an LTE network (i.e. there is no need for the terminals to be under network coverage). To distinguish between UEs intended to be used in the commercial or the PPDR domain, the latter are explicitly referred to as 'Public Safety ProSe-enabled UE' along the 3GPP specifications. Other ProSe capabilities introduced specifically to be supported in Public Safety ProSe-enabled UE are:

- Support for **ProSe Group Communication** and **ProSe Broadcast Communication** among a number of Public Safety ProSe-enabled UEs (these features rely on the use of a common ProSe E-UTRA Communication path).
- Ability for a UE to function as **ProSe UE-to-Network Relay** between E-UTRAN and UEs not served by E-UTRAN. This capability allows an out-of-coverage area Public Safety ProSe-enabled UE to have access to network services and applications through another Public Safety ProSe-enabled UE that supports the ProSe UE-to-Network Relay function.
- Ability for a UE to function as **ProSe UE-to-UE Relay** between two other UEs that are out of direct communication with each other. This relay function allows communications between these Public Safety ProSe-enabled UEs without the communications media (e.g. voice, data) being transported via the infrastructure of the PPDR network.

The support of the two abovementioned relay functionalities is not included in Release 12 but expected to be completed within Release 13, together with other requirements for service continuity and QoS priority/pre-emption of ProSe Direct Communication sessions. Further details of the Discovery and Communication features are addressed in the following.

4.4.2.1 ProSe Discovery

ProSe Discovery identifies that ProSe-enabled UEs are in proximity of each other, using E-UTRA (with or without E-UTRAN) or EPC when permission, authorization and proximity criteria are fulfilled. The proximity criteria can be configured by the operator.

The use of ProSe Discovery must be authorized by the operator, and the authorization can be on a 'per-UE' basis or a 'per-UE per-application' basis. An authorized application can interact with the ProSe Discovery feature to request the use of certain ProSe Discovery preferences.

The network controls the use of E-UTRAN resources used for ProSe Discovery for a ProSe-enabled UE served by E-UTRAN.

A plausible realization of the ProSe Direct Discovery capability is illustrated in Figure 4.19. It is based on the following premises:

- ProSe operates in uplink spectrum (in the case of FDD) or uplink subframes of the cell providing coverage (in the case of TDD).
- ProSe Discovery resources are configured by the RAN. Resources are semi-statically allocated and their amount does not practically impact on the LTE capacity (e.g. capacity reduction can be <1%). The eNBs assign part of discovery resources via broadcast control signalling (i.e. System Information Blocks (SIBs)) to authorized ProSe Discovery devices.

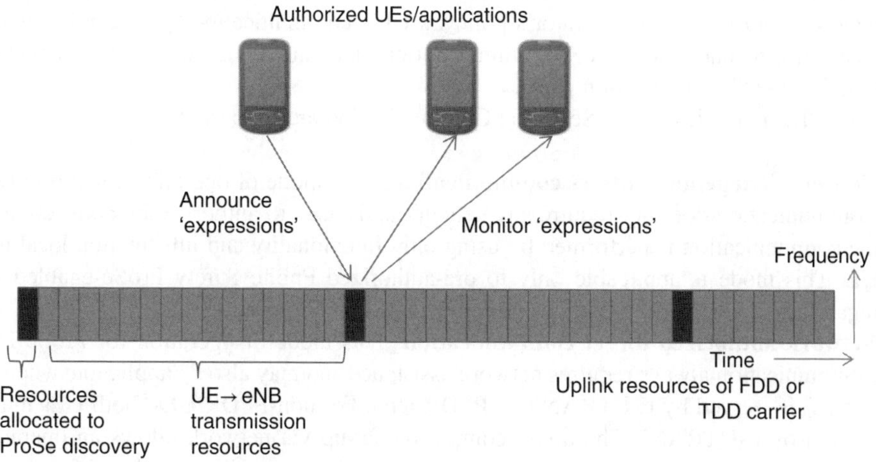

Figure 4.19 Operation of ProSe Discovery.

- All UEs can use the allocated resources either to broadcast their needs/services or listen to others. To that end, UEs broadcast 64- or 128-bit service identifiers named *Expressions* for each service they offer. An expression is used by peers in the discovery of proximate services, apps and context. Expressions can also be used by proximate peers to establish direct communications. The operator or some other entity has to manage a database of expressions and services they represent; UEs transmit and receive expressions within the discovery resources.
- UEs in the proximity read 'expressions' to determine relevance. UEs wake up periodically and synchronously to discover all UEs within range.

ProSe Discovery can be used as a stand-alone process (i.e. it is not necessarily followed by ProSe Communication) or as an enabler for other services. The UE-to-UE interface for ProSe Direct Discovery (and also for ProSe Direct Communication) is denoted as sidelink and specified in 3GPP TS 36.300 [20].

4.4.2.2 ProSe Communication

ProSe Communication enables the establishment of new communications paths between two or more ProSe-enabled UEs that are in ProSe communications range. The ProSe Communication path could use E-UTRA or WLAN.

In the case of WLAN, only ProSe-assisted WLAN Direct Communication (i.e. when ProSe assists with connection establishment management and service continuity) is considered part of ProSe Communication. Indeed, direct control of the WLAN link is outside the scope of 3GPP. However, service requirements that enable the 3GPP EPC to provide network support for connection establishment, maintenance and service continuity for WLAN Direct Communication are currently addressed by the 3GPP solution.

In the case of E-UTRA, the network controls the radio resources associated with the E-UTRA ProSe Communication path. The use of ProSe Communication must be authorized by the operator. The operator shall be able to dynamically control the proximity criteria for ProSe Communication. Examples of the criteria include range, channel conditions and achievable QoS. According to the operator's policy, a UE's communications path can be switched between an EPC path and a ProSe Communication path, and a UE can also have concurrent EPC and ProSe Communication paths.

Two different modes for ProSe Direct Communication are supported:

1. **Network-independent direct communication**. This mode of operation for ProSe Direct Communication does not require any network assistance to authorize the connection, and the communication is performed by using only functionality and information local to the UE. This mode is applicable only to pre-authorized Public Safety ProSe-enabled UEs, regardless of whether the UEs are served by E-UTRAN or not.
2. **Network-authorized direct communication**. This mode of operation for ProSe Direct Communication always requires network assistance and may also be applicable when only one UE is 'served by E-UTRAN' for PPDR UEs. For non-PPDR UEs, both UEs must be 'served by E-UTRAN'. The direct connection set-up via network allows an operator to authorize and control the direct connection set-up between UEs and determine the user traffic routing between direct and network-based paths.

Moreover, in the case of Public Safety ProSe-enabled UEs:

- ProSe Communication can start without the use of ProSe Discovery if the Public Safety ProSe-enabled UEs are in ProSe communications range.
- Public Safety ProSe-enabled UEs must be able to establish the communications path directly between Public Safety ProSe-enabled UEs, regardless of whether the Public Safety ProSe-enabled UE is served by E-UTRAN, as well as being able to participate in ProSe Group Communication or ProSe Broadcast Communication between two or more Public Safety ProSe-enabled UEs which are in proximity. Any of the involved Public Safety ProSe-enabled UEs need to have authorization from the operator.
- ProSe Communication is also facilitated by the use of a ProSe UE-to-Network Relay, which acts as a relay between E-UTRAN and UEs not served by E-UTRAN. The use of this relay function is controlled by the operator.
- In addition, ProSe Communication can also take place over a ProSe UE-to-UE Relay, a form of relay in which a Public Safety ProSe-enabled UE acts as a ProSe E-UTRA Communication relay between two other Public Safety ProSe-enabled UEs.

A summary of the possible configurations of ProSe Communication for PPDR and non-PPDR users is depicted in Figure 4.20:

- **Direct configuration, under network coverage and control**. Applicable to both PPDR and non-PPDR use cases. ProSe data plane is established directly between the communicating devices, while part of the control signalling is still performed through the involvement of the network.
- **Locally routed configuration**. Applicable to both PPDR and non-PPDR use cases. Neither data nor control plane is supported directly between devices, but data traffic is routed locally at the eNB, in contrast to the common case where data plane for each terminal is terminated at the P-GW of the LTE network.
- **Off-network operation**. Only applicable to PPDR. Public Safety ProSE-enabled UEs may communicate in case of absence of E-UTRAN coverage, subject to regional regulation and operator policy, and limited to specific public safety designated frequency bands and terminals.
- **ProSe UE-to-Network Relay**. Only applicable to PPDR. Under network control or off-network.
- **ProSe UE-to-UE Relay**. Only applicable to PPDR. Under network control or off-network.
- **ProSe Group Communication** (one to many) and **ProSe Broadcast Communication** (one to all). Only applicable to PPDR. Under network control or off-network. Communication is realized by means of a common ProSe E-UTRA Communication path established between the Public Safety ProSe-enabled UEs.

It is assumed that ProSe Communication transmission/reception does not use full duplex on a given carrier. Moreover, from the individual UE's perspective, on a given carrier, ProSe communication signal reception and cellular uplink transmission do not use full duplex either and TDM can be used. This includes a mechanism for handling/avoiding collisions. All data carrying physical channels is being assumed to use SC-FDMA for communication. In scenarios where a number of device-to-device communications links coexist, distributed link scheduling

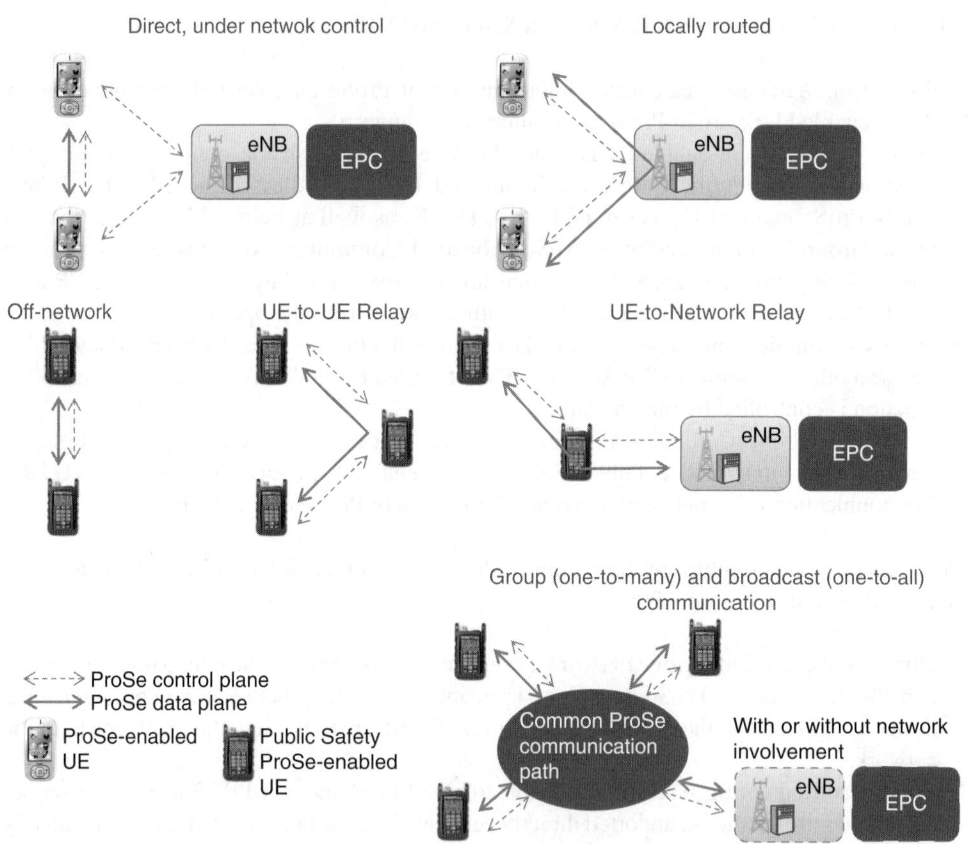

Figure 4.20 Possible configurations of ProSe Communication for PPDR and non-PPDR users.

can help to improve resource utilization by enabling maximal spatial reuse depending on the interference environment. Priority mechanisms can also be in place so that a transmitter will not transmit if this action will result in degrading higher-priority scheduled links [52].

4.4.3 ProSe Functional Architecture

The functional architecture to support ProSe features in EPS is illustrated in Figure 4.21. The new functional elements and reference points introduced are:

- **ProSe Application** is the application in the UE side that uses the features provided by ProSe. A ProSe Application is given a globally unique identifier (application ID). The UE hosting the ProSe Application is required to support procedures for ProSe Direct Discovery of other ProSe-enabled UEs (over PC5 reference point) as well as procedures for the exchange of ProSe control information between ProSe-enabled UE and the ProSe Function over the user plane (over PC3 reference point). In the case of Public Safety ProSe-enabled UE, additional functions may be included to support procedures associated with one-to-many ProSe Direct

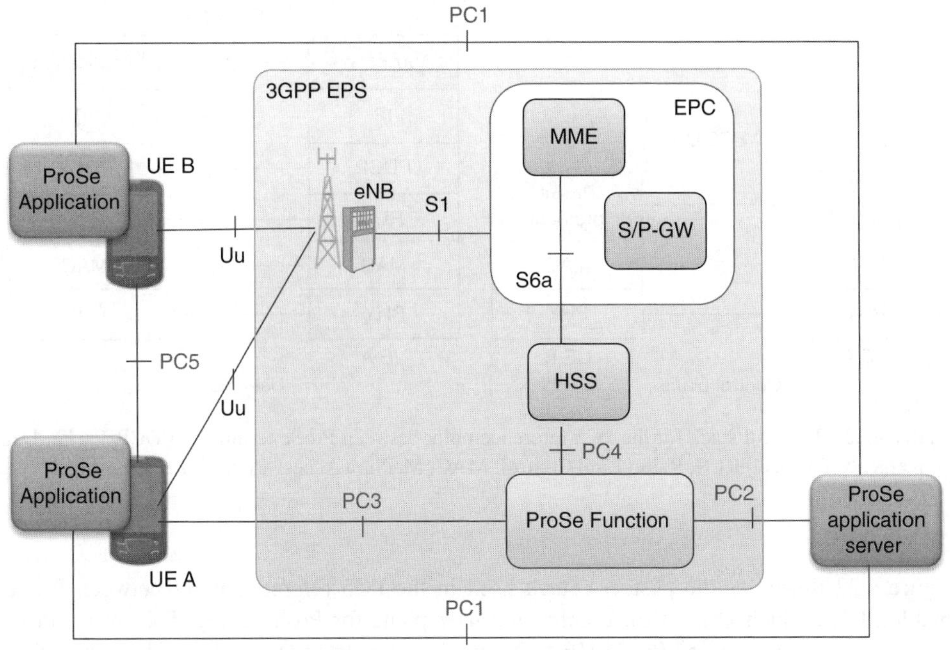

Figure 4.21 Functional architecture for ProSe.

Communication and UE-to-Network Relay. The configuration of parameters (e.g. including IP addresses, group security material, radio resource parameters) can be preconfigured in the UE or, if in coverage, provisioned by signalling (over the PC3 reference point).

- **ProSe Application Server** supports capabilities for storage and mapping of application and user identifiers. Specific application level signalling between the ProSe Application Server and the ProSe Application is done over PC1. The ProSe Application Server also interacts with the ProSe Function over the PC2 reference point.
- **ProSe Function** is the logical function that is used for network-related actions required for ProSe. It consists of three main sub-functions that perform different roles depending on the ProSe feature:

1. Direct Provisioning Function is used to provision the UE with necessary parameters in order use ProSe Direct Discovery and Prose Direct Communication.
2. Direct Discovery Name Management Function is used for open Prose Direct Discovery to allocate and process the mapping of ProSe Application IDs and other identifiers used in ProSe Direct Discovery. It uses ProSe-related subscriber data stored in HSS for authorization for each discovery request (retrieved over PC4 reference point). It also provides the UE with the necessary security material in order to protect discovery messages transmitted over the air.
3. EPC-level Discovery ProSe Function that includes storage of ProSe-related subscriber data and/or retrieval of ProSe-related subscriber data from the HSS, storage of a list of applications that are authorized to use ProSe EPC-level Discovery and EPC-assisted WLAN direct discovery and communication, etc.

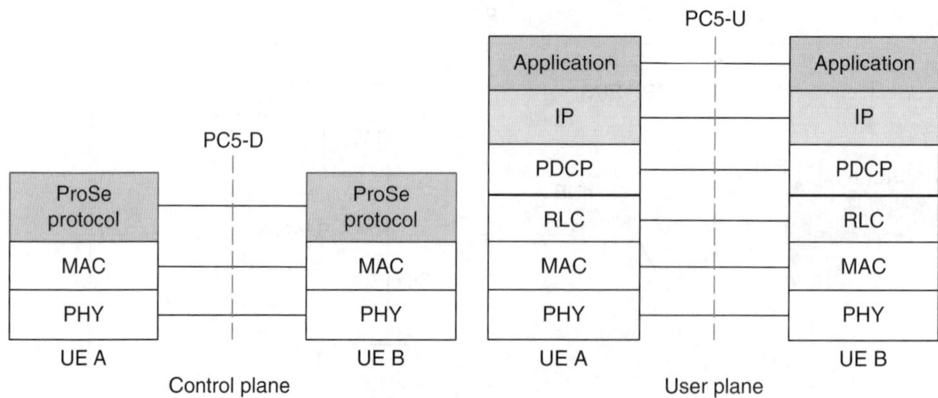

Figure 4.22 Protocol stack for the PC5 reference point between ProSe terminals. PDCP, Packet Data Convergence Protocol; RLC, Radio Link Control; MAC, Medium Access Control; PHY, Physical layer.

Figure 4.22 illustrates the protocol stack used in the PC5 reference point between ProSe-enabled UEs, which is used for control and user plane for ProSe Direct Discovery, ProSe Direct Communication and ProSe UE-to-Network Relay. In the user plane, IP packets from upper layer applications are exchanged between the ProSe UE on top of the conventional PDCP/RLC/MAC/PHY layers enhanced with the *sidelink* features defined in TS 36.300 [20]. On the control plane, a new 'ProSe Protocol' has been defined to support the associated procedures for ProSe Service Authorization, ProSe Direct Discovery and ProSe EPC-level Discovery [65]. For further details on the functional architecture and related procedures, the reader is referred to 3GPP TS 23.303 [63].

4.5 Prioritization and QoS Control for PPDR

Prioritization capabilities are essential elements of a mission-critical system at times of emergency and network congestion. Prioritization is the network's ability to determine which connections have priority over others and to allocate network resources accordingly.

Whenever a user attempts to establish a connection in a cellular network, certain administrative actions take place. In addition to authentication, authorization and some other administrative procedures, the network shall also determine through an admission control function whether it has sufficient resources to accept a new connection or not. These resources include bandwidth, processing power and other operational elements within the system. Prioritization within a cellular network ensures that users of 'high priority' can establish connections with higher level of certainty relative to users of 'low priority'. Hence, when multiple connection requests are being served, a network with prioritization capabilities will allocate resources to connections in an order dictated by the prioritization level. Prioritization capabilities may also include pre-emption of ongoing connections, so that the network could terminate or downgrade low-priority connections to free up resources and allocate a higher-priority connection. In addition to admission control, prioritization levels can also be taken into account by other resource management functions when handling network overload situations (e.g. load/congestion control mechanisms).

In general, priority levels for connections can be defined and assigned based on various criteria including user's role (or user priority), user application types, incident type in case of emergency-triggered prioritization schemes, etc. As a matter of principle, for a given application type, connections initiated by users with higher user priority take priority over the connections initiated by users with lower user priority. However, such priority may not hold if the application types are different. For example, a priority scheme may choose not to provide a connection priority to a higher-priority user with video application rather than to a lower-priority user with voice application. The determination of connection priority levels and its mapping to user priority, application type and other attributes is a matter that hinges upon both the public safety needs and the technology supporting it. Prioritization schemes used in a network can also be set out to distinguish between home and roaming users. For a prioritization scheme to be effective, any network or system involved in the provision of the 'end-to-end' call or session has to be either aware of the prioritization scheme or dimensioned not to be the communications bottleneck.

In LTE networks, prioritization capabilities are closely connected to QoS control capabilities, both of them being essential attributes of a mission-critical system [66, 67]. Indeed, QoS control is the network's ability to assign classes to different applications based on certain performance attributes and objectives and maintain the network performance for the application (e.g. packet loss, delay and throughput) within the acceptable range. On the other hand, prioritization capabilities are more related to the treatment received when establishing, modifying or releasing a connection. Therefore, during network congestion, a user with high priority shall be able to access his/her applications with no degradation in QoS, whereas a user with lower access priority may not be served or experience degraded QoS of his/her applications. It is worth noting that priority mechanisms should only act during those instances when the demand for bandwidth exceeds the available bandwidth. At all other times, priority mechanisms should not interfere with the transmission of information or access to network resources.

In networks used for PPDR applications, prioritization and QoS control can be exploited to discriminate:

- Among PPDR traffic in a public or private LTE network
- Among commercial and PPDR traffic sharing a public LTE network
- Among PPDR and other potential (secondary) users sharing a private LTE network

Prioritization and QoS control management in LTE networks can be realized through three constituent features that were already introduced in the first release of the LTE specifications (Release 8) and improved over the subsequent releases (illustrated in Figure 4.23):

1. **Access priority**. LTE provides a number of features for overload control of the radio signalling channels used for network access. This is necessary to protect the network from excessive signalling from large numbers of devices trying to get access to the network. Notice that access priority is key to avoid high-priority connections being blocked at this stage, when the importance of this communication has not been indicated to the network yet. Access priority is managed through (i) access control capabilities and (ii) priority signalling fields within the RRC protocol.

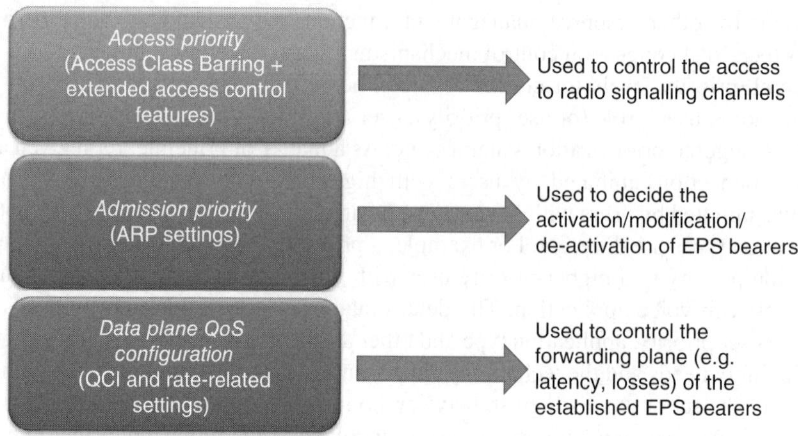

Figure 4.23 Constituent features for prioritization and QoS control in LTE.

2. **Admission priority**. This refers to the decision about the activation/modification/deactivation of EPS bearer services. Admission priority is managed through the ARP settings.
3. **Data plane QoS configuration**. This refers to the configuration of the user plane of the established bearers in terms of, for example, throughput, latency and packet loss. It is managed through QCI, GBR/MBR and AMBR settings.

The way that LTE priority and QoS control capabilities can best serve PPDR needs shall be eventually determined by PPDR users. In this regard, organizations such as NPSTC have outlined requirements for these capabilities to be deployed in the nationwide PPDR LTE network in the United States [66].

The next subsections cover in more detail each of these constituent features for prioritization and QoS control in LTE. In addition, in a last subsection, the MPS is described. MPS is a service already specified by 3GPP that could be implemented on top of the previously discussed prioritization features.

4.5.1 Access Priority

Access priority is concerned with the initial access to the system. Access priority shall allow the network operator to prevent normal users from making connection attempts or reducing its frequency in specified cells, so that priority users' connection attempts do not get blocked before being able to initiate the control procedures to set up a priority call/session.

3GPP networks in general and LTE in particular support two mechanisms that allow an operator to impose cell reservations or access restrictions:

1. The first mechanism uses an indication of cell status and special reservations for control of UE cell selection and reselection procedures. This mechanism allows the barring of a cell, so that users are not pexrmitted to camp on that cell. Cell reservation also considers a situation in which camping on the cell is only allowed for operator UEs [68].

2. The second mechanism, referred to as Access Class Barring (ACB), does not influence in selecting the cell to camp on, but the related cell access restrictions shall be checked by the UE when initiating any access attempt (i.e. RRC connection establishment procedure).

Access control allows the network operator to prevent overload of the access channel under critical conditions, though it is not intended that access control be used under normal operating conditions [69]. In 3GPP networks, all UEs are members of 1 out of 10 randomly allocated mobile populations, defined as access classes (ACs) 0–9. In addition, UEs may be members of one or more out of five special categories (ACs 11–15):

Class 15: Public land mobile network (PLMN) staff
Class 14: Emergency services
Class 13: Public utilities (e.g. water/gas suppliers)
Class 12: Security services
Class 11: For PLMN use

AC 10 is also defined, but its use is only intended to control emergency calls (e.g. 911 in the United States, 112 in Europe) attempts. AC information is stored in the SIM/USIM of the UE and in the subscription profile within the network. Over-the-air modification of AC settings is supported.

The information about the permitted ACs is signalled over broadcast channels in the air interface on a cell-by-cell basis. If the UE is a member of at least one of the permitted classes and the AC is applicable in the Serving Network (SN), access attempts are allowed in that cell. Any number of these classes may be barred at any one time. ACs 0–9 are valid for use in both Home and Visited PLMNs, whereas ACs 12, 13 and 14 are only valid for use in networks of the home country[4] and ACs 11 and 15 are only valid for use in the HPLMN or any Equivalent HPLMN (EHPLMN).[5] For the establishment of the RRC connection, the 'establishmentCause' field of the RRC can be used to indicate 'highPriorityAccess' when ACs 11–15 are used [70].

Additional features that complement and/or extend the basic access control capabilities in E-UTRAN are the following [69]:

- **Enhanced Access Control**. Barring status is not limited to 'barred/unbarred'. The SN broadcasts 'barring rates' (e.g. percentage value) and 'mean durations of access control' information parameters. These parameters are provided for different types of access attempts (i.e. mobile originating data or mobile originating signalling). With this information, the UE determines the barring status by drawing a uniform random number between 0 and 1. Access is allowed when the uniform random number is less than the current barring rate. Otherwise, the access attempt is not allowed, and further access attempts of the same type are then barred for a time period that is calculated based on the 'mean duration of access control'.
- **Service Specific Access Control** (SSAC). SSAC allows E-UTRAN to apply independent access control for multimedia telephony services for mobile originating session requests (e.g. a 'barring rate' and 'mean duration of access control' are broadcasted for voice and video services).

[4] Home country is defined as the country of the mobile country code (MCC) part of the IMSI.
[5] An Equivalent HPLMN (EHPLMN) is any PLMN to be declared as the HPLMN regarding PLMN selection. A list of EHPLMN can be stored in the UE.

- **Access Control for Circuit-Switched Fallback** (CSFB). Similar to SSAC, access control support for CSFB provides a mechanism to regulate the access to E-UTRAN to perform CSFB calls. It minimizes service availability degradation (i.e. radio resource shortage, congestion of fallback network) caused by mass simultaneous mobile originating requests for CSFB and increases the availability of the E-UTRAN resources for UEs accessing other services.
- **Extended Access Barring** (EAB). It is a mechanism to restrict network access for low-priority devices. Low-priority access configuration was introduced in Release 10 to aid with congestion and overload control in scenarios where a high number of devices can be connected to the eNB for delay-tolerant low-bit-rate data services (e.g. machine-to-machine (M2M) scenarios). A network operator can restrict network access for UE(s) configured for EAB in addition to the common access control and domain-specific access control when network is congested while permitting access from other UEs. The UE can be configured for EAB in the USIM or in the mobile equipment. EAB can be initiated when MMEs connected to eNB(s) request to restrict the load for UEs configured for low access priority or if requested by the network management system. When EAB is activated and UE is configured for EAB, it is not allowed to access the network. When the UE is accessing the network with a special AC (ACs 11–15) and that special AC is not barred, the UE can ignore EAB. Also, if it is initiating an emergency call and an emergency call is allowed in the cell, it can ignore EAB. During the work of 3GPP on System Improvements to Machine-Type Communication (SIMTC) Release 11, dual priority access was also introduced to allow for devices to hold dual-priority applications that will need 'normal' (default) priority access (e.g. in order to send infrequent service alerts/alarms) in addition to the 'low-priority/delay-tolerant' access (e.g. that will be used for the vast majority of their connection establishments). In addition to access restrictions, a 'low access priority' indicator is included in the signalling from the device to the RAN and to the core network. In the LTE and UMTS RAN specifications, the indicator is named 'delay tolerant'.

The access control information broadcasted in the air interface and the expected behaviour of UEs are specified in 3GPP TS 36.331 [70] and TS 22.011 [69]. In the case of multiple core networks sharing the same access network, the access network shall be able to apply ACB for the different core networks individually.

Using an administrative terminal or over-the-air configuration mechanisms, a PPDR operator should be able to assign a UE to one or more prioritized AC, which will determine the preferential initial access to the network. As well, the PPDR operator should be able to dynamically control which ACs are able to utilize the network in the event of congestion. It should be emphasized that once a UE has been admitted to the network and the UE remains active (i.e. RRC connected states) on the system, a change in the responder's AC will not discontinue (pre-empt) the UE's service. However, should the UE become idle, the UE will be required to once again pass the AC criteria. As well, because AC values (0–15) are stored in the UE's USIM, the UE's assigned AC(s) would be the same numerical value(s) regardless of the network that the UE is trying to get access to (e.g. dedicated network or commercial network(s)). Therefore, in the case of scenarios where the roaming of PPDR users to commercial networks is supported, as most 'average' commercial users will be assigned ACs 0–9, only having an AC between 0 and 9 should be avoided by critical PPDR UEs.

4.5.2 *Admission Priority*

Admission decision is based on the ARP parameter (15 priority levels + pre-emption and pre-emptability flags). This is enforced by each eNB based on the ARP settings obtained from the core network for the activation of the default and dedicated EPS bearer services.

3GPP specifications [22] point out that the ARP priority levels 1–8 should only be assigned to resources for services that are authorized to receive prioritized treatment within an operator domain (i.e. that are authorized by the SN). The ARP priority levels 9–15 may be assigned to resources that are authorized by the home network and thus applicable when a UE is roaming. This ensures that future releases may use ARP priority level 1–8 to indicate, for example, emergency and other priority services within an operator domain in a backward compatible manner. This does not prevent the use of ARP priority level 1–8 in roaming situation in case that appropriate roaming agreements exist that ensure a compatible use of these priority levels.

In the case of using a commercial network for the delivery of PPDR communications, the support of pre-emption is of special interest for PPDR practitioners. Pre-emption could be applied when there is contention for radio resources such that the resources requested by a PPDR user/application are not available. In that case, the network could pre-empt commercial users until sufficient resources become available to permit the PPDR responder to use his/her applications. In any case, the use of the pre-emption capability needs to be properly engineered. For instance, an active ‘911’ or ‘112’ session shall not be pre-empted. Moreover, pre-emption mechanisms should first attempt to free up resources by limiting the choices and QoS of the commercial users’ applications (e.g. throttling down the AMBR or GBR parameters of established EPS bearers) while not depriving these users from minimum connectivity (e.g. voice calls, SMS). A proposal for the mapping of commercial services and PPDR services onto the set of ARP values is provided in the work by R. Hallahan and J.M. Peha [71].

The allocation of ARPs has to be defined by the system administrator as part of the subscription profiles and/or PCC rules. An LTE network may support:

- **Default admission priority settings**. Default priority should be thought of as the day-to-day prioritization settings that the network will automatically use most of the time but special incidents or needs. Because congestion can occur at any moment, the default priority framework must be carefully designed to accommodate the widest range of responder activities. Default priority levels may be set in the user profiles.
- **Dynamic admission priority settings**. Dynamic priority refers to the ability of an authorized responder or administrator to override the default priority assigned automatically by the network and to be able to dynamically set or modify the priorities assigned to users. Typically, human intervention is required to trigger a dynamic priority change, such as pressing the UE’s emergency button. As well, users’ priorities may be assignable by the incident commander in real time (e.g. an incident commander should have the capabilities with support to influence network priority rather than having this responsibility fall to an overwhelmed public safety dispatcher or to the individual radio users). The user may be made aware of the default and incident-specific settings of his/her access priority via the UE human interface.

The following parameters are expected to be considered when defining priorities [66, 72]:

- Role of the user involved in the response to an incident in accordance with incident management systems or incident command systems
- User operational state (e.g. immediate peril, responder emergency, etc.)
- Location of the user (users that are too far from the incident or unable to intervene effectively should be assigned a lower priority)
- Application category (e.g. mission-critical/non-mission-critical categories)
- Application type (e.g. latency-sensitive real-time, video streaming, latency-tolerant M2M, client–server database queries, web browsing, etc.)

The work by R. Hallahan and J.M. Peha [71] provides some considerations on operational policies and arrangements to deploy PPDR priority access over public access cellular networks. Recently, the TIA in the United States, which develops standards for the information and communications technology industry, released the document TIA-4973.211 that describes requirements for a mission-critical priority and QoS control service for a wireless broadband network. It includes requirements to determine a user's default priority on the broadband network and provides requirements for dynamic prioritization changes to meet situational needs. The requirements allow an operator to define consistent and deterministic policies to moderate usage of a shared wireless broadband network.

4.5.3 Data Plane QoS Configuration

The treatment of the user plane (e.g. delay, packet loss, scheduler priority) is based on the QCI parameter, complemented with the GBR/MBR parameters for GBR bearers.

Like admission priority, data plane QoS configuration is typically assigned by an authorized administrator, and it is enforced on a per-eNB basis. As well, all the discussion around default/dynamic settings and the type of parameters considered when defining admission priorities is extensible to the configuration of the user plane.

With regard to QCI, as described in Section 4.2.2, standard QCI values can be suitable for PPDR applications insofar as they already consider some values that may be exploited in PPDR scenarios. Adopting the industry standard QCI definitions may be beneficial in terms of interoperability (e.g. PPDR scenario has the ability to roam for added coverage and capacity to commercial LTE systems). Should the need arise, LTE does allow custom QCIs to be created.

Complementary to QCI, GBR/MBR parameters for GBR bearers and AMBR for non-GBR bearers provide LTE with the rate limiting and bandwidth control capabilities over the amount of radio resources that are made available to a given user. Details on the different rate-related parameters are provided in Section 4.2.2. These parameters would allow, for example, limiting the maximum bit rate for general data services such as using Internet access. This would prevent a user from dominating non-GBR resources at an eNB. Standards/profiles could be created to consistently apply rate limits per UE across the entire network. In the presence of congestion, the network may further provide a guaranteed minimum bandwidth for a UE's non-GBR traffic in order to prevent starvation.

When configuring a new streaming voice or video application for use, the minimum and maximum bandwidth needs of the application are usually well known (e.g. codec bandwidth needs). Real-time voice and video applications typically require dedicated resources. Therefore, user entities could have the ability to configure application minimum and maximum bandwidth needs when commissioning new applications for use on the network. In any case, real-time adjustment of network bandwidth controls is to be strongly avoided for both UEs and applications because of the high complexity involved [66].

4.5.4 MPS

MPS [73] is a service already standardized within 3GPP that is realized through the access control, admission priority and data plane QoS configuration features described in the previous sections.

MPS is intended to be utilized for voice, video and data bearer services in the packet-switched domain and the IMS of 3GPP networks. MPS complements another service called Priority Service [74], specified for establishing voice calls through the 3GPP CS domain.

An MPS user is an individual who has received a priority level assignment from a regional/national authority (i.e. an agency authorized to issue priority assignments) and has a subscription to a mobile network operator that supports the MPS feature (i.e. MPS subscription). Upon MPS invocation, the calling MPS user's priority level is used to identify the priority to be used for the session being established. Both on-demand MPS invocation (i.e. priority treatment is explicitly requested by MPS user) and always-on MPS subscription (priority treatment is provided by default to all packet-switched sessions) are considered. Pre-emption of active sessions shall be subject to regional/national regulatory requirements. Also, subject to regional/national regulatory policy, a mobile network should have the capability to retain public access as a fundamental function. Therefore, MPS traffic volumes should be limited (e.g. not to exceed a regional/national specified percentage of any concentrated network resource, such as eNB capacity), so as not to compromise this function.

A feasibility study to support MPS over 3GPP networks was first addressed under Release 7 [75], and as a result, MPS Stage 1 requirements were developed within Release 8 in TS 22.153 [73]. Within Release 10, TR 23.854 [76] addressed some enhancements for MPS related to priority aspects of EPS packet bearer services and priority-related interworking between IMS and EPS packet bearer services. These enhancements enable the network to support end-to-end priority treatment for MPS call/session origination/termination, including the NAS and access-stratum (AS) signalling establishment procedures at originating/terminating network side as well as resource allocation in the core and radio networks for bearers. The basic MPS considering priority handling of IMS-based multimedia service, EPS bearer services and CSFB were completed in Release 10. In Release 11, SRVCC from LTE to UTRAN/GERAN/1xCS was also addressed based on Release 10 SRVCC specification [76].

As a result, MPS treatment can be applied to:

- **IMS-based multimedia services**. An MPS session (e.g. a voice/video call with priority treatment) can be activated through the IP Multimedia Subsystem (IMS) platform. A session request shall include an MPS code/identifier followed by the destination address (e.g. SIP URI, Tel URI). Consistency of the priority indication and treatment between the IMS service layer and signalling and data bearers established through the mobile network is achieved through Policy Control and Charging functionality [21]. Thus, PCC functionality

determines the QoS profile (i.e. ARP and QCI parameters) to be assigned to the corresponding bearer services. Besides user prioritization, service prioritization is also possible with this model.
- **Priority EPS bearer services**. The activation of MPS does not require the use of IMS. Hence, when IMS is not used, on-demand MPS can also be activated/deactivated via, for example, HTTPS server that would interact with the operator PCC infrastructure (the HTTPS server will act as an AF within the PCC model).
- **Circuit-switched fallback (CSFB)**. This functionality is needed to allow the initiation or termination of CS services (e.g. voice service) from/to an UE while the UE is attached to an E-UTRAN that does not provide direct access to CS services. CSFB instructs the terminal to move to GERAN/UTRAN to handle the service. In this context, priority treatment for calls initiated/terminated using the CSFB functionality will also be permitted.

4.6 Isolated E-UTRAN Operation

A high availability network shall implement features that provide increased robustness and alternate radio path establishment in the case of degraded network due to the failure of one or more infrastructure nodes or network connectivity. In many critical incident scenarios, the benefit of ensuring the ability to communicate between PPDR officers on the ground will be of utmost importance, even though they may be moving in and out of the LTE network coverage or following the loss of backhaul communications.

The support of the ProSe capabilities described in Section 4.4 are already central for ensuring a high availability of PPDR communications services as they can provide off-network operation in the case of complete network failure. However, even in the case of a major disaster, a likely situation is that the backhaul connectivity of a base station could be lost but the base station itself is still operational. In such cases, any features enabling the base station to act alone, isolated from the network, can be still of great support for increased resilience of PPDR communications in the affected area.

Both commercial and PPDR systems need to be able to survive network equipment failures and overload situations, though the requirements for public safety are more rigorous. In June 2013, 3GPP agreed to study how to enhance the resilience of LTE networks for public safety applications. In this context, 3GPP established the WI 'Feasibility Study on Isolated E-UTRAN Operation for Public Safety (FS_IOPS)' as part of Release 13 [77]. The WI FS_IOPS is intended to address a feasibility study to define use cases and identify potential requirements for isolated E-UTRAN operation in support of mission-critical network operations to be reported in 3GPP TR 22.897 [78]. Stage 1 specifications have been reported in Ref. [79], and a study on architecture enhancements to support these features has started (to be reported in 3GPP TR 23.797).

Two main scenarios, illustrated in Figure 4.24, can benefit for the addition of isolated E-UTRAN operation features:

1. Infrastructure failures. In the event of an interruption to normal backhaul connectivity, isolated E-UTRAN operation aims to adapt to the failure and maintain an acceptable level of network operation in the isolated E-UTRAN. The restoration of service is the eventual goal.

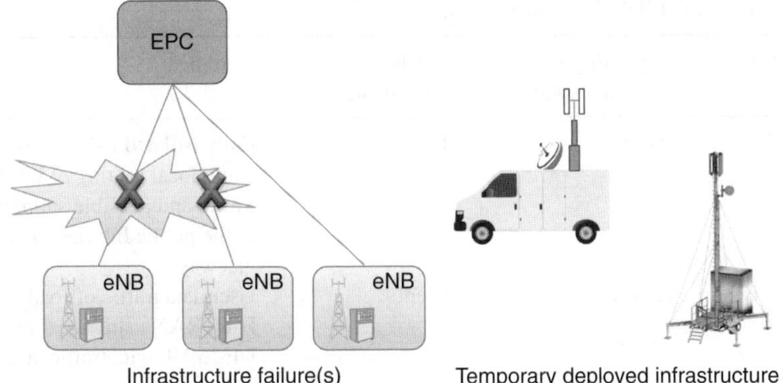

Figure 4.24 Scenarios under scope for the isolated E-UTRAN operation features.

2. Deployment of infrastructure with no connectivity to the network core. To provide voice, video and data communications service for PPDR officers who are out of LTE network coverage, the PPDR authorities may deploy a mobile command post equipped with one or a set of nomadic eNB (NeNB). A NeNB may consist of base station, antennas, microwave backhaul and support for local services. The NeNB can provide coverage or additional capacity wherever coverage was never present (e.g. forest fire or underground rescue) or wherever coverage is no longer present, for example, due to natural disaster.

In this context, isolated E-UTRAN can be created:

- Following an event isolating the E-UTRAN from normal connectivity with the EPC
- Following deployment of stand-alone E-UTRAN NeNBs

Isolated E-UTRAN can comprise:

- Operation with no connection to the EPC
- One or multiple eNBs
- Interconnection between eNBs
- Limited backhaul capability to the EPC
- The services required to support local operation (e.g. group communication)

The features to be developed for isolated E-UTRAN are closely linked to ProSe and GCSE features. As addressed in previous sections, ProSe and GCSE have defined requirements for public safety discovery and communications (including group communications) in the cases of no network coverage and of full (E-UTRAN and EPC) network coverage. Therefore, the need for discovery and group communications has to be considered in the case of isolated E-UTRAN. Isolated eNB(s) can bring benefits from exploiting locally routed communications for PPDR UEs such as the following:

- The communications range achievable between PPDR UEs may be enhanced compared with direct communications using ProSe.

Table 4.6 Isolated E-UTRAN scenarios [78].

IOPS scenario	Signalling backhaul status	User data backhaul status	Comment
No backhaul	Absent	Absent	Fully isolated E-UTRAN operation using local routing of UE-to-UE data traffic and possible support for access to the public Internet via a local gateway
Signalling-only backhaul	Limited	Absent	User data traffic offload at the E-UTRAN using local routing of UE-to-UE data traffic and possible support for access to the public Internet via a local gateway
Limited backhaul	Limited	Limited	Selective user data traffic offload at the E-UTRAN using local routing of UE-to-UE data traffic and possible support for access to the public Internet via a local gateway
Normal backhaul	Normal	Normal	Normal EPC-connected operation

- PPDR eNB(s) permanently or temporarily without backhaul can act as a radio resource manager for ProSe communications between PPDR UEs to reduce interference and increase system capacity.
- Isolated eNB could offer additional benefits by extending the network architecture, for example, with the use of Local IP Access (LIPA)-like features [19] that enable access for IP-capable UEs, connected via the isolated eNB, to other IP-capable entities (e.g. application servers) attached directly to that eNB.

Table 4.6 illustrates the features expected to be supported by the eNB in scenarios with different backhaul limitations.

4.7 High-Power UE

In coverage-limited deployments, the maximum transmit power of the device (i.e. uplink maximum transmit power) could be a bottleneck to achieve high data rates at long distances from the eNB. At the initial stage, LTE specifications only considered support for one UE power class (Class 3) with a maximum transmit power of 23 dBm for any of the supported operating bands. In Release 11, an additional power class (Class 1) was defined only for Band 14 devices, that is, the band allocated in the United States for PPDR. Therefore, LTE currently specifies the following UE power classes [18]:

Class 3: defined for all supported bands with a maximum transmit power of 23 dBm, though the tolerance is not exactly the same in all the bands

Class 1: defined only for Band 14, with a maximum transmit power of 31 dBm

The specification of Class 1 devices in a PPDR network would allow for a better coverage and availability/throughput performance than provided by commercial systems, particularly in rural areas. This can be eventually achieved by using higher-power UE(s) for vehicular mobile applications. An illustrative example of the coverage range increase that can be achieved by Class 1 in front of Class 3 devices is given in the following. Assuming the Hata rural path loss model [18], a carrier frequency of 790 MHz within Band 14 and an eNB antenna height of 45 m above the ground, the difference in maximum transmit power of 8 dB between Class 1 and Class 3 will result in the same increase of compensable propagation loss 8 dB, which in turn equates to around 70% increase in the cell radius and over 180% in cell coverage area.

The support of high-power class devices is mainly limited by the coexistence requirements with systems operating in adjacent bands. The approach followed in the specification of Class 1 for Band 14 was to maintain the same coexistence impact in terms of throughput/out-of-band emissions from the Band 14 Class 1 UE to base station receivers in Band 13 through tighter RF requirements for the high-power UE where applicable. These coexistence studies are reported in TR 36.837 [80].

4.8 RAN Sharing Enhancements

3GPP specifications add support for RAN sharing among several core network operators. In this way, different core network operators are allowed to connect to a shared RAN. The operators do not only share the radio network elements but may also share the radio resources themselves (i.e. radio spectrum).

In the commercial domain, RAN sharing is perceived as a method for QoS improvement in congested areas, reduction of expenditures or coverage improvement. RAN sharing is expected to allow operators for a higher flexibility in pursuing different network deployment strategies [81, 82]:

- **A greenfield deployment**. Two operators jointly agree to build out a new technology. At the outset, the new shared network infrastructure and operations can be based on capacity and coverage requirements of both operators. The operator can fund the built-on 50:50 or according to their expected needs.
- **Buy-in**. One of the sharing operators has already built a RAN and is looking for another operator to share this network. In this case, the second operator would either pay a capacity usage fee or upfront fee to acquire in the network.
- **Consolidation**. Either 2G, 3G or 4G networks, which have already built out by each of the sharing operators, need to be consolidated into one joint network. This type of network sharing usually holds significant cost advantages, but it also presents substantial design challenges.

In the PPDR domain, RAN sharing could be explored as a way to reduce the high upfront costs associated with the deployment of a stand-alone, dedicated PPDR network by partnering with commercial network operators to share the RAN equipment. Indeed, a majority of the upfront costs are related to establishing coverage, with approximately 70% of the capital expenditures (CAPEX) involving site acquisition, access equipment, civil works (i.e. construction of the site, installation of the equipment) and laying cables for electricity and backhaul.

The existing LTE 3GPP-compliant technology makes different RAN sharing options technologically possible today. These options are analysed in Chapter 6, while the description in this section centres on the technical features supported by LTE specifications as to the RAN sharing implementation.

The support for RAN sharing in 3GPP was already introduced under Release 6 for GERAN and UMTS, being later updated for LTE. The details of network sharing for GERAN, UTRAN and E-UTRAN have been defined in 3GPP TS 23.251 [82]. The arrangements for network sharing between the involved entities can vary widely, being influenced by a number of factors including business, technical, network deployment and regulatory conditions. Within all of this variation, there is a set of common roles centred on connecting network facilities between the parties participating in a network sharing agreement:

- **Hosting RAN provider**. The hosting RAN provider is identified as sharing a hosting RAN with one or more participating operators. The hosting RAN provider has deployed and operates a RAN in a specific geographic region covered under the network sharing arrangement. It has primary operational access to a particular licensed spectrum that is part of the network sharing arrangement, though it does not necessarily own licensed spectrum but has agreement to operate it. The hosting RAN provider can be a mobile network operator, though other entities can be involved through outsourcing, joint ventures or leasing agreements for operating and owning the RAN infrastructure or managing the sharing agreements.
- **Participating operator**. The participating operator is identified as using shared RAN facilities provided by a hosting RAN provider, possibly alongside other participating operators. The participating operator uses a portion of the particular shared licensed spectrum to provide communications services under its own control to its own subscribers. The sharing agreement between the hosting RAN provider and the participating operators may or may not include sharing of a part of the radio spectrum of the hosting RAN provider (e.g. a mobile virtual network operator (MVNO) as a participating operator would use the spectrum provided by the hosting RAN provider). Mobile network operators can take on the role of participating operators, as well as other entities such as outsourcing, joint ventures or leasing agreements for operating or owning the service infrastructure.

There are two identified architectures to be supported by network sharing (illustrated in Figure 4.25 for the case of an LTE network):

1. **Multi-Operator Core Network** (MOCN). Multiple core network nodes are connected to the same RAN node (i.e. RNC in UTRAN, BSC in GERAN and eNB in LTE). The core network nodes are operated by different participating operators.
2. **Gateway Core Network** (GWCN). Besides shared RAN nodes, the participating operators also share some core network nodes (i.e. MSC/SGSN for UMTS and GSM and MME for LTE).

The UE behaviour in both of these configurations shall be the same. No information concerning the configuration of a shared network shall be indicated to the UE. At the end, network sharing is an agreement between operators and shall be transparent to the user. This implies that a supporting UE needs to be able to discriminate between core network operators available in a

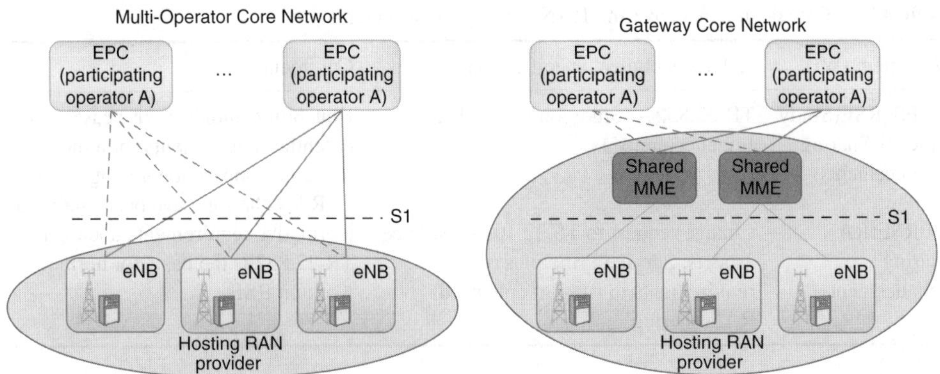

Figure 4.25 RAN sharing architectures: Multi-Operator Core Network (MOCN) and Gateway Core Network (GWCN).

shared RAN and that these operators can be handled in the same way as operators in non-shared networks. To that end, each LTE cell in a shared RAN shall include information concerning the available core network operators in the broadcast system information so that LTE UEs can take that information into account in the network and cell (re)selection procedures.

The scenarios allowed from Release 6 specifications are rather limited and do not cover more complex scenarios that arise due to recent needs for more dynamic cooperation among operators. Importantly, current 3GPP RAN sharing specifications do not cover specific mechanisms to distribute radio access capacity between core network operators attending to some specific needs/conditions. In practical situations, such functionality is achieved based on inter-operator cooperation and requires substantial network reconfiguration effort.

Two main WIs have been established within 3GPP to improve the support for RAN sharing:

1. 'Study on RAN Sharing Enhancements' [83], initiated in Release 12
2. 'RAN Sharing Enhancements (RSE)' [84], initiated in Release 13

Table 4.7 lists the main 3GPP technical specifications and reports that are related to both WIs.

Regarding more advanced scenarios for RAN sharing, 3GPP TR 22.852 [85] provides a study on scenarios of multiple operators sharing radio network resources and creates potential requirements that complement the existing system capabilities for sharing common E-UTRAN resources. The scenarios illustrate:

- Means for efficiently sharing common E-UTRAN resources according to identified RAN sharing scenarios (e.g. pooling of unallocated radio resources)
- Means to verify that the shared network elements provide allocated E-UTRAN resources according to sharing agreements/policies
- Indication of and potential actions upon overload situation in consideration of sharing agreements/policies
- Means to flexibly and dynamically allocate RAN resources on demand at smaller timescales than those supported today

Table 4.7 3GPP documents covering RAN sharing enhancements.

Work Item (WI)	Related technical specifications/reports	Comments
WI FS_RSE(Study on RAN Sharing Enhancements)	TR 22.852 – 'Study on RAN Sharing Enhancements'	Feasibility studies on use cases and potential requirements for a more dynamic cooperation among operators on RAN sharing have been analysed
WI RSE(RAN Sharing Enhancements)	Change request to TS 22.101 – 'Service aspects; Service Principles on requirements to support enhanced RAN sharing'	Normative requirements document. TR 22.852 is the basis for normative work on RSE

As a result of the analysis of the new scenarios, means that complement the existing system capabilities for sharing common RAN resources are addressed in Release 13:

- **Flexible allocation of shared RAN resources**. The hosting RAN shall be able to allocate RAN resource capacity to each of the participating operators through different approaches such as fixed allocation (i.e. guaranteeing a minimum allocation and limiting to a maximum allocation), fixed allocation for a specified period of time and/or specific cells/sectors and first-come/first-served allocation (i.e. on demand). A hosting RAN provider shall be able to define in the hosting RAN the allocated share of RAN capacity for each participating operator and differentiate traffic associated with individual participating operators. Admission control shall be conducted based on the proportion of assigned RAN usage for each participating operator, and the shared RAN shall be capable of applying differentiated QoS attributes per participating operator.
- **On-demand capacity negotiation**. The hosting RAN shall be able to offer by automatic means sharable E-UTRAN resources as on-demand capacity to participating operator's networks. The participating operator's networks shall be able to request offered on-demand resources. The hosting RAN provider shall be able to allow a participating operator to request the cancellation of granted on-demand requests. The hosting RAN provider shall be able to withdraw a granted request (within SLA/business agreement).
- **Selective access to operation, administration and maintenance (OAM) functions**. The hosting RAN shall be able to provide and control selective OAM access (e.g. to allow link test in the base station and provide fault reports) to each participating operator to perform OAM tasks supporting the participating operator's use of the hosting RAN. The hosting RAN shall be able to allow participating operators to retrieve selective OAM status information to the same level of detail as would be available from a non-shared E-UTRAN.
- **Load balancing while respecting the agreed shares of RAN resources**. The hosting RAN shall be able to support load balancing within a shared RAN while respecting the agreed shares of RAN resources based on the whole cell load level and the load level for each participating operator. The hosting RAN shall be able to perform load balancing per participating operator.
- **Generation and retrieval of usage and accounting information**. A hosting RAN shall report events supporting the accounting of network resource usage separately for each participating operator.

- **Handover functionality due to dynamic RAN sharing agreements**. At the beginning of RAN sharing, participating operators shall be able to direct both connected and idle UEs towards the hosting RAN, and the hosting RAN provider shall be able to direct both connected and idle UEs away from the hosting RAN at the end of the RAN sharing period. If required by the on-demand granted RAN sharing agreements, participating operators shall be involved in the decision of where to drive connected and idle UEs to when multiple options are available at the end of the RAN sharing period.
- **Public Warning System (PWS) in shared RAN**. The hosting RAN shall be able to broadcast PWS messages originated from the core networks of all participating operators.

Though work on enhancements to RAN sharing in Release 13 initially focused on E-UTRAN, the work is going to be extended to GERAN and UMTS. The main motivation for this is that many operators already share RAN resources on GERAN and UTRAN, and it would be beneficial in terms of efficiency and costs to provide similar enhancements for sharing using those RATs also. Also, for GERAN especially, it is likely that some operators will come to a point where there will be a single (2G GSM) network to support legacy traffic from all the operators in a particular country/region. Therefore, it will be important to ensure that GERAN-based networks can be effectively shared with the aim of reducing costs. Virtualization techniques are gaining attraction as potential means to implement some of the abovementioned RAN sharing capabilities [86].

References

[1] 3rd Generation Partnership Project (3GPP) official website. Available online at http://www.3gpp.org/LTE (accessed 28 March 2015).
[2] Cisco Visual Networking Index: Global Mobile Data Traffic Forecast Update, 2013–2018.
[3] TCCA Critical Communications Broadband Group, 'Mission Critical Mobile Broadband: practical standardisation and roadmap considerations', White Paper, February 2013.
[4] Iain Sharp (Netovate), 'Delivering public safety communications with LTE', White Paper on behalf of 3GPP, September 2013. Available online at http://www.3gpp.org/IMG/pdf/130902_lte_for_public_safety_rev2_1.pdf (accessed 28 March 2015).
[5] Balazs Bertenyi, Chairman of 3GPP TSG-SA, 'Developments in 3GPP – Release 12 and beyond', 20 May 2014. Available online at http://www.3gpp.org/ftp/Information/presentations/presentations_2014/2014_05_bertenyi_3GPP_Rel12_beyond.pdf (accessed 28 March 2015).
[6] ETSI TR 103 269-1 V1.1.1, 'TETRA and Critical Communications Evolution (TCCE); Critical Communications Architecture; Part 1: Critical Communications architecture reference model', July 2014.
[7] Draft TS 103 269-2 V0.0.2, 'TETRA and Critical Communications Evolution (TCCE); Critical Communications Architecture; Part 2: Critical Communications application mobile to network interface architecture', December 2014.
[8] Open Mobile Alliance, 'OMA overview', NPSTC Governing Board Meeting, San Antonio, TX, November 2014.
[9] 3GPP SA6 Working Group, 'Mission critical applications'. Available online at http://www.3gpp.org/specifications-groups/sa-plenary/sa6-mission-critical-applications (accessed 28 March 2015).
[10] R. Ferrús and O. Sallent, 'Extending the LTE/LTE-A Business Case: Mission- and Business-Critical Mobile Broadband Communications', Vehicular Technology Magazine, IEEE, vol.9, no.3, pp.47, 55, September 2014.
[11] Balazs Bertenyi, 'LTE Standards for Public Safety – 3GPP view', Critical Communications World, 21–24 May 2013.
[12] Donny Jackson, 'Congress told of "significant progress" toward mission-critical voice over LTE', Urgent Communications, December 2013.
[13] Erik Dahlman, Stefan Parkvall, Johan Skold and Per Beming, '3G Evolution: HSPA and LTE for Mobile Broadband', Amsterdam: Academic Press, 2009.

[14] Erik Dahlman, Stefan Parkvall and Johan Sköld, '4G: LTE/LTE-Advanced for Mobile Broadband', Amsterdam: Academic Press, 2013.
[15] Magnus Olsson, Stefan Rommer, Catherine Mulligan, Shabnam Sultana and Lars Frid, 'SAE and the Evolved Packet Core', Amsterdam: Academic Press, 2009.
[16] Stefania Sesia, Issam Toufik and Matthew Baker, 'LTE – The UMTS Long Term Evolution: From Theory to Practice', Chichester: John Wiley & Sons, Ltd, 2009.
[17] Gonzalo Camarillo and Miguel-Angel Garcia-Martin, 'The 3G IP Multimedia Subsystem (IMS): Merging the Internet and the Cellular Worlds', 3rd Edition, Chichester: John Wiley & Sons, Ltd, September 2008.
[18] 3GPP TS 36.101, 'User Equipment (UE) radio transmission and reception (Release 12)', March 2014.
[19] 4G Americas, '4G mobile broadband evolution: 3GPP Release 11 & Release 12 and beyond', White Paper, February 2014.
[20] 3GPP TS 36.300, 'Evolved Universal Terrestrial Radio Access (E-UTRA) and Evolved Universal Terrestrial Radio Access Network (E-UTRAN); Overall description; Stage 2 (Release 12)', December 2014.
[21] 3GPP TS 23.203, 'Policy and charging control architecture'. Available online at http://www.3gpp.org/DynaReport/23203.htm (accessed 18 April 2015).
[22] 3GPP TS 23.401, 'General Packet Radio Service (GPRS) enhancements for Evolved Universal Terrestrial Radio Access Network (E-UTRAN) access'.
[23] Analysis Mason, 'Creating a real-time infrastructure for BSS and revenue management', November 2013.
[24] 3GPP TS 33.401, '3GPP System Architecture Evolution (SAE); Security architecture'.
[25] 3GPP TS 33.102, '3G Security; Security architecture'.
[26] 3GPP TS 33.402, '3GPP System Architecture Evolution (SAE); Security aspects of non-3GPP accesses'.
[27] 3GPP TS 33.210, '3G security; Network Domain Security (NDS); IP network layer security'.
[28] IETF RFC-4301, 'Security Architecture for the Internet Protocol', December 2005.
[29] 3GPP TS 33.310, 'Network Domain Security (NDS); Authentication Framework (AF)'.
[30] GSMA IR.88, 'LTE and EPC roaming guidelines', Version 10.0, July 2013.
[31] GSM Association, 'Official Document IR.92 – IMS profile for voice and SMS', Version 7.0, 3 March 2013.
[32] GSM Association, 'Official Document IR.94 – IMS profile for conversational video service', Version 5.0, 4 March 2013.
[33] Miikka Poikselkä, Harri Holma, Jukka Hongisto, Juha Kallio and Antti Toskala, 'Voice over LTE (VoLTE)', Chichester: John Wiley & Sons, Ltd, February 2012.
[34] APCO Global Alliance, 'Updated Policy Statement – 4th Generation (4G) Broadband Technologies for Emergency Services', October 2013. Available online at http://www.apcoglobalalliance.org/4g.html (accessed 28 March 2015).
[35] NPSTC Public Safety Communications Report, 'Push-to-Talk over Long Term Evolution Requirements', July 2013. Available online at http://pdf.911dispatch.com.s3.amazonaws.com/npstc_push-to-talk_report_july2013.pdf (accessed 28 March 2015).
[36] Andrew M. Seybold, 'Voice over Public Safety Broadband', Public Safety Advocate e-newsletter, October 2012.
[37] 3GPP SP-130326, 'Revised WID on Group Communication System Enablers for LTE (GCSE_LTE)', 3GPP TSG SA Meeting #60, Oranjestad, Aruba, 17–19 June 2013.
[38] 3GPP TD SP-140228, 'New WID on Service Requirements Maintenance for Group Communication System Enablers for LTE (SRM_GCSE_LTE)', 3GPP TSG SA Meeting #64, Sophia-Antipolis, France, 16–18 June 2014.
[39] 3GPP SP-130728, 'New WID on Mission Critical Push To Talk over LTE (MCPTT)', 3GPP TSG SA Meeting #62, Busan, Korea, 9–11 December 2013.
[40] 3GPP TS 22.468, 'Group Communication System Enablers for LTE (GCSE_LTE)'.
[41] 3GPP TS 23.468 V12.0.0, 'Group Communication System Enablers for LTE (GCSE_LTE); Stage 2 (Release 12)', February 2012.
[42] 3GPP TS 23.246, 'Multimedia Broadcast/Multicast Service (MBMS); Architecture and functional description'.
[43] D. Lecompte and F. Gabin, 'Evolved Multimedia Broadcast/Multicast Service (eMBMS) in LTE-Advanced: Overview and Rel-11 Enhancements', Communications Magazine, IEEE, vol.**50**, no.11, pp.68, 74, November 2012.
[44] 3GPP TS 22.179, 'Mission Critical Push to Talk (MCPTT) (Release 13)', December 2014.
[45] 3GPP TS 22.779, 'Study on architectural enhancements to support Mission Critical Push To Talk over LTE (MCPTT) services (Release 13)', November 2014.

[46] Open Mobile Alliance, 'Push to talk over Cellular 2.1 Requirements', OMA-RD-PoC-V2_1-20110802-A, Version 2.1, August 2011.
[47] 3GPP TR 23.979, '3GPP enablers for Open Mobile Alliance (OMA) Push-to-talk over Cellular (PoC) services; Stage 2 (Release 7)', June 2007.
[48] Open Mobile Alliance, 'Push to talk over Cellular (PoC) – Architecture', OMA-AD-PoC-V2_1-20110802-A, Version 2.1, August 2011.
[49] Jerry Shih and Bryan Sullivan (AT&T), 'Comparison of MCPTT Requirements and PCPS 1.0', OMA-TP-2014-0177, Critical Communications Workshop, 15 August 2014.
[50] CEPT ECC Report 199, 'User requirements and spectrum needs for future European broadband PPDR systems (Wide Area Networks)', May 2013.
[51] Qualcomm, 'LTE Direct: operator enabled proximity services', March 2013.
[52] Sajith Balraj (Qualcomm Research), 'LTE Direct overview', 2012. Available online at http://s3.amazonaws.com/sdieee/205-LTE+Direct+IEEE+VTC+San+Diego.pdf (accessed 28 March 2015).
[53] 3GPP SP-110638, 'WID on Proposal for a study on Proximity-based Services', 3GPP TSG SA Plenary Meeting #53, Fukuoka, Japan, 19–21 September 2011.
[54] 3GPP TD SP-130605, 'Update of ProSe WID to get SA2 TS and SA3 TR numbers', 3GPP TSG SA Meeting #62, Busan, Korea, 9–11 December 2013.
[55] 3GPP RP-122009, 'Study on LTE device to device proximity services', 3GPP TSG RAN Meeting #58, Barcelona, Spain, December 2012.
[56] 3GPP SP-140386, 'Editorial update of SA1 ProSe phase 2 WID', 3GPP TSG SA Meeting #64, Sophia Antipolis, France, 16–18 June 2014.
[57] 3GPP TD SP-140573, 'Revised Rel-13 WID for Proximity-based Services', 3GPP TSG SA Meeting #65, Edinburgh, Scotland, 15–17 September 2014.
[58] 3GPP SP-140629, 'New Study to create a dedicated SA3 TR on Security for Proximity-based Services', 3GPP TSG SA Meeting #65, Edinburgh, Scotland, 15–17 September 2014.
[59] 3GPP TR 22.803 V12.2.0 (2013-06), 'Feasibility study for proximity services (ProSe) (Release 12)', June 2013.
[60] 3GPP TS 22.115, 'Service aspects; Charging and billing'.
[61] 3GPP TS 22.278, 'Service requirements for the Evolved Packet System (EPS)'.
[62] 3GPP TR 23.703, 'Study on architecture enhancements to support proximity services (ProSe) (Release 12)', February 2014.
[63] 3GPP TS 23.303 V12.3.0, 'Proximity-based Services (ProSe); Stage 2 (Release 12)', December 2014.
[64] 3GPP TR 36.843 V12.0.1, 'Feasibility Study on LTE device to device proximity services; Radio aspects', March 2014.
[65] 3GPP TS 24.334, 'Proximity-services (ProSe) User Equipment (UE) to ProSe Function protocol aspects; Stage 3 (Release 12)', December 2014.
[66] NPSTC Broadband Working Group, 'Priority and QoS in the Nationwide Public Safety Broadband Network, Rev 1.0', April 2012. Available online at http://www.npstc.org/download.jsp?tableId=37&column=217&id=2304&file=PriorityAndQoSDefinition_v1_0_clean.pdf (accessed 28 March 2015).
[67] National Public Safety Telecommunications Council (NPSTC), Broadband Working Group, 'Mission critical voice communications requirements for public safety', September 2011. Available online at http://npstc.org/download.jsp?tableId=37&column=217&id=1911&file=FunctionalDescripton (accessed 28 March 2015).
[68] 3GPP TS 36. 304, 'User Equipment (UE) procedures in idle mode'.
[69] 3GPP TS 22.011, 'Service accessibility'.
[70] 3GPP TS 36.331, 'Radio Resource Control (RRC); Protocol specification'.
[71] Ryan Hallahan and Jon M. Peha, 'Policies for public safety use of commercial wireless networks', 38th Telecommunications Policy Research Conference, October 2010.
[72] Mobile Broadband for Public Safety – Technology Advisory Group Public Security Science and Technology, 'Public Safety 700 MHz Mobile Broadband Communications Network: Operational Requirements', February 2012.
[73] 3GPP TS 22.153, 'Multimedia priority service'.
[74] 3GPP TR 22.950, 'Priority service feasibility study'.
[75] 3GPP TR 22.953, 'Multimedia priority service feasibility study'.
[76] 3GPP TR 23.854, 'Enhancements for Multimedia Priority Service'.
[77] 3GPP SP-130596, 'Updates to the WID on Feasibility Study on Study on Isolated (was "Resilient") E-UTRAN Operation for Public Safety (FS_IOPS, was FS_REOPS)', Busan, Korea, 9–11 December 2013.

[78] 3GPP TR 22.897 V13.0.0, 'Study on Isolated E-UTRAN operation for public safety', June 2014.
[79] 3GPP TS 22.346 V13.0.0, 'Isolated Evolved Universal Terrestrial Radio Access Network (E-UTRAN) operation for public safety; Stage 1 (Release 13)', September 2009.
[80] 3GPP TR 36.837 V11.0.0 (2012-12), 'Public safety broadband high power User Equipment (UE) for band 14 (Release 11)', December 2012.
[81] GSMA White Paper, 'Mobile infrastructure sharing'. Available online at http://www.gsma.com/publicpolicy/wp-content/uploads/2012/09/Mobile-Infrastructure-sharing.pdf (accessed 18 April 2015).
[82] 3GPP TS 23.251, 'Network sharing; Architecture and functional description'.
[83] SP-110820, 'Proposed WID on Study on RAN Sharing Enhancements', 3GPP TSG SA Plenary Meeting #54, Berlin, Germany, 12–14 December 2011.
[84] TD SP-130330, 'Proposed update to New WID on RAN Sharing Enhancements (RSE)', 3GPP TSG SA Meeting #60, Oranjestad, Aruba, 17–19 June 2013.
[85] 3GPP TR 22.852 V13.1.0 (2014-09), 'Study on Radio Access Network (RAN) sharing enhancements (Release 13)', September 2014.
[86] X. Costa-Perez, J. Swetina, T. Guo, R. Mahindra and S. Rangarajan, 'Radio Access Network Virtualization for Future Mobile Carrier Networks', Communications Magazine, IEEE, vol.**51**, no.7, pp.27, 35, July 2013.

5

LTE Networks for PPDR Communications

5.1 Introduction

Various arrangements are currently in use by governments and PPDR organizations for the procurement of PPDR communications services over dedicated narrowband PMR networks. The defining elements commonly used to characterize such arrangements include:

- **Ownership of the infrastructure**. The network used for PPDR communications is typically owned by the PPDR agency in the case of systems only used by that organization. In the case of networks shared by several PPDR organizations, the network assets can be owned by the government itself or by a public entity or department created by the government for such purpose. These options are typically referred to as government-owned (GO) arrangements. Furthermore, in some arrangements, commercial service providers like mobile network operators (MNOs), infrastructure vendors or system integrators can also be the owners of the network assets. This case is typically referred to as contractor owned or commercial owned (CO).
- **Operator of the infrastructure**. Running and maintaining the network can be conducted by the owner itself or outsourced to a third-party company that provides specialized managed network services. The combination of ownership and network operation mainly leads to three different models: government owned and government operated (GO–GO), contractor owned and contractor operated (CO–CO) and government owned and contractor operated (GO–CO).
- **Users admitted in the network**. The network can be exclusively used to serve PPDR agencies or be shared with other users. In the case of GO–GO and GO–CO arrangements, potential users that could share a GO network for PPDR could be restricted to public or private organizations that may have to deal with public safety-related problems as part of their operations, such as hospitals, ambulances, public transport firms, water and energy distribution companies, money transportation companies, security firms, etc. In the case of

Mobile Broadband Communications for Public Safety: The Road Ahead Through LTE Technology, First Edition.
Ramon Ferrús and Oriol Sallent.

a CO–CO arrangement, sharing options could be more diverse depending on the contract conditions agreed upon between the government and/or PPDR agencies and the commercial entity that provisions the PPDR network and services.

- **Designation of spectrum for PPDR use**. The spectrum used in PPDR communications can be specifically designated for this purpose (i.e. dedicated spectrum assigned for PPDR use). Alternatively, there is the option of procuring commercial services from an MNO to provide PPDR services without designating dedicated spectrum for PPDR. Intermediate options are also possible such as reserving some spectrum for PPDR but sharing this resource with an MNO in a hybrid arrangement.

In European countries, the GO–GO and GO–CO models along with the designation of some amount of spectrum for dedicated PPDR use are the most common arrangements for current voice-centric PPDR networks. In these arrangements, the government, directly or via a controlled entity, procures equipment, deploys and operates a closed network dedicated to the PPDR users. These networks are typically shared by several PPDR organizations such as police, firefighters, medical services and civil protection of the same country or region. In many countries, the government creates a 'pseudo-independent' operator, often state-owned, in order to take responsibility for the build, management and financing of the network. Examples of GO–GO models are the ASTRID network in Belgium and VIRVE network in Finland. Examples of GO–CO networks are BDBOS in Germany and BOS in Austria. To a lesser extent, the CO–CO model is also deployed in some European countries. As in the case of GO–GO and GO–CO models, the designation of dedicated spectrum for PPDR is the prevailing approach in these CO–CO arrangements. Examples of CO–CO networks are Airwave in the United Kingdom and SINE in Austria.

In the United States, while the prevailing model is still largely based in disparate and separate analogue or P25 radiocommunications system owned and operated by individual PPDR agencies, many efforts have been conducted in the last years towards the realization of regional (multi-county) or state-wide shared networks that could provide service to multiple agencies. In this regard, different models are followed for the governance structure of these shared networks that range from direct contractual agreements among the participating PPDR agencies to governance models enacted by (state) legislative actions that usually also include funding mechanisms for construction [1]. A common denominator of these arrangements is that the network remains under the ownership of government entities. Nevertheless, another PPDR communications delivery model in which the services are provided by a private company (e.g. a telecommunications carrier) is also gaining momentum in the United States. This model is commonly referred to as carrier-hosted 'P25 as a service', which draws many similarities with the CO–CO concept previously discussed. This service model has been offered by Bell Canada in Canada for years, while its adoption in the United States is still at an incipient stage [2].

5.1.1 Separation of Service and Network Layers in PPDR Communications Delivery

The adoption of the LTE standard brings up a major change in the way that PPDR services can be provisioned with respect to current narrowband voice-centric PMR systems. The use of LTE allows the service layer, which comprises the specific applications and service delivery platforms (SDPs) needed for PPDR, to be decoupled from the underlying network layer, whose

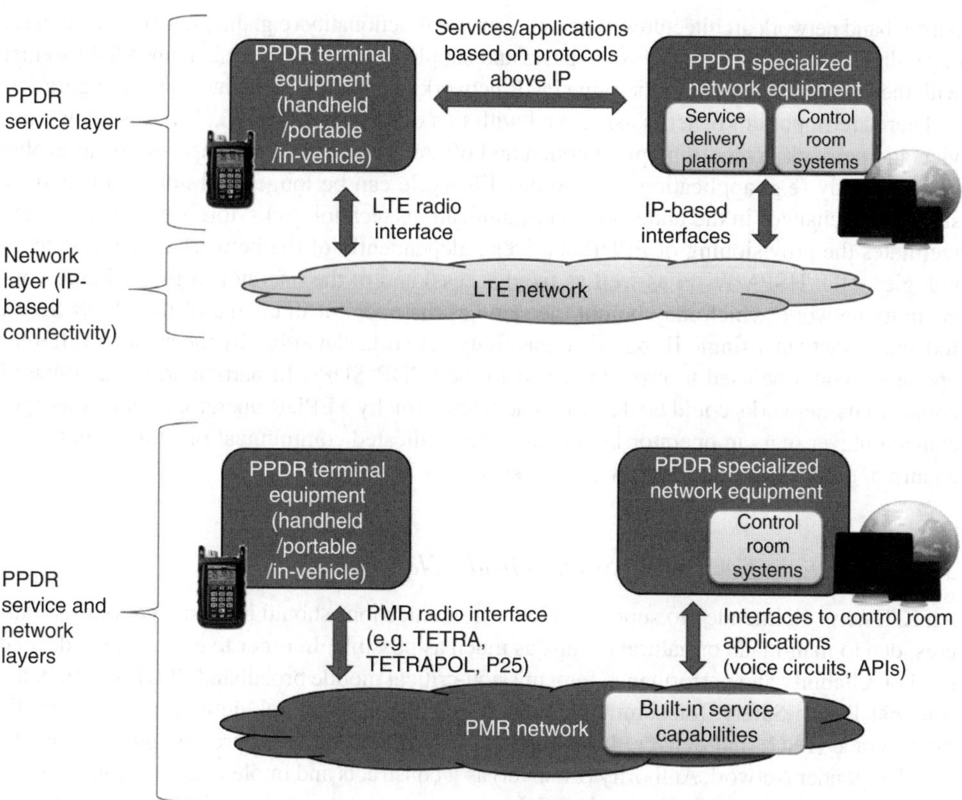

Figure 5.1 Separation of the PPDR service layer from the underlying IP network layer.

main task becomes the delivery of IP-based connectivity to the service layer components. This approach, illustrated in Figure 5.1, follows the principles and general reference model for next-generation networks (NGNs) that establishes a functional division between [3]:

- **Transport functions**. These are solely concerned with conveyance of digital information, of any kind, between separate network points. Transport functions mainly encompass the network and lower layers of the Open Systems Interconnection (OSI) 7-layer basic reference model. Transport functions may deliver user-to-user connectivity, user-to-SDP connectivity and SDP-to-SDP connectivity.
- **Service functions**. These provide the user services, such as a voice, data and video services or some combination thereof. Service functions may be organized in a complex set of geographically distributed SDPs and applications, or in a simple case, just involve a set of service functions and applications located in two terminating entities.

According to this model, the 'PPDR service layer' can be implemented mainly as application clients within the PPDR terminals and application servers within the PPDR SDPs and control room systems (CRS). This 'PPDR service layer' is then deployed over the LTE connectivity services, including individual EPS bearer services, MBMS bearer services and ProSe communications services. It is worth highlighting that such a decoupling does not exist in current narrowband network architectures, where the service functionality (e.g. the protocols and control capabilities used to support the voice basic and supplementary services) is embedded together with the transport functions in the same PMR network platform, as illustrated also in Figure 5.1.

There are important benefits associated with this decoupling. Remarkably, this allows service and network layers to be provisioned and offered separately and, importantly, to evolve independently (e.g. applications technology life cycle can be longer and not be affected by subsequent changes in the underlying communications technology). Moreover, this approach facilitates the provisioning of PPDR services independently of the network and access technologies (LTE/HSPA/Wi-Fi as well as fixed access) and of the operator of the IP-based connectivity network, which may be not the same as the operator in charge of the PPDR SDPs. Indeed, more than a single IP-based connectivity network, run either by the same or different operators, could be used to provide access to the PPDR SDPs. In particular, these IP-based connectivity networks could be dedicated networks, run by a PPDR operator or any other specialized player (e.g. an operator in charge of the dedicated communications infrastructure in an airport) as well as public access networks, run by MNOs.

5.1.2 Design of a 'Public Safety Grade' Network

An LTE network intended to support PPDR communications should be designed to resist failures, due to man-made or natural events, as much as possible. In order to establish the desired level of reliability and performance for a mission-critical mobile broadband PPDR solution, the National Public Safety Telecommunications Council (NPSTC) has defined the term public safety grade (PSG) and delivered a number of recommendations to provide guidance to the First Responder Network Authority (FirstNet) as it constructs and implements a US nationwide PPDR LTE network [4]. Qualitatively, PSG communications are defined simply as the effect of reliable and resilient characteristics of a communications system. On this basis, the NPSTC's report provides measurable characteristics that would differentiate a mission-critical communications system from a standard or commercial-grade network. The report covers in detail environmental considerations, service-level agreements (SLAs), reliability and resiliency elements, coverage design details, push-to-talk (PTT) support, applications, site hardening, installation and operations and maintenance. A network or system intended to be considered PSG must address these topics during its design and later in its implementation.

In addition to the NPSTC recommendations, there are other documents that develop network-related requirements and identify the expected key characteristics for an LTE network to deliver mobile broadband PPDR communications [5–13]. Based on these, Table 5.1 gathers some of the key characteristics that illustrate the expected performance and capabilities of an LTE network designed for PPDR use.

5.2 Delivery Options for Mobile Broadband PPDR Networks and Services

The definition of the right deployment scenario and associated business model for the delivery of mobile broadband PPDR communications is currently a matter of strong interest for PPDR agencies and governments around the world. As described in the previous section, the ownership of the new infrastructure, the management model for such infrastructure, the eligible

Table 5.1 Key characteristics expected in an LTE network designed for PPDR use.

Areas	Characteristics
Coverage and capacity	Geography-based coverage planning – as opposed to population coverage Symmetric UL/DL usage pattern – as opposed to DL-dominated traffic for commercial users Strongly varying cell load, with limited possibilities to predict (stochastic) Ubiquitous coverage, not just outdoors but inside buildings and other hard-to-reach locations (e.g. subways) Off-network operation, to enable direct communication between PPDR users outside network coverage Deployable systems, to secure wide area PPDR communication also when users are outside normal network reach Support for air–ground–air (AGA) communication
Network availability and resilience	Robust network sites with enhanced physical protection and battery backup (site hardening) Geo-redundant network functions (intra-network redundancy) Fallback to other networks at network failure (inter-network redundancy) Multiple backup options Hardening should not be a one-size-fits-all approach, but particularized to specific conditions of the sites and facilities locations Network availability at least 99.99% of the time (i.e. <50 min/year of unplanned outages)
Security	User data encryption end to end Elevated link security, protecting both user plane and control plane Elevated O&M security, securing node and network configurations and stored user data Enhanced identity management, provisioning users with 'right to use' of network resources Mutual authentication of infrastructure and terminals Methods for temporarily and permanently disabling terminals and smart cards Functions to detect and compensate for jamming at the air interface
Priority control	Differentiated priority classes so that high-priority communications are never blocked Dynamic control (real time) of priority and resource management. Also applicable in a visited network Emergency calls
Accountability and service assurance	Real-time KPI monitoring to enforce service-level agreements between PPDR users and network operators (PPDR operators and/or MNO) Quality-of-service (QoS)-dependent charging to enable charging of PPDR services based on priority and QoS
Functionality and performance	Data and voice services and features (e.g. group communications, PTT) in need for PPDR operations (initial deployment may be only for data services) Reliability and adequate QoS unaffected by location (fast call set-up, reliable delivery of data)

(*continued*)

Table 5.1 (*continued*)

Areas	Characteristics
Interoperability	Ability of communications equipment to comply to the same technical standard and to operate within the tuning range of the frequency bands used by these systems (this is especially relevant for European PPDR organizations) Seamless operation of broadband equipment across borders in Europe, including ad hoc deployments of extra capacity coverage when and where needed
Hierarchical control levels in a multi-jurisdictional shared network	Ability of operations centres at differed jurisdictional levels (e.g. national/regional/local) to exercise control on the network use Ability of PPDR agencies to have local network control to secure capacity and ensure their public safety priorities are met

users to be served and the potential designation of some amount of dedicated spectrum are among the central defining elements that have to be fixed. In addition, the formulation of a service delivery model for mobile broadband communications based on the LTE technology has to take into consideration the following elements:

- **Role of the commercial MNOs**. Due to the distinct technological platforms in use in the PPDR and commercial domains at the time that the current voice-centric service delivery models were formulated, options such as roaming to commercial networks were not feasible. Nevertheless, this situation has radically changed with the adoption of a common radio access technology such as LTE (as described in Chapter 4) and the possibility to decouple the network and service layers for PPDR service delivery (as described in Section 5.1.1). In this context, synergies and cost-sharing approaches between the PPDR and the commercial domains are expected to be central in the network delivery options for next-generation mobile broadband PPDR communications. Indeed, relying on commercial network's capacity for the delivery of PPDR data services can be a first step for a reduced time-to-market solution with affordable investment levels, while dedicated capacity for PPDR can be progressively deployed in specific areas and within a longer time frame.
- **Coexistence roadmap with narrowband systems**. While LTE is expected to become a full replacement of the current voice-centric networks in the long term, an integration of both types of networks is seen as the most likely scenario to deliver mission-critical voice with mission-critical data in the short to medium term. Augmenting the current infrastructures to support mobile broadband could be considered as the starting point for many administrations.
- **Possible joint efforts and synergies between the PPDR sector and other mission/business critical users**. The increasing need for mobile broadband in other sectors such as utilities and transportation motivates the need to consider possible joint efforts and synergies among them for the potential sharing of a common or complementary mission-critical communications delivery platform.

Bearing in mind the above considerations, Table 5.2 compiles an overview of different service delivery options for mobile broadband PPDR communications that have been analysed so far [14–20].

Table 5.2 Studies that cover the analysis of delivery options for mobile broadband PPDR networks and services.

Study	Delivery options for mobile broadband PPDR networks and services[a]
TCCA, 'A Review of Delivery Options for Delivering Mission Critical Solutions' [14]	• (Option A) Take service from standard commercial networks • (Option B) Operate as a mobile virtual network operator (MVNO) • (Option C) Take service from commercially owned dedicated network • (Option D) Build own dedicated network • (Option E) Combination of the above models (e.g. to solve timing issues with regard to the availability of dedicated networks and/or to complement coverage and functionality issues) • Other hybrid approaches: o (Option F.1) Take service from standard commercial networks but fund network enhancements on coverage and resilience o (Option F.2) Take service from commercially owned dedicated network but allow operator to sell excess capacity to commercial users o (Option F.3) Build own dedicated network but outsource operation of the network o (Option F.4) Build dedicated network but share network elements with commercial operators
CEPT ECC Report 218 [15]	• Dedicated network infrastructure for PPDR o (Option A.1) Mobile broadband network planned, built, run and owned by the authority o (Option A.2) Mobile broadband service provided through service offering • Commercial network(s) infrastructure providing broadband services to PPDR users o (Option B.1) Same mobile broadband services to PPDR as to public customers o (Option B.2) Mobile broadband services to PPDR with special requirements • Hybrid solutions with partly dedicated and partly commercial network infrastructure o (Option C.1) Geographical split between dedicated and commercial network infrastructure o (Option C.2) MVNO model where PPDR users share RAN with the public users o (Option C.3) MVNO model with partly dedicated/partly shared RAN network o (Option C.4) Extended MVNO model where PPDR have dedicated core and service nodes and dedicated carriers in the radio transmitters/receivers in the RAN part of the commercial mobile broadband network
Simon SCF Associates, 'Study on Use of Commercial Mobile Networks and Equipment for "Mission-Critical" High-Speed Broadband Communications in Specific Sectors' [16]	• (Option A) Dedicated specialized networks using specialized equipment only. TETRA for voice and narrowband data in dedicated bands, government owned/government operated (GO–GO). This option is actually a continuation of the current model in most European countries and is mainly considered in the study for comparative purposes with the other options • (Option B) Commercial networks using commercial equipment only. Resilient LTE on one or several commercial networks, no extra spectrum, owned/operated by MNOs (CO–CO model)

(*continued*)

Table 5.2 *(continued)*

Study	Delivery options for mobile broadband PPDR networks and services[a]
	• (Option C) Dedicated specialized networks using commercial equipment. Dedicated/hardened LTE network in dedicated spectrum, owned/operated by government, public enterprise or public–private partnership (GO–GO or GO–CO models) • (Option D) Hybrid solutions involving dedicated specialized and commercial networks. Voice on TETRA (GO–GO) plus broadband data on resilient LTE for broadband capability (GO–CO or CO–CO) • (Option E) Common multipurpose network for use by the PPDR, transportation and energy sectors simultaneously. Resilient LTE on one or more commercial networks, with extra spectrum, government owned but MNO operated (GO–CO)
German Federal Ministry of the Interior, 'On the Future Architecture of Mission Critical Mobile Broadband PPDR Networks' [17]	• (Option A) Service provider scenario. This scenario is derived from services that commercial mobile radio network operators offer for business customers. For these services, a special Access Point Name (APN) is configured in the PPDR terminal equipment • (Option B) MVNO scenario. This scenario is based on roaming. If PPDR organizations have an own broadband mission-critical network and use this scenario to provide coverage in additional areas, then this scenario is similar to a national roaming agreement between two commercial mobile radio networks. If PPDR organizations have no broadband network, then this scenario is similar to an MVNO of a commercial mobile radio network • (Option C) Radio access network (RAN) sharing scenario. This scenario assumes that a user-owned mission-critical mobile broadband PPDR core network is set up. For the RAN, a RAN sharing agreement with a commercial mobile network operator reduces the investment and operational cost • (Option D) Dedicated RAN scenario. This scenario assumes that a user-owned mission-critical mobile broadband PPDR RAN based on LTE and a user-owned mission-critical mobile broadband PPDR core network are set up
APT Report on 'PPDR Applications Using IMT-based Technologies and Networks' [18]	• (Option A) Deployment of a dedicated PPDR network owned and operated by the PPDR agency or controlling entity, based on IMT technology • (Option B) A combination of a dedicated PPDR network owned and operated by the PPDR agency (or entity) and commercial network services, based on a common IMT technology, to facilitate roaming where the PPDR agency as a preferential subscriber with suitable assigned priority, under negotiated contract arrangements • (Option C) Similar to the two options above but with the dedicated PPDR network being owned and operated by a commercial entity under negotiated contract arrangements • (Option D) Sharing the commercial IMT network operator's IMT infrastructure as a closed/private subnetwork under specific contract arrangements (e.g. as a virtual private network (VPN)) or as a preferential subscriber with suitable assigned priority

[a] The tagging of the options shown in the table does not obey to any order or preference expressed in the reference studies but mainly added here for the sake of clarity.

Regardless of some slight differences in terminology and focus of the different studies, the potential delivery models under consideration can be broadly categorized as follows:

- Options centred on the deployment of new mobile broadband **dedicated, specialized networks** purposely built for mission-critical services, either GO or CO
- Options centred on the use of (enhanced) **public access commercial networks** for the delivery of mission-critical services, either with ordinary or specific PPDR contracts (i.e. covering special requirements for the delivery of the PPDR services)
- Options formulated as some sort of **hybrid combination** involving both dedicated and commercial infrastructures

A description of the different options, with an analysis of its advantages, limitations and specific factors to be considered, is addressed in the following sections.

5.3 Dedicated Networks

A dedicated mobile broadband network for PPDR can be specifically designed and built to meet PPDR users' requirements. This approach is often regarded as the best solution from the PPDR users' point of view. The primary advantage of a dedicated network is that the government and PPDR users deploy and manage their own broadband network, having full control and guaranteed network access in mission-critical situations and without commercial users to take capacity during an incident. This approach would be the logical continuation of the present scenario with dedicated TETRA and TETRAPOL networks in Europe. Obviously, the main disadvantage of this approach is the high cost (capital and operational) of coping with such a new dedicated infrastructure. Moreover, the procurement process and network buildout could typically take 3–5 years to deliver the first service to users, thereby delaying the start of broadband services for the users of that network. The designation of some dedicated spectrum is a valuable asset that the government can bring up as part of the funding scheme for a dedicated network. However, the designation and clearance of new dedicated spectrum for mobile broadband PPDR can also take some years in those countries that have not initiated this process yet.

Within the dedicated network option, two main alternatives regarding the network's ownership can be distinguished:

1. **GO network**. In this case, a governmental agency owns and finances the equipment and supporting infrastructure, the operation and maintenance support systems and pays for the running cost. This option gives the government and PPDR users full control over the procurement process, the network procured and the operating assets. The control is in PPDR users' hands over the network management subsystem, which also allows the network capacity to be tailored to changing needs and circumstances. The operation of the network can be carried out by the governmental agency itself (i.e. GO–GO model) or outsourced to an external, more specialized and capable organization (i.e. GO–CO model).
2. **CO network**. In this case, the network is deployed and operated by a commercial company or a consortium of investors/industrial partners (e.g. MNOs). The network is purposely built to offer services to the PPDR users. Therefore, this is a CO–CO model in which the government contracts the mobile broadband services from a commercial company

according to a set of negotiated technical requirements. PPDR requirements (e.g. in terms of coverage, availability and resilience) are stipulated in the contract, as well as the obligations and SLAs that are binding to the network provider. Long-term contracts of at least 10–15 years may be required to ensure business viability. In this option, the network provider finances the most part of the upfront costs, including equipment and supporting infrastructure. On the other hand, the government pays, for example, an agreed monthly recurring fee for the provisioning of the mobile broadband service to the PPDR users.

On the basis that the high costs and funding schemes are likely to be the biggest hurdles to overcome in any form of dedicated network, cost-efficient network footprints specifically tailored to fulfil PPDR particular needs are needed (i.e. the design of a dedicated network may not necessarily need to mimic that of a commercial network). In addition, enlarging the user base served through the dedicated infrastructure and finding ways of monetizing the excess bandwidth are some of the key dimensions that deserve careful consideration. These two aspects are further discussed in the following subsections.

5.3.1 Cost-Efficient Network Footprints

Traffic demand in PPDR wireless communications networks is much less predictable, both geographically and temporally, than in commercial networks. Major incidents can happen anytime and anywhere. When such incidents do arise, communications needs are substantial and tend to be concentrated around a relatively small area. It would be impractical, both in economic and engineering terms, to plan a conventional wireless network based on such eventualities, since much of the capacity would never be used.

A more cost-efficient approach is to plan a network to provide a basic minimum level of wireless connectivity at all locations that can be rapidly expanded on an ad hoc basis to provide additional capacity to cater for unforeseen incidents. The rationale of this approach lays on the fact that two types of traffic can be distinguished in a PPDR network: the light traffic induced by routine activities, such as patrols and surveillance, and the heavier traffic due to a large number of PPDR personnel at a major incident scene. The network architecture must be designed to meet the peak capacity requirement due to the sum of both types of traffic.

In the conventional architecture adopted in commercial networks, the radio access infrastructure mainly consists of stationary base stations (BSs) connected spread at fixed locations over the territory. If such conventional architecture is adopted to deploy a wide area PPDR network, the placement of BS needs to be dense enough to meet the peak demand. An alternative approach is to design the radio access infrastructure based on a reduced number of more sparsely deployed stationary BSs for supporting light routine traffic complemented with a distributed set of **mobile BSs ready to be quickly deployed** to any incident scene by vehicle or helicopter [21]. A premise of this lighter architecture is that a mobile BS can be dispatched to the incident scene and can be set up and operational in a very short time. This imposes a requirement on the density and placement of mobile BSs, as well as on the technologies used to link and integrate the mobile BSs with the operation of the fixed infrastructure (e.g. fast settable wireless backhaul solutions, self-organizing features to automatically configure the BS settings). The analysis conducted in Ref. [21] shows that an architecture partially based on mobile BSs can potentially offer over 75% reduction in terms of the total number of fixed BSs that are necessary to cope with the same traffic demand. The analysis of a range of traditional, cellular-based solutions all the way down to a solution more dependent on mobile deployables is also under consideration by FirstNet in the United States [22]. Cellular designs using 35 000, 24 000 and 14 000 sites have been considered, the latter being a hybrid design with a very thin network and much more heavily reliance on transportable systems that can allow PPDR personnel to bring the network with them for those events that occur in areas where it doesn't make sense to have a site around the clock. Solutions based on all-deployable public safety LTE networks are also being trialled in the United States [23].

Another compelling approach is the use of **extended cell range** solutions, also known as **boomer cells**, to reduce the permanent footprint of a dedicated PPDR mobile broadband network [24]. The conventional wisdom is that PPDR LTE could make economic sense in metropolitan, urban and even some suburban areas, where the capacity demand justifies the establishment of many cell sites located relatively close together, with coverage typically extending less than 4–5 km from the site. However, the case of rural areas raises concern in purely economic terms. For this reason, the use of cells with much larger cell radius for coverage in rural and remote areas can be a relevant capability to consider in the design of a dedicated PPDR network.

Extended cell range or boomer cells have been used for years in various cellular technologies to provide wide area coverage. Testing and evaluation of boomer cells able to extend an LTE cell radius to range of up to tenths of kilometres have been conducted since 2014 by the Public Safety Communications Research (PSCR) in the United States [25]. In particular, initial trials considered an LTE eNB antenna deployed at 85 m height. The transmit power of the eNB was set to 40 W, which is a typical transmit power for a macrocell. PRACH Preamble[1] Format 1 was used to allow extended cell range up to 77 km (48 miles). On the terminal side, an LTE vehicular modem with 200 mW (23 dBm) transmit power (i.e. the same as a commercial cell phone) and two external antennas were tested, one omnidirectional car roof antenna with a gain of 3 dB as a practical application for mobile use and another 16 dB gain directional antenna as a practical application for fixed situations. In the case of the rooftop antenna, coverage ranges of around 40 km were demonstrated. In the case of the directional antenna, transmission at a distance of 77 km was possible, achieving a data rate of a few megabits per second. Additional experiments are planned for 2015 considering the eNB antenna placed at 280 m height and a PRACH Preamble Format 3, targeting the extension of the coverage up to 100 km. Besides the increased coverage, raising the antenna height is expected to improve the performance at a fixed location. This is particularly beneficial for the uplink, which is typically the limiting factor in the link budget. The use of high-power devices, which have already been specified by 3GPP for LTE Band 14, is especially relevant in this scenario.

Another big challenge for PPDR communications that could importantly affect the placement and density of the network sites is the capability to provide **in-building coverage** in large structures, such as office buildings, apartment buildings, warehouses, parking structures, tunnels and basements. In these scenarios, distributed antenna systems (DAS) and signal boosters are typically used to enhance coverage in commercial mobile communications. Deploying this type of solutions to improve indoor coverage in some critical locations could

[1] 3GPP has specified four random access preamble formats for FDD in TS 36.211 that allow for different maximum round-trip propagation delays in large cells. In particular, Format 0 allows up to 14 km, Format 1 up to 77 km, Format 2 up to 29 km and Format 3 up to 100 km.

ultimately reduce the number of outdoor macrocells. Regulation could be a key factor to favour this case by issuing rules mandating, for example, the support of in-building DAS for public safety in new constructions. Indeed, installing an in-building system for cellular users to have good connectivity is almost a necessity for building owners in today's environment. The relatively low cost associated with adding public safety support to the cellular support that the market demands and the maturity of the DAS technology for 700/800 MHz make this a compelling approach [26, 27]. However, special attention should be paid to the requirements on the PPDR DAS (e.g. UPS requirements, percentage of coverage in critical areas, components enclosures requirements, certification and periodical testing). Much more stringent requirements for PPDR DAS compared to carrier-grade DAS could be a deterrent for the adoption of a common solution for both domains [28, 29].

Finally, yet importantly, the adoption of **small cells** is also expected to have a huge impact on the architecture of next-generation mobile broadband PPDR networks [30]. As mobile operators roll out small cells to enhance LTE coverage, the technology is likely to expand to mission-critical broadband networks as well. 'Small cells' is an umbrella term for operator-controlled, low-powered radio access nodes, including those that operate in licensed spectrum and unlicensed carrier-grade Wi-Fi. Small cells typically have a range from 10 to several hundred metres [31]. Types of small cells include femtocells, picocells and microcells – broadly increasing in size from femtocells (the smallest) to microcells (the largest). These small cells, along with the mainstay macrocellular tower and rooftop antennas, make up what is known as a heterogeneous network (HetNet). Placing small cells in key locations such as police stations, fire stations and major government buildings will allow public safety an opportunity to cost-effectively add the coverage and capacity they need. Furthermore, advances in small cell technologies will also undoubtedly benefit the development of the mobile BS solutions discussed previously in this subsection.

5.3.2 Expanding the User Base beyond PPDR Responders

Enlarging the user base of a dedicated infrastructure can greatly contribute to its economical sustainability. An expanded user base beyond the PPDR responders group would help to achieve higher economies of scale and monetize the excess capacity that a private LTE network only serving PPDR traffic may have. On a day-to-day basis, not all of the broadband capacity available on a dedicated network would be certainly used. Therefore, leveraging the priority features supported in the LTE standard, part of the network capacity could be shared with non-first-responder users to generate revenue that could be used to fund the operations and upgrades of the broadband network.

Besides the PPDR community, a dedicated network purposely built to support mission-critical communications can also become a compelling business proposition for other users who depend on efficient mobile communications to carry out their job and whose daily activities can also be fundamental to the health, safety and well-being of the citizens [20]. They comprise critical infrastructure services and other public and private entities such as [14, 16]:

- **Utilities** (electricity, gas, water). Nowadays, utilities use radiocommunications systems mainly for the routine but business critical demands of managing their distribution networks and supply chains. Repairs and monitoring are typical tasks. Nevertheless, when managing serious disruptions in their supplies, the use of the radiocommunications systems can be considered 'mission critical' because of the potential scale of its economic impact. As utilities become more dependent on sensors, automated switches and automated metering and consider implementing other smart-grid capabilities, the need for reliable, low-latency data connectivity becomes increasingly important. Utilities are particularly interested in deploying sensor technologies that can alert them of weaknesses in their infrastructures before it becomes noticeable to customers. Many of the most critical technologies that utilities use do not require much bandwidth, but they cannot be encumbered by latency issues and must be reliable. For instance, everyday teleprotection for electric utilities requires continuity of service with reliability levels of up to 99.999% and latency below 5 ms. On the other hand, the use of applications such as video surveillance of remote sites could drive bandwidth requirements higher in the future. In addition, workers in the field are increasingly equipped with portable data devices embedding professional applications, so that the use of broadband connectivity is an emerging requirement in many phases of day-to-day work such as fleet management, damage assessment, customer relations, fault mapping and diagnosis. Even though utilities are relying on commercial MNOs to support their workers in the field, an own dedicated communications network would be preferred for the most critical operations because service disruptions generate legal liabilities. Telecom service providers specialized in the utility sector have already started to announce plans to deploy private LTE networks to meet the increasing data demand of its utilities and its commercial customers [32].

- **Transportation** (buses and trams, trains and metro, ports and airports). There are many applications in use nowadays in the transport sector that rely to some extent on radiocommunications (traffic signals, variable message signs, vehicle detection systems, access control schemes, security cameras, real-time passenger information, etc.). While most of these are short-range communications or do not need much bandwidth, broadband is necessary for applications such as the transmission of real-time video to monitoring and evaluation centres. For example, roadside cameras for traffic flow monitoring are crucial for early detection of accidents and delays, to enforce speed limits, lane access restrictions and so on. Furthermore, real-time video from buses and trains enhances public safety and security and ensures that emergencies can be dealt with quickly and effectively. In addition, obtaining up-to-date information on the location of buses and trains enables passengers to be informed of waiting time while supervisory staff can allocate additional capacity when necessary. In the European rail sector, a replacement for GSM-R and an affordable migration time frame will need to be found in the future. This replacement is mainly motivated by the cost increase that is expected sometime in the next decade, when the production of GSM equipment will end. Applications such as train control, line side signalling and others, which are increasingly complex applications, will be more demanded. Fast and reliable communications are also needed in airports and ports in order to deal with the rapid turnaround of aircraft and ships and ensure that operations are conducted in a safe and secure manner.
- Other potential users, such as customs, coast and border guards, military and paramilitary as well as governments and their administrations, including public work departments responsible for maintaining city's infrastructure (storm water collection systems, streets, trees and parks, etc.). In the commercial sector, mining, fuel and petrochemical, manufacturing and other forms of industry require group-based resilient communications, and all of these sectors are likely to take advantage of broadband services.

A detailed compilation of the requirements, operational procedures and functional and safety needs for the utility and transportation sectors can be found in Ref. [16], together with a description of the wireless equipment and networks currently in use in these sectors.

There are strong arguments for these potential users having access to the same reliable network as PPDR:

- **Operational benefits in emergency response**. The ability of users such as utilities and transportation to perform their functions effectively can have a significant impact on public safety. The loss of electrical power is one of the most disruptive side effects of a disaster, often cascading into the breakdown of other services needed for normal life (e.g. communications networks can be impaired by power cuts even when not damaged by the disaster itself). When a major incident occurs, one of the first logistical issues that must be determined is whether these critical infrastructure systems are operating properly or not. In fact, in some cases, traditional public safety is virtually helpless to take action until a utility is able to address a problem at a scene. One illustrative example is an incident involving a fire associated with a gas leak: coordination between the first responders and the gas company staff might be critical to shut off the leak and allow the firefighters to do their job [33]. As another example, if electrical and gas power can be leveraged and drinking water is readily available, the response effort to a disaster situation is much simpler than if one of these critical components is not operational or has been compromised. Therefore, a tight coordination between PPDR and critical infrastructure services is essential and can be facilitated by the use of the same highly reliable communications infrastructure.
- **Business opportunities**. Besides increasing the user base and thus bringing new revenue streams for the self-sustainability of the network, these user groups potentially could also bring valuable assets that can be usable for the PPDR network. For example, electric utilities have high-voltage electrical lines and towers crossing rural, remote, forested, mountainous and similar areas. Such lines also carry fibre-optic cable and such towers can support network sites. Similarly, railroads often cross rural or mountainous areas, providing both assets and potential users.

Nevertheless, there are also some challenges that could prevent the sharing of a common dedicated infrastructure for various critical communications if these issues are not properly addressed:

- **Competition with commercial providers**. Critical infrastructure entities and the other types of user groups listed above are also commonly considered as prime enterprise customers for most commercial network operators. Unlike the PPDR community, which is mainly comprised of government entities and not-for-profit corporations, companies such as utilities are for-profit enterprises that commercial providers would like to retain as customers. Hence, commercial service providers may oppose a government-funded infrastructure or assets (read dedicated spectrum) that could be used to serve part of this market. In this regard, a proper regulation to determine who are the eligible users of such an infrastructure is key to ensure a level playing field for all of the involved players. Indeed, the notion of government competing with commercial business is always a controversial subject, with potentially significant legal, economic and political implications.

- **Low commonality in specific sector requirements**. Differences in, for example, coverage and functional requirements among the different types of users may render useless the option of using a common dedicated network infrastructure to cover all of critical users' needs. For instance, utilities might not need most of the PPDR-specific features sought from LTE even though their repair crews work in teams. In the context of intelligent transport applications, many have only local connectivity needs (e.g. vehicle to roadside, vehicle to approaching vehicle) so that a cellular network infrastructure is unnecessary for these applications.
- **Preferential access in times of congestion**. In times of crisis or major incident, the network operator can ensure that those who need the service most will not be blocked by overload. While the fact that PPDR shall have priority and pre-emptive access to the available capacity is not questioned, one key point in any negotiation with other user groups such as utilities could be to provide also priority access for a handful of their key applications [33]. On a normal day, utilities would expect to use considerable excess capacity, but that usage can be reduced dramatically during an emergency, when applications that are not particularly time sensitive can be throttled down or stopped. For example, automated meter reading can be turned off to make more bandwidth available for PPDR. However, certain applications that are critical to letting utilities know whether their systems are functioning properly or not should not be turned off at any time. If this is not assured, utilities would have little incentive to invest in a dedicated network. Fortunately, these most critical applications require relatively small amounts of bandwidth.

In the United States, the possibility of collaborating with utilities and transportation on the build-out of the FirstNet broadband network is being considered [33, 34]. Indeed, in the law enacted in February 2012 that created FirstNet, language was included that allows a broader definition of the first responders, opening the door to embrace critical infrastructure users. Just for illustrative purposes, US commercial carriers like Verizon and AT&T are able to spread the capital costs of their nationwide networks across more than 100M subscribers. In contrast, the dedicated broadband system being pursued by FirstNet, which is expected to be more reliable and with greater coverage than the commercial counterparts, could count with a user base roughly estimated to be as high as 5M subscribers if only first responders are allowed access [33].

In a similar matter, the Utilities Telecom Council (UTC) in Canada has requested Industry Canada for the reclassification of some utilities with regard to their eligibility and priority in using the PPDR spectrum that has been allocated for the deployment of a dedicated network. In Canada, a public safety hierarchy has been defined with three categories. Category one users are the traditional first responders such as police, fire and medical services. Category two users include forestry, public works, public transit, hazardous materials clean-up, border protection and 'other agencies contributing to public safety'. Category three users include 'other government agencies and certain non-governmental agencies or entities'. Currently, hydro and gas utilities are classified as category three users so that they could only get access to this capacity in emergencies. According to UTC Canada, for utilities to see the worth of the network, they should be given day-to-day access to the spectrum for certain mission-critical applications.

Besides enlarging the user base of a dedicated network with critical communications users, there are other more controversial propositions that even consider supporting some citizen's communications over the dedicated network. One of these proposals is the so-called Dynamic

Spectrum Arbitrage (DSA) concept, proposed by a private company in the United States as a way to enable real-time auctions for bandwidth at any given time or location to monetize the excess throughput in the FirstNet's network [35]. The rationale behind this approach is that the FirstNet system is expected to have a significant amount of excess bandwidth available during routine times that can be sold to operators to generate revenue that can be used to fund further FirstNet's deployment and operation. Under the DSA concept, a DSA provider would finance the buildout of the PSG broadband network and would be given the right to sell excess bandwidth to operators, on the basis that first responders would get all of the bandwidth during an emergency response. A Dynamic Spectrum Arbitrage Tiered Priority Access (DSATPA) engine has been designed (and patented). The engine is intended to allow network operators to bid for access to unused broadband capacity on the FirstNet system on a nearly real-time basis, similar to the approach used by utilities in the energy markets. The use of this capacity is expected to be enforced through real-time modifications in the network operators' subscriber databases, adapting setting such as subscribed quality-of-service (QoS) profiles, access restrictions for roaming and packet data networks to which the wireless devices are allowed to connect. In this way, MNOs' subscribers can be properly steered between the networks. This flexibility in the management of the excess capacity is expected to be valuable for commercial services that are delay tolerant and can be accommodated easily over time such as software upgrades, M2M traffic, massive data backups, etc.

Another controversial proposal is to use the PPDR dedicated infrastructure also to support emergency communications with the citizens, in particular for the deployment of reliable alerting systems [36]. In this way, every commercial device could be mandated to support the frequency band allocated for PPDR to have such an alerting capability. According to Ref. [36], this approach could, on the one hand, make the chipsets supporting the PPDR spectrum band to become less expensive because of the economies of scale and, on the other hand, justify an additional charge to consumers and business users for the financing of this alerting capability (e.g. if a fee of 50 cents per month were established in the United States across 300M commercial devices, this could generate almost $2B funds each year for the sustainability of the FirstNet network).

5.4 Commercial Networks

As long as commercial broadband networks are becoming an important part of a society's infrastructure, there is also an increasing consensus that these infrastructures undoubtedly will play a role in the delivery of critical communications solutions. However, there are important differences between the PPDR and commercial network models that should be taken into account when assessing the specific role that commercial networks could have in the overall delivery solution for mobile broadband PPDR communications. A compilation of some key differences is provided in Table 5.3 [16], which shows divergences in many fundamental aspects such as goals, capacity and coverage planning approaches, availability standards, types of communications services in demand, control over subscriber information, etc.

The delivery of PPDR mobile broadband services over commercial networks has to be necessarily formulated via contractual agreements between the individual PPDR agencies, or a public entity on behalf of a group of PPDR agencies, and one or more MNOs. In some countries, it may be possible to have a national roaming agreement in place, allowing the PPDR users to have a single subscription with a particular MNO but to get services from

Table 5.3 Main differences between the PPDR network and commercial network models.

Issues	Commercial network operator model	PPDR network model
Goals	Maximize revenue and profit	Protect life, property and the state
Capacity	Defined by 'busy hour' on a typical day	Defined by 'worst-case' scenario
Coverage	Population density	Territorial, focused on whatever may need protection across a country geography
Availability	Outages undesirable (revenue loss/customer loss)	Outages unacceptable (live lost or threatened)
Communications	One to one	Dynamic groups, one to many, field crews/ control centre
Broadband data traffic	Internet access (mainly downloads)	Traffic mainly within agency (more uploads than downloads)
Subscriber information	Owned by carrier	Owned by agency
Prioritization	Minimal differentiation, by subscription level or application	Significant differentiation, by role and incident level (dynamic)
Authentication	Carrier controlled, device authentication only	Agency controlled, user authentication
Preferred charging method	Per minute for voice, per GB for data, per message for SMS	Quarterly or annual subscription with unmetered use

From Ref. [16].

other networks if that particular network fails or if the user moves out of the coverage of the currently used network.

One of the key advantages of a PPDR service delivery model based on commercial mobile broadband networks is the fact that 3G and 4G data networks are already deployed and becoming increasingly ubiquitous. Therefore, PPDR agencies can implement mobile broadband solutions early and explore some of the possibilities that broadband data offers without significant upfront investment. While early implementations may be limited to data services, including video delivery, it may be possible to obtain new capabilities as soon as the specific features that 3GPP is developing for critical communications are released and deployed by the MNOs (e.g. ProSe, group communications system enablers). Clearly, the use of commercial networks can result in a very cost-effective solution, since there is limited infrastructure CAPEX from the PPDR side (mainly limited to specific PPDR SDPs), and since the network is supporting many more commercial users than PPDR users, there are significant economies of scale. Needless to say, the applications considered as non-critical (such as administrative applications) can always be well handled via a commercial infrastructure.

On the downside, one of the main disadvantages compared to a dedicated network for PPDR is the less control that the PPDR agencies may have on the footprint and performance of the network (e.g. coverage of low populated areas, response times of public operators in the case of network disturbance, prioritization of PPDR users, security, control of subscribers' profiles, etc.). Remarkably, special consideration should be given to the lower network availability standards typically in place in the commercial networks. In this regard, it is a fact that during emergencies, public mobile networks are more prone to suffer from congestion and even suffer important shutdowns during disasters, power outages and other events. On the contrary, many of today's PPDR narrowband networks have been built to provide service availability close to 99.999%, which means less than 5 min of downtime per year [4]. Various system design elements are used to create this performance standard in current PPDR network, including redundant radio BSs, the use of self-healing backhaul transmission networks, backup power supplies and automatic failover to redundant critical components. Obviously, the achievement of this service availability level has a direct translation into increased network costs, which may not be considered as economically viable or desirable by a commercial network operator. Consequently, the fear that the network might fail due to congestion during a crisis is often cited as the most compelling reason for emergency responders not to rely on commercial networks for their communications needs [16]. The 2013 annual report on major communications network outages by the European Union Agency for Network and Information Security (ENISA) found that 'Overload was the cause affecting by far most user connections, more than 9M connections on average per incident'. Even if the commercial network does not fail, unless PPDR users have been able to negotiate guaranteed capacity on the network, there is the danger that access will become restricted, or lost altogether, during times of peak demand. Experience shows that such peaks often occur during major incidents when critical communications users often need the service most [14].

These disadvantages can be modulated to some extent depending on the nature of the contract established between the PPDR entities and MNOs. This view leads to define two major types of arrangements for the delivery of mobile broadband services to PPDR users:

1. **Ordinary contracts**, similar to those established with other corporate and business users. In this case, PPDR users get the same service as the general public. No special requirements to the service offering or to service priority exist, with the exception of potential national roaming if it has been agreed. Governments or PPDR agencies pay an agreed recurring (monthly) fee for the mobile broadband service. This option is currently in use by many PPDR agencies that use mobile data communications in smartphones/tablets/laptops for routine and mainly administrative tasks. However, this is not considered as an option for mission-critical activities.
2. **Specific PPDR contracts**, covering special requirements for the delivery of the PPDR services. These special requirements are negotiated and formalized through specific SLAs that may include conditions on the use of priority access in critical incidents, a response time for network outages, a target for coverage and a target for latency, among others [14]. In addition to recurring fees for the payment of the service, specific public funding schemes might be in place to cope with (part of) the extra costs incurred by the MNOs for the fulfilment of the specific PPDR contract.

Regarding the specific PPDR contracts, hardening and disaster recovery plans for improved resilience are likely to be among the most critical extra costs that arise. In this respect, government paying for improved resilience is an option to be considered, though this may be classified as state aid as long as it may turn into a competitive advantage for a given MNO. Another critical extra cost can be associated with coverage extension requirements, which may also include specific areas such as tunnels or underground facilities. This can also be achieved through public funding or by establishing coverage obligations on the MNOs for the land-mass geographic coverage, data speed and building penetration.

Even considering the added cost of resilience and improved coverage, a recent study [16] on the use of commercial mobile networks and equipment for mission-critical high-speed broadband communications found that the PPDR communications delivery model based on commercial LTE operation was the cheapest option in simple financial terms. However, the same study pointed out that the main issue with this option is not the technology challenge of building a resilient network, but the regulatory, legal and contractual context.

In this regard, the next subsection provides some considerations on the organizational and contractual aspects between the PPDR users and MNOs for the delivery of mission-critical communications over the commercial networks. After that, some key conclusions from the study [16] are echoed with respect to the conditions and regulatory changes that would help in removing the obstacles of using commercial mobile broadband networks for mission-critical purposes. Finally, some insight into the operation and status of existing prioritization systems supported in some commercial networks and currently used by PPDR users for privileged access to voice communications services is provided in the last subsection.

5.4.1 Organizational and Contractual Aspects

The establishment of a central public entity as the contracting authority with MNOs rather than the individual PPDR agencies can help in leveraging the buying power of the entire PPDR sector by pooling the demand in one or more contracts and executing long-term service procurement with the MNOs.

This central role can be played by a new entity established for that purpose. This entity could be a kind of 'delivery partner' or 'programme manager' as envisioned in the UK contractual structure for the delivery of the ECS (described in Chapter 3). Another approach is to establish an MVNO as the entity responsible for the delivery of mobile broadband PPDR services, managing some network and service aspects and taking care of the arrangements with the MNOs and system integrators. An example of this approach is the Belgian network operator ASTRID with the Blue Light Mobile MVNO service for data communications (see Chapter 3 for further details on ASTRID's approach and Section 5.5.2 in this chapter for a detailed description of the applicability of the MVNO model for PPDR communications).

Establishing contracts with multiple MNOs would also bring in some advantages. The most obvious is an increased availability level, as there is no a single network being the point of failure. In addition, the ability to roam across multiple MNOs would enable PPDR users to get connection through the most appropriate network at any situation. Hence, PPDR terminals would be connected to, for example, the nearest BS among those from different service providers, thus exploiting the diversity of BS sites and frequencies (multi-band and multi-operator site diversity). This ability can be expected to turn in better spectrum efficiency and achievable data rates for PPDR users. Moreover, involving multiple MNOs can also allow a selective hardening of sites across the multiple operators to ensure the coverage as needed in each location. This could reduce the costs of hardening considerably since only a fraction of the MNO's sites would be hardened. This approach would also satisfy state aid rules by providing equal advantage to all the MNOs.

When using commercial networks for mission-critical purposes, the governments are expected to write specific legal requirements on the operator down in a contract. These contracts may include clauses on everything from government step-in upon failure, SLAs on minimum availability, contract transfer limits, limits on force majeure claims and so on. While these conditions are essential for a proper service assurance, there is the risk that the demand of stringent SLAs (e.g. very high availability levels, penalties for SLA infringements with liability for damages due to service interruptions) may ultimately hinder the incentives for mobile operators to offer such specialized services to the mission-critical sectors, especially for the PPDR sector. Examples of legal requirements imposed on the operator by the governmental authority are provided in Ref. [15] and captured in Table 5.4. Therefore, the ability to draw up balanced contractual and/or financial measurements is essential for both the PPDR community and MNOs. Anyway, contracts are not always 100% solid in comparison to the guarantees associated with a government-controlled network (e.g. in an extreme case, the operator business may go bankrupt or be sold).

5.4.2 Commercial Networks' Readiness to Provide Mission-Critical PPDR Services

While it is a fact that public mobile networks are not yet engineered for resilience and advanced voice features that critical communications users require, the challenge for MNOs is to decide whether they can recoup or not the additional investment that would be required from what is a niche market in comparison to their consumer customer base. In the context of an increasing pressure on public budgets and the introduction of a new and more efficient generation of mobile broadband technology, European research project HELP [37] considered strengthening the role and commitment of commercial wireless infrastructures in the provision of PPDR communications by exploiting network and spectrum sharing concepts (more details on the PPDR communications framework developed in this project are given later on in Section 5.6.7). More recently, the European Commission requested an in-depth independent study on the costs and benefits of using commercial networks for mission-critical purposes [16]. The study explored the communications requirements and options in three sectors – PPDR, utilities and transportation – and conducted a cost analysis of five network deployment scenarios. Among the analysed options (listed in Table 5.2), the use of commercial LTE networks was found to be the most attractive in terms of its value for money. The study concluded that it could be possible for commercial mobile broadband networks to be used for mission-critical purposes, though only if the following five conditions were fully met:

1. The behaviour of commercial MNOs must be constrained to provide the services needed by mission-critical users while preventing the use of 'lock-in' techniques to take unfair advantage of this expansion of the MNOs' market power and social responsibility. Such changes include not just stronger commitments to network resilience but the acceptance of limits on price increases and contract condition revisions, ownership continuity assurances and a focus on QoS for priority mission-critical traffic. Equally important for long-term relationships will be the mission-critical services' perception of MNO behaviour and performance. For that, measures will be needed that go beyond SLAs at a commercial contract level, and new regulations regarding commercial MNO services must be enforced by national regulatory agencies.
2. Commercial networks have to be 'hardened' from radio access network (RAN) to core and modified to provide over 99% availability – with a target of 'five nines'. Geographic

Table 5.4 Examples of legal requirements that might be contractually imposed on the operator by the governmental authority.

Level	Legal requirements that might be imposed on the MNO by the governmental authority in the contract for PPDR services
Most stringent	**Parent company guarantee (PCG)/performance bond (PB)**: (i) PCG – The supplier's parent company agrees to meet the supplier's financial and/or performance obligations should the supplier fail to do so. (ii) PB – The supplier provides the government customer with a PB usually valued at between 5 and 10% of the contract price. The government customer can redeem the bond if the supplier fails to meet its contractual obligations (even if the financial costs of the failure are lower than the value of the bond). Both are standard in government customer contracts
	Intellectual property: The government customer is to own the rights in any new, project-specific intellectual property developed by the supplier. Often, a non-negotiable requirement of the customer. This is a standard approach
	Liability: Government customer contracts may specifically exclude a waiver of consequential and indirect damages, and the overall liability cap may be in excess of contract value. Losses for breach of the confidentiality provisions may be uncapped altogether. Light breaches will in some countries result in liabilities in the order of €500M and in other cases be fully uncapped. Standard provision in many jurisdictions
	Open book accounting: The government customer has access to the supplier's financial records in order to see any reduction in the supplier's costs in performing the contract. If costs have reduced, the supplier and government customer will split the 'profit'. Sometimes, the split is 50/50 although it is not uncommon for the government customer to receive the majority of any such profit. Provision is becoming more common
	Most favoured customer: The government customer must have the best price. The supplier cannot sell the same (or similar) products and services to another customer at prices lower than those paid by the government customer. Reasonably frequent, but not a standard provision
Moderately stringent	**Step-in**: The government customer has the right to take over the performance of the contract in certain circumstances. For example, where the supplier suffers an insolvency event (e.g. insolvency, arrangement with creditors, etc.) or commits a material breach of the contract. The supplier is not paid during step-in and may also have to meet the government customer's additional costs associated with step-in. The government customer may hand back the services to the supplier or terminate the contract. Provision is becoming more common
	Termination: The government customer has extensive rights to terminate, often including termination for convenience. Whereas the supplier will only be permitted to terminate in very limited circumstances (e.g. protracted failure to pay undisputed fees). Standard provision
	Change of control: A change of control of the supplier will be subject to the government customer's approval, which often may be withheld at the customer's absolute discretion. In some instances, changes of control are prohibited altogether. Standard provision
	Financial strength: The supplier is required to show financial strength on a regular, ongoing basis. If the supplier's financial strength diminishes, the government customer may terminate the contract. Provision is becoming more common

(*continued*)

Table 5.4 (*continued*)

Level	Legal requirements that might be imposed on the MNO by the governmental authority in the contract for PPDR services
	Control over performance: This is typically very stringent in government contracts – the government customer takes a more involved role than is usual in other contracts, for example, in testing and acceptance procedures. In our experience, government customers are slower and less inclined to approve and accept elements of a project than commercial customers, resulting in delayed payment to the supplier. Standard approach
	Liquidated damages/service credits (LD/SC): Although they can vary from contract to contract, LD/SC regimes are, in our experience, often more onerous with government contracts as the government cannot allow or afford the project to fail or be delayed or services to be compromised. Also see comments regarding force majeure events. Standard provision
	Force majeure: Force majeure events are often defined much more narrowly than in commercial contracts. For example, industrial action is usually excluded from government customer contracts (although sometimes permitted if it is nationwide or industry-wide). Force majeure clauses may also include a proviso that, due to the very purpose of the contract (i.e. public safety), a circumstance will not be considered a force majeure event if the party invoking that event reasonably ought to have taken into account when the contract was signed. Standard provision
	Export control: A US company is responsible for ensuring that its products are not exported to prohibited countries. This responsibility extends to onward sale by the company's customers. Government customers will not accept such restrictions imposed on it by another government. Accordingly, standard export control provisions are routinely excluded from government customer contracts. Standard provision
	Source code escrow: The supplier must place the system source code into escrow with a third party, at the supplier's cost. The source code can be released to the government customer in specified events such as insolvency of the supplier, breach of contract by supplier, etc. Standard provision
Least stringent	**Assignment**: The supplier is not usually allowed to assign the contract without the government customer's prior consent, which may be withheld at the customer's discretion. Standard provision
	Security clearance: The government customer may require certain supplier employees (e.g. those who have access to certain customer sites) to undergo national security clearance. Standard provision
	Data: Recording and retention obligations for data processed under government customer contracts may be subject to specific data protection legislation, which can be quite burdensome. Provision is becoming more common
	Continuous improvement: The supplier must improve the operation of the system over time at no additional cost to the government customer. Reasonably frequent, but not a standard provision
	Taxes: Contractual obligation on the supplier to regularly pay its taxes. Failure to do so would amount to breach of contract by the supplier. Standard provision
	Confidentiality: Government customers are usually reluctant to agree to standard confidentiality provision, preferring to use their own. Standard provision

From Ref. [15].

coverage must also be extended as needed for mission-critical purposes and indoor signal penetration improved at agreed locations.

3. All this network hardening and extended coverage, along with the addition of essential mission-critical functions and resilience, must be accomplished at reasonable cost. No more should be spent on the selective expansion and hardening of commercial networks for mission-critical use than it would cost to build a dedicated national LTE network for that purpose.
4. Hardened LTE networks must be able to provide the different types of service required by each of the three sectors. Each sector uses broadband in quite different ways. That is, not just for streaming video, image services and database access, as in PPDR, but for very-low-latency telemetry and real-time control for utilities and transport.
5. Overcoming ingrained preferences that some countries could have for state-controlled networks for applications that implicate PPDR. This is not simply a legal, regulatory or economic question. Some countries have specific histories of state control as part of their culture, traditions and politics, not to mention investments in current technologies with long payback cycles. Thus, some countries may want to continue using dedicated networks in the short and medium term even if they cost more (examples are Germany, Italy and France for PPDR).

A crucial barrier pointed out by Forge et al. [16] for the use of commercial networks is the current MNO's mass-market business model, which needs to be suitably amended to provide the appropriate levels of service to priority clients with special needs. In this regard, the study concludes that specific regulatory measures may be needed to reassure the three sectors (particularly those with regulatory obligations on continuity of service, such as the utilities) and ensure that the MNO's performance levels are maintained over decades. These measures are necessary to gain the trust of the user communities that MNO commercial behaviour will never disrupt mission-critical services. In particular, measures needed to build the confidence of these users in the MNOs would be:

- Being prepared to upgrade to high standards of reliability and correct service failures as quickly as possible, without any degradation in that commitment over several decades
- Acceptance of long-term (15–30-year) contract commitments to mission-critical customers, with stable conditions and agreed rates
- Providing priority access to mission-critical services, especially when emergencies create the risk of network overload
- Providing geographic coverage to meet the needs of mission-critical users
- Willingness to cooperate with other MNOs and MVNOs – for instance, in handing over a mission-critical call to another operator with a better local signal
- Keeping to the spirit and letter of long-term contracts for mission-critical services without arbitrary changes in technical features, tariffs or service conditions
- Readiness to submit cost-based pricing analyses of tariffs with full open book accounting for national regulatory authorities (NRAs) and government clients
- Willingness to offer new charging regimes and metering procedures
- Removal of excessive charges for international roaming across the European Union (EU) and avoidance of 'surprise charges' for previously agreed services

In this regard, a number of measures that give NRAs specific new powers to cope with this situation on behalf of mission-critical services are proposed [16]:

- MNOs should be mandated to support mission-critical services. There are two possible ways to do this. One possibility is that to operate as an MNO or an MVNO, licensees must agree to provide mission-critical services. That may entail extended geographic coverage, hardening to meet minimum standards of availability and resilience and designating an overall programme manager for mission-critical services. It is effectively an additional set of licence conditions to operate a public mobile service. The other possibility is that any purchase or exercise of a mobile spectrum licence brings with it the obligation to support mission-critical services for as long as the licence is valid. Note that the grant of spectrum conditioned on mission-critical provisions offers to the NRA the power to reassign the spectrum to a new operator, if the original one fails to perform. This mechanism gives effective control to the NRA through the potential for spectrum reassignment, such that the spectrum is not lost. The first alternative would begin with new licences; the second would begin with the transfer of an existing licence. This would not be a universal service obligation but a specific service obligation that extends MNO/MVNOs' responsibilities for the long term. On the positive side, it would be an obligation that brings with it a new revenue stream, accompanied by government investments in resilience that would benefit all the users of the network.
- NRAs should have the power to introduce regulations that support and enforce the provisions of long-term MNOs' contracts with mission-critical users.
- NRAs should be authorized to grant priority access to commercial mobile network services for mission-critical communications when justified by circumstances, including the handover of calls between MNOs when required. This may require the amendment of existing guidelines, statutes or regulations.
- NRAs should support the governments in setting tariffs for mission-critical services by researching into the true costs of MNO operation. That may require forensic accounting and suitable preparations of the cost base declarations by MNOs.

5.4.3 Current Support of Priority Services over Commercial Networks

The ability to prioritize users is a key component in any model for the delivery of mission-critical communications services over commercial networks. Prioritization may also entail the ability to pre-empt or degrade some users or services during times of emergencies.

While prioritization capabilities are part of the LTE standard (see Chapter 4 for a detailed description), they have not been tested much in a real-world environment to date. In the context of the dedicated network being built in the United States by FirstNet, testing activities are being addressed by the PSCR programme, a joint effort of the Department of Homeland Security and NTIA/FirstNet, which is targeted to advance on public safety communications interoperability. The goal is to verify that QoS, priority and pre-emption features function correctly before first responders use them in a real situation [38]. This provides an opportunity for network equipment and UE vendors to debug such features and verify that pieces of equipment from various vendors interoperate properly. Early reports [39] indicate that the PSCR testing has already validated some basic features such as priority pre-emption of bearers, ARP and QCI configuration, admission control and packet scheduling. In addition, further development and testing are still required for advanced features such as Multimedia Priority Service (MPS), RRC Establishment Cause support and SIP priority. While the work at PSCR is focused on the support of priority and pre-emption on a dedicated LTE network, the LTE features tested could also be applicable to commercial networks. Besides the validation of the technical capabilities, the deployment of priority and pre-emption in LTE commercial networks requires the establishment and validation of clear policies to ensure that such actions do not have negative unintended consequences. An example that is often cited when it comes to the applicability of pre-emption in a commercial network is how to deal with the case of an 'ordinary' call between a medical specialist and somebody providing advice for a cardiac arrest that is just happening in the same area where there is a sudden increase of higher-priority PPDR traffic.

Thus far, schemes for privileged access to commercial mobile networks have only been adopted for voice communications in some countries. The current situation is that there is not a global prioritization system, though some efforts have been conducted at international organizations such as ITU-T by developing recommendations to facilitate the use of public telecommunications services during emergency, disaster relief and mitigation operations (e.g. ITU-T Rec. E.106 [40]).

The United States has been a pioneer in the development of this type of systems, initially implementing the prioritization of voice calls in the wireline network, known as Government Emergency Telecommunications Service (GETS) [41], followed by its extension to commercial mobile networks, known as Wireless Priority Service (WPS) [42]. This system gives priority to calls made by phones subscribed to WPS after dialling a specific prefix. Canada and Australia have also implemented a system based on the US WPS [43].

In the EU, only the United Kingdom possesses a prioritization system, known as Mobile Telecommunication Privileged Access Scheme (MTPAS), which substituted the old Access Overload Control (ACCOLC) system [44]. Upon the Police Incident Commander request, MTPAS restricts the access to the cell sites in the area where an emergency situation has been declared. In turn, Sweden is developing standards for the implementation of priority services but lacks of a system at the moment [45].

Following a brief discussion about the technical foundations of priority support in commercial networks for voice services, the next subsections cover the scope and capabilities of the WPS used in the United States and the MTPAS used in the United Kingdom as the two more representative initiatives for priority services on public mobile telephone networks. Further information on other initiatives related to the delivery of priority services on commercial networks can be found in Ref. [46]. Recently, some patents to enable prioritization and ruthless pre-emption to authorized users have been filled [47].

5.4.3.1 Technical Foundation: Priority Service

Priority Service [48] was first introduced in 3GPP Release 6 specifications and is applicable to voice calls offered through both GERAN and UTRAN. A user of the Priority Service is assigned access class(es) in the range 11–15 to receive priority access to the network. In addition to access attempt priority provided by access control mechanisms, a Priority Service call receives end-to-end priority treatment, including priority access to traffic channels, priority call progression and call completion. Priority Service is activated on a per-call basis using specific dialling procedures: the user dials a given service code (SC) after the destination number. It is worth noting that Priority Service supported by the 3GPP systems is one element in the ability to deliver calls of a high-priority nature from mobile-to-mobile networks, mobile-to-fixed networks and fixed-to-mobile networks. Hence, in order to effectively provide an 'end-to-end' solution, all Priority Service providers (mobile networks, transit networks, fixed networks) should adhere to uniform, nationwide operating access procedures.

3GPP technical report TR 22.950 [48] establishes the high-level requirements for Priority Service, and TR 22.952 [49] provides a 'guide' that describes how existing 3GPP specifications support such requirements. In fact, Priority Service is supported by relying on a service known as Enhanced Multi-Level Precedence and Pre-emption (eMLPP) that was already specified in GSM Release 98 and updated within GSM/UMTS Release 6 to be compatible with the Priority Service requirements. eMLPP service is specified in Refs [50–52].

The eMLPP service is offered to a subscriber as a network operator's option for all basic services subscribed to and for which eMLPP applies. The eMLPP service supports two capabilities: precedence and pre-emption. Precedence involves the assignment of a priority level to a call. eMLPP supports a maximum of seven priority levels as defined in Ref. [50]. The highest level (A) is reserved for network internal use. The second highest level (B) may be used for network internal use or, optionally, depending on regional requirements, for subscription. These two levels (A and B) may only be used locally, that is, in the domain of one mobile switching centre (MSC). The other five priority levels (0–4) are offered for subscription and may be applied globally, for example, on inter-switch trunks, if supported by all related network elements, and also for interworking with ISDN networks providing the MLPP service. Levels A and B shall be mapped to level 0 for priority treatment outside of the MSC area in which they are applied. For each of the seven priority levels, the network operator can administer the parameters that control the treatment of that priority within its domain. This treatment includes the selection of a target call set-up time and whether pre-emption is allowed for each priority level or not. An example for an eMLPP configuration is given in Ref. [50]. 3GPP specifications do not define specific mechanisms to achieve the target set-up times as defined by the service provider. The use of pre-emption is an operator choice. In case pre-emption is not provided, priority levels may be associated with different queuing priorities for call establishment.

The eMLPP priority level for a given call depends on the calling subscriber. The maximum precedence level for each subscriber is set at subscription time. Information concerning the maximum priority level that a subscriber is entitled to use at call establishment is stored in the Home Subscriber Server (HSS). On the UE side, SIM/Universal Subscriber Identity Module (USIM) also stores eMLPP subscribed levels and related data (e.g. fast call set-up conditions). The priority level can be selected by the user on a per-call basis (up to and including their maximum authorized precedence level). The selection of the priority level is done via man–machine interface (MMI) at the UE in the case of an eMLPP subscription including priority levels above 4. For mobile-terminated calls, the priority level is established based on the priority of the calling party, and it is applied at the terminating end (presuming the call's priority is passed via signalling between the originating and terminating networks). Interworking with ISDN MLPP is required. The eMLPP service also applies to roaming scenarios, if eMLPP is supported by the related networks.

eMLPP also supports automatic answering or called party pre-emption. Hence, if the called mobile subscriber is busy and automatic answering applies for calls with a sufficient priority level, the existing call may be released (if pre-emption applies) or may be placed on hold in order to accept an incoming call of higher priority.

With all the above mechanisms defined, no further development of the eMLPP service is envisaged in ongoing 3GPP releases.

5.4.3.2 WPS

WPS [42] is a method of improving connection capabilities for a limited number of authorized national security and emergency preparedness (NS/EP) cell phone users on commercial wireless networks. WPS was mandated by the FCC in the United States. In the event of congestion in the commercial network, an emergency call using WPS will wait in queue for the next available channel. WPS calls do not pre-empt calls in progress or deny the general public's use of the radio spectrum. WPS is available to key NS/EP personnel with leadership responsibilities at the federal, state, local and tribal levels of government and in critical private industries (e.g. finance, telecommunications, energy, transportation, etc.). These users can range from senior members of the presidential administration to emergency managers and fire and police chiefs at the local level and to critical technicians in wireline and wireless carriers, banking, nuclear facilities and other vital national infrastructures.

The implementation of WPS in commercial 3GPP networks is based on the Priority Service described in the previous section. WPS actually uses five priority levels, and a Priority Service call is invoked by dialling *SC+destination number, with an SC of '272'. Within US networks, a WPS user is assigned one or more access classes in the range 12–14 to receive priority access to the network, in addition to an assigned access class in the range 0–9. Pre-emption capability is not required by WPS. Additional details of the implementation of the WPS in the US networks can be found in Ref. [49]. The WPS system places a limit on the number of voice channels that can be used for priority calls at any given time (25% of voice channels on each cell site [53]). This ensures commercial users not being completely starved of access to the network.

5.4.3.3 MTPAS

MTPAS [44] is a service launched in the United Kingdom in September 2009 that aims to provide privileged access to the mobile networks by installing a special SIM card in the telephone handset.

The service is only available to some categories of emergency responders who are entitled to get privileged access SIMs in their staff's mobile phones. The activation of privileged access in SIMs does not necessarily require the physical change or acquisition of a SIM. Mobile operators can change the access of a SIM (from normal public access to privileged access and vice versa) remotely using SIM over-the-air (OTA) technology.

The MTPAS is activated upon request in an emergency situation. Thus, using an agreed protocol, the Police Gold Commander, in charge of the response to a major incident, notifies all MNOs that a major incident has been declared. In such situation, if networks become congested, handsets installed with a privileged access SIM will stand a much higher likelihood of being able to connect to their network and make calls than other customers.

MTPAS supersedes ACCOLC, the former scheme for managing mobile privileged access. SIMs issued under the old scheme will continue to work under the MTPAS.

5.5 Hybrid Solutions

Hybrid solutions, used here as an umbrella term to categorize those PPDR service delivery options that are based on a combination to different extent of dedicated and commercial mobile broadband network infrastructures, can help to strike the right balance between the approaches presented in the previous sections. In some cases, hybrid solutions can be considered as interim solutions in the medium run that could pave the way for smoothly transition towards the desirable long-term solution. For example, commercial networks can be exploited until a dedicated network become available. In other cases, hybrid solutions can be considered as the most valuable option from a cost–benefit perspective even for the long term.

Hybrid solutions can be formulated around one or a combination of the following approaches:

- **Support of national roaming for PPDR users over commercial networks**. This model can be pursued when the approach is to build a dedicated mobile broadband network for PPDR in some parts of a country (e.g. in the most populated areas and most important roads) and rely on broadband services provided by one or more commercial MNOs for the remaining part of the country. Therefore, the PPDR operator of the dedicated network establishes roaming agreements with the national commercial carriers to get access to their networks and so complement the coverage and capacity of the dedicated infrastructure.
- **Deployment of a PPDR MVNO**. In this model, a PPDR MVNO is set up as a global provider for PPDR communications services to multiple PPDR agencies, avoiding each individual PPDR organization to make specific arrangements with the commercial MNOs. The PPDR MVNO does not have its own RAN but only relies on those deployed by the MNOs. This centralized approach facilitates the possibility to build out a dedicated core and service infrastructure for the PPDR users, enabling the PPDR MVNO instead of the MNOs to have full control of critical capabilities for the PPDR users such as subscriptions, service profiles and service offerings.
- **RAN sharing with MNOs**. In this model, a dedicated RAN is deployed through infrastructure sharing agreements or partnerships with MNOs. This approach could importantly minimize the initial investment required to launch service while providing the PPDR users with wide geographic service availability based on existing 3G/4G commercial footprints.
- **Network sharing of critical and professional networks**. In this model, the dedicated communications systems deployed in specific areas or facilities by other public or private companies are leveraged or integrated as part of a global, interoperable PPDR communications system. For example, critical infrastructures such as airports and transportation hubs have typically their own dedicated communications systems adapted to its specific industrial and professional environments and used to support their diverse day-to-day activities (maintenance, diagnostics, security, crisis situation). Priority management within these networks could allow for sharing critical and non-critical services on the same network, including the operation of PPDR teams displaced there when needed. The vital condition for this type of model remains the introduction of a highly secure network that is available under all circumstances, comprising a developed arbitration system between users. This scheme would enable that absolute access is guaranteed to the service for security forces and priority services, particularly in a crisis period, while finding an economic equation via the commercial uses [54].

A further insight into the above hybrid solutions is provided in the following subsections.

5.5.1 *National Roaming for PPDR Users*

Roaming is defined as the use of mobile services from another operator, which is not the home operator. The most well-known form of roaming is international roaming, which allows users to use their mobile devices when abroad. Nevertheless, national roaming is also quite common. National roaming is concerned with networks of operators within the same country. In some countries, national roaming is imposed by the NRAs with the objective to promote and stimulate competition by facilitating the entrance of new actors in the market. Sometimes, national roaming is implemented on a voluntary basis between providers (as part of business agreements), without the intervention or request from the NRA. Different types of existing national roaming schemes are the following [55]:

- **New entrants**. This national roaming scheme aims to facilitate new entrants in the market, with the goal to improve competition. The new entrant makes a national roaming agreement to have an immediate full geographic coverage without high initial investments. Such agreements are usually temporary.
- **Coverage in rural areas**. In this scheme, a provider extends its coverage to scarcely populated (rural) areas using a national roaming agreement. This national roaming scheme aims to facilitate smaller operators who may not be able to sustain the costs of covering a large territory with a low density of population.
- **Regional licences**. In countries with regional spectrum licences, national roaming can be used by operators to offer services in other regions. This allows smaller operators to provide a national service.
- **In-flight/at-sea mobile services**. Some operators provide roaming services to allow customers to use mobile services in flight or at sea, for instance, using roaming agreements with satellite communications providers.
- **Emergency roaming**. The roaming agreement can be used for resilience purpose, supporting the traffic of customers affected by an outage. This scenario has been investigated by the ENISA as a solution for mitigating outages in order to foster security and resilience of European communications networks and ensure that European citizens can communicate also in the case of major outages [55].

National roaming for PPDR is needed to extend the coverage of the PPDR services to the areas not covered by the dedicated network. National roaming with multiple MNOs enables higher coverage and capacity as well as improved resilience, since most areas are covered by BSs of several operators.

The selection of the serving network would be controlled by the PPDR network operator. Indeed, in current commercial solutions, network selection relies on information that is directly stored in the SIM card of the mobile terminal:

- Home network and equivalent home networks list (an equivalent home network defines a network that should be treated as it was the home network) in priority order.
- Operator-controlled networks list including the codes for networks preferred by the operator in priority order. The different access technologies for each network are also reported.
- User-controlled networks list including the code for networks preferred by the user in priority order. The different access technologies for each network are also reported.
- Forbidden networks list including the codes of networks with access denied to the device with a reject message.

- Equivalent networks list including list of equivalent networks codes as downloaded by the actual registered network. This list is replaced for each new location registration procedure. These networks are equivalent to the current network in the network selection.

Based on the previous information, network selection can be automatic or manual. In automatic mode, the mobile device scans the spectrum and finds all available networks. The UE then chooses a network in priority order based on the following list:

1. Home network or equivalents list in priority order.
2. User-controlled networks list in priority order.
3. Operator-controlled networks list in priority order.
4. Other networks with received high-quality signal in random order.
5. Other networks in order of decreasing signal quality.

In this way, the UE requests first to the home network or equivalent as described in the first list. If none of them is found, the device selects a network in the next list and so on. In manual mode, the device researches and displays all available networks including the networks present in the forbidden networks list (in opposition to the automatic mode). The networks are presented to the user in the same priority order as do the automatic mode. Then, the user selects arbitrary a network from the networks list.

Solutions currently in use by MNOs or roaming hub operators to control the distribution of roamers in scenarios with multiple candidate networks (i.e. dynamic roaming steering) can find its applicability in the PPDR national roaming scenario. For instance, solutions for OTA provisioning of SIM card settings [56] can be used to organize the roaming priority lists within the PPDR SIMs and to specify a given roaming behaviour in a dynamic manner according to specific PPDR needs.

5.5.2 Deployment of an MVNO for PPDR

MVNO models have existed for more than 10 years in the commercial domain, and therefore, the principles are well established. Nevertheless, there are many different ways of implementing the MVNO model depending on how far the MVNO wants to control its services and how much control an MNO is willing to offer.

A standard definition of MVNO does not exist. Ofcom, the UK regulator, offers probably the most general definition of MVNO as 'an organisation that provides a mobile (sometimes called wireless or cellular) service to its customers but does not have an allocation of spectrum'. Indeed, an MVNO is a business model that has emerged by the rupture of the traditional mobile value chain of an MNO. This has allowed new players to participate in the mobile value chain and extract value leveraging their valuable assets. The traditional mobile value chain can be separated into two main areas [57]:

1. The RAN, which is exclusively exploited by MNOs and requires a licence granted by the regulatory authority to use the spectrum.
2. The rest of the elements required to deliver the service to the customers. As it is shown in the upper part of Figure 5.2, this area of the value chain includes the operation of the core network (e.g. switching, backbone, transportation, etc.), the operation of the service platforms and value-added services (e.g. SMS, voicemail, etc.), the operation of the back-office process

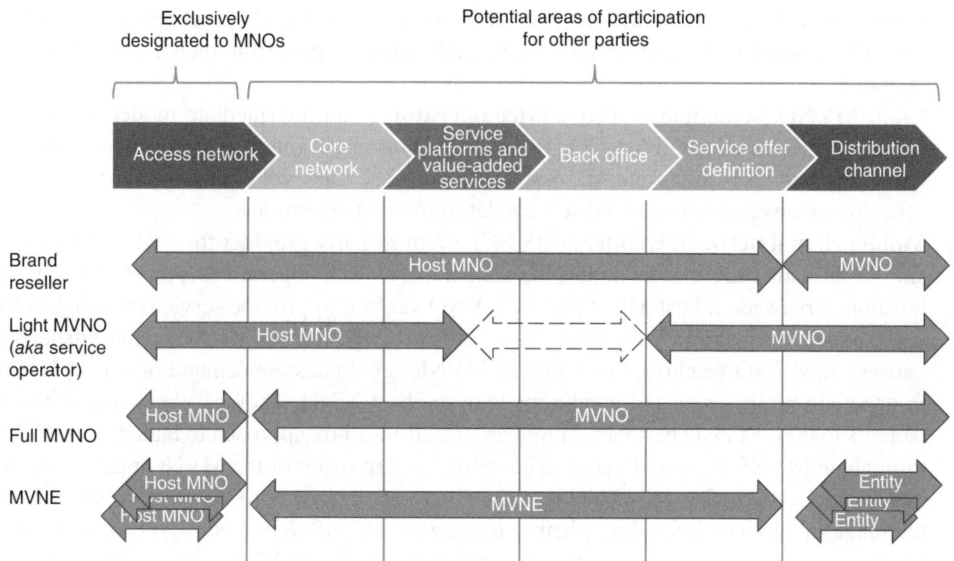

Figure 5.2 Different MVNO business models.

to support business processes (e.g. subscriber registration, handset and SIM logistics, billing, balance check, customer care, etc.), the definition of a mobile value offer and the final delivery of the products and services to the client through the distribution channel.

It is in this second area of the value chain where other parties can participate by innovating, operating or selling mobile services. Both MNOs and MVNOs can take advantage of this business model. MNOs can exploit its network capacity, IT infrastructure and service and product portfolio to acquire untargeted segments, add a new revenue stream from wholesale business and reduce spare capacity and cost to serve per user. In turn, an MVNO can exploit its brand awareness, distribution channels and customer base to provide customized value proposition and complementary products and services to its customers. An MVNO venture brings multiple benefits to a company such as a new revenue stream, a low-cost entry strategy to the mobile market, a new vehicle to strengthen the value proposition and an opportunity to increase customer acquisition and/or retention.

Based on how the value chain is restructured between MNO and MVNO, the four main business models illustrated in Figure 5.2 have emerged in the MVNO market:

1. **Branded reseller** is the lightest MVNO business model, where the venture just provides its brand and, sometimes, its distribution channels. The MNO provides the rest of the business, from access network to the definition of the mobile service offer. This model requires the lowest investment for a new venture, and it is the fastest to implement. However, most of the business levers remain with the MNO. Therefore, the new venture has a very limited control of the business levers and value proposition of the service.
2. **Full MVNO** is the most complete model for a new venture, where the MNO just provides the access network infrastructure and, sometimes, part of the core network, while the new

venture provides the rest of the elements of the value chain. This MVNO business model is typically adopted by telecom players that could gain synergies from their current business operation.
3. **Light MVNO**, sometimes called **service operator**, is an intermediate model between a branded reseller and a full MVNO. This model allows new ventures to take control of the marketing and sales areas and, in some cases, increase the level of control over the back-office processes and value-added service definition and operations.
4. **Mobile virtual network enablers** (MVNE) is a third-party provider focused on the provision of infrastructure that facilitates the launch of MVNO's operations. An MVNE can be positioned between a host MNO and an MVNO venture to provide services ranging from value-added services and back-office processes to offer definition. MVNEs reduce the entry barriers of MVNO ventures, given that an MVNE aggregates the demand of small players to negotiate better terms and conditions with the host MNO. They pass on some of these benefits to their MVNO partners. Moreover, the all-in-a-box approach to launch an MVNO through an MVNE has accelerated, even more, the explosion of the MVNO market. Some MVNE models are also called mobile virtual network aggregator (MVNA), depending on the range of services offered or whether they aggregate different host MNOs or only one. MVNE models range from telco-in-a-box offering, where the MVNE just offers core network, value-added services and back-office services, to full MVNE as shown in Figure 5.2.

The success of the business model is dependent on creating a win–win situation between the MVNO and the MNO, where the MVNO has more knowledge than the MNO of a specific market segment that the MNO may not wish to develop itself. This may be due to particular market circumstances and/or difficulties in penetrating that market segment. The commercial relation between MVNO and MNO is based on relatively simple SLAs for the MVNO to obtain bulk access to network services at wholesale rates. Often, the MVNO bulk-buys minutes or data, based on coarse MVNO usage level predictions (for multiple months) and corresponding spare capacity predictions by the MNO. Typically, there is no differentiation between the MVNO and MNO users on the RAN. Thus far, there have been three main groups of players that have taken advantage of the MVNO business model:

1. Telecom companies. MNOs can benefit from the MVNO model by serving untapped segments that their current value proposition is unable to attract. Additionally, telecom operators in general (fixed and/or mobile) can use the MVNO opportunity to enter to new geographies through a wholesale business based on an MVNE model. Finally, fixed telecom operators with no mobile offering can strengthen its value proposition providing a 4-play offer including mobile services.
2. Non-telecom companies such as retailers, media and content generators, financial institutions, travel and leisure, postal services, sports clubs and food and beverage, among other types of companies, can take advantage of their current assets (brand, customer base, channels, content, etc.) to exploit a new business or to strengthen their current value proposition and customer loyalty.
3. Investors can take advantage of the MVNO model by investing in new opportunities to create value by participating in the telecom industry.

In this context, the applicability of an MVNO model to cope with the market segment of 'PPDR communications', or more broadly that of 'critical communications', is gaining momentum. From an MNO's standpoint, this market can be considered as a small, but very demanding, market in comparison to their mainstream large consumer markets. This is where an MVNO model is expected to fill the gap and bring value.

In the case of an MVNO model targeting the delivery of PPDR communications, the goal of the MVNO is to leverage the existing commercial mobile broadband radio infrastructure to create and operate dedicated services for the critical users. The MVNO stands between the user organization(s) and the MNOs, and it manages all the services for the users, such as provisioning, monitoring and managing all operational processes including incident, problem, change, configuration and release management that are needed to control the quality of the services. Therefore, out of the MVNO business models illustrated in Figure 5.2, a full MVNO is the most suitable approach to follow [58]. Alternatively, a light MVNO without core network equipment (i.e. P-GW and/or GGSN) but with control at least over the service platforms and subscriber management would also be feasible. Clearly, dedicated services can deliver added value including better availability, security, quality control and better customer care than can be delivered by the commercial MNOs individually.

The MVNO requires deep expertise in the relevant mobile communications technologies and a thorough understanding of the critical user requirements in order to make the best use of the existing commercial cellular radio infrastructures if it is to create and operate services dedicated to critical users. The MVNO has to negotiate special contractual arrangements with the MNO and a detailed SLA to reach its objectives. The MVNO will also manage the funding and financial aspects of the project. In order to spread the risk of network failure and exceptional traffic conditions as well as improving coverage, it may be advantageous to negotiate capacity with more than one MNO.

Better availability can be achieved by allowing roaming to any of the national MNOs. This also results in achieving the best coverage possible as well as improved resilience, provided the infrastructures are not shared. However, the implementation of national roaming could require (too) large investments by the providers if not national roaming practices are already in place for other purposes. As a way to address this, PPDR MVNO operators could initially leverage on existing international roaming infrastructure (i.e. use of international roaming hubs).

Reliability and service availability can be increased for PPDR users by granting higher access priority onto the commercial infrastructures based on special agreements with the MNOs. Security and confidentiality of the data flows can be managed by the MVNO (e.g. end-to-end security mechanisms between PPDR terminal equipment and MVNO's service platforms). A schematic view of the MVNO model for PPDR mobile broadband communications is depicted in Figure 5.3.

The key advantage of an MVNO model is that it represents a relatively low-risk and limited investment option that would allow PPDR agencies to incorporate mobile broadband data into their daily operations in the short term, retaining the control on key aspects such as subscriber management, service profiles and service offerings. The implementation of an MVNO model could enable users to gain experience in the use of mobile broadband data and assist in the development of future plans. Additionally, an MVNO model can become very efficient especially looking for a 'shared' network/solution between different critical user organizations, as it enables them to reach an economy of scale through a common approach rather than each organization setting up its own MVNO with the MNOs. Indeed, a compelling reason for having a PPDR MVNO model is to negotiate and follow up the requirements and service commitments instead of having a situation where all the individual PPDR organizations

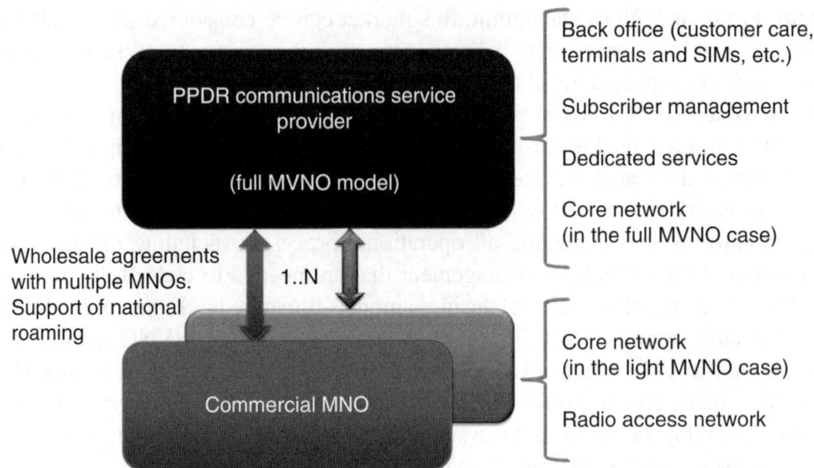

Figure 5.3 MVNO model for PPDR mobile broadband communications.

make a contract with commercial operators themselves. A shared solution, through an MVNO, will also create an environment that will facilitate the exchange of information between user organizations when needed.

An MVNO can also function as a platform for PPDR application innovation from a very early stage and help to adapt services to the evolving requirements of the user's mobile data applications. Critical communications users are now at the early stage of a new era regarding mobile data applications, and they will face growing and changing expectations towards the mobile data services. For existing PMR operators (i.e. TETRA, TETRAPOL, etc.), the MVNO model can provide a first step towards an eventual dedicated broadband network in the future while enabling the operator and users to better understand the benefits of mobile broadband. This gives also the advantage of a '1-stop' service for different types of communications (i.e. PPDR users can go to the same entity for as both voice + data). The adoption of this model also represents a compelling opportunity for MNOs to build the confidence of the PPDR users in the MNOs as reliable and necessary players in the delivery of mission-critical data services.

On the downside, the disadvantages of this model mainly come from the limitations in terms of footprint and performance of the commercial MNO's infrastructures, as discussed in Section 5.4. Hence, similarly to the case of standard commercial networks, the MVNO solution will only provide service where the operators have installed coverage and network resilience levels will only be those considered necessary by the commercial provider. However, as also discussed in Section 5.4, special arrangements can be established between the PPDR MNVO and the MNOs for increased levels of availability through increased robustness in the network design, extended coverage and priority access privileges for the PPDR users, especially in relation to emergencies and disaster events.

The adoption of an MNVO model for PPDR to use commercial capacity does not prevent the future deployment of private LTE networks. Indeed, the MVNO and dedicated network models are considered complementary [58, 59]. This model has been already put in place in Europe by the Belgian PPDR national network operator (ASTRID), which offers the Blue

Light Mobile MVNO service for the delivery of data communications over the national commercial networks in Belgium. MSB in Sweden also plans to deploy a similar model to provide mobile broadband to their PPDR users, currently using a TETRA network for mission-critical voice. The MVNO model was also at the core of the PPDR communications framework developed in the European research project HELP [37], which is addressed in further detail in Section 5.6.7.

5.5.3 RAN Sharing with MNOs

The deployment of a dedicated network may require a wide range of infrastructure and spectrum sharing mechanisms through public–public and public–private partnerships to be able to achieve its goals in a cost-efficient manner. Commercial MNOs are among those potential infrastructure and spectrum sharing partners. Commercial carriers can bring some key attributes to the table that other partners cannot, such as expertise in deploying a nationwide LTE network, back-office support systems and access to backhaul assets. Sharing agreements or partnerships with MNOs can minimize the initial investment required to launch service while providing the PPDR users with wide geographic service availability based on existing 3G/4G commercial cell sites. Indeed, it is a fact that many national TETRA and TETRAPOL networks in Europe already share some buildings, power, air condition and transmission with commercial operators [16, 13].

Network sharing can be viewed from two perspectives: geographical and technical [60]. From the geographical point of view, network sharing can take various aspects depending on the business model intended and the areas already covered by each involved mobile operator. The dimensions of network sharing can be characterized as:

- **Stand-alone**. Two or more network operators own and control their respective physical network infrastructure. No network sharing or interaction between operators exists (outside roaming agreements).
- **Full split**. Each operator covers a specific non-overlapping geographical region, and each operator extends its own area of operation by using the other operator's network.
- **Unilateral shared region**. One mobile operator has full control over the network infrastructure in one region, and a new greenfield operator enters the market by utilizing this existing infrastructure. This model lowers the barrier for new business entrants and allows existing operators to better capitalize on their installed resources.
- **Common shared region**. Two operators of similar size want to have a presence in the same region, and they decide to share the infrastructure. This scenario is mostly attractive in rural areas, where the estimated revenues are low and operators need to carefully plan their investment.
- **Full sharing**. Two or more mobile operators decide to fully share their network, either access, core or both, in order to render the operations and management of the network more efficiently.

From the technical perspective, the range of solutions encompass:

- **Passive RAN sharing**. This is usually defined as the sharing of space or physical supporting infrastructure that does not require active operational coordination between network operators. Therefore, equipment from different mobile operators is installed at the same sites, and operators share the costs for site construction and renting. The construction of the tower site and its operation may be outsourced to third-party companies. Site and mast sharing are the two main forms of passive sharing. **Site sharing**, involving the co-location of sites, is perhaps the easiest and most commonly implemented form of sharing. Operators share the same physical compound but install separate site masts, antennas, cabinets and backhaul. **Mast sharing**, or tower sharing, is a step up from operators simply co-locating their sites and involves sharing the same mast, antenna frame or rooftop.

- **Active RAN sharing**. This implies two or more MNOs sharing one or several elements such as antennas, radio equipment (e.g. LTE eNBs) and/or backhaul transmission equipment. The MNOs can keep using separate frequency channels or pool together their spectrum resources and operate them in parallel according to predefined resource allocation agreements. While active sharing can be limited to individual elements, 3GPP networks such as LTE offer support for what is known as (full) RAN sharing. This is the most comprehensive form of access network sharing: each of the individual RAN access networks is incorporated into a single network, which is then split into separate networks at the point of connection to the core. MNOs continue to keep separate logical networks and spectrum, and the degree of operational coordination is less than for other types of active sharing. The 3GPP standardizes the necessary procedures for (full) RAN Sharing. There are two supported architectures: Multi-Operator Core Network (MOCN) and Gateway Core Network (GWCN). Further details about LTE RAN sharing in 3GPP are provided in Chapter 4.
- **Core network sharing**. The core network may be shared at one of two basic levels, namely, the transmission network backbone and core network logical entities. With regard to transmission network, it may be feasible to share spare capacity that one operator may have with another operator. The situation may be particularly attractive to new entrants who are lacking in time or resources (or desire) to build their own ring. Therefore, they may purchase capacity, often in the form of leased lines, from already established MNOs. On the other hand, core network logical entity sharing represents a much deeper form of sharing infrastructure and refers to permitting a partner operator the access to certain or all parts of the core network. This could be implemented at different levels depending on what platforms the MNOs wish to share (e.g. service platforms and value-added systems).
- **Network roaming and MVNO.**[2] Sometimes, network roaming and MVNO are also considered forms of network sharing. However, there are no requirements for any common network elements for these types of sharing to occur. As long as a roaming and/or wholesale agreement between the two operators exists, then the visited network (in case of roaming) or host network (in the case of MVNO) can be used to serve the other operator's traffic. For this reason, some operators may not classify roaming as a form of sharing since it does not require any shared investment in infrastructure [61]. At this point, it is also worth noting the difference between roaming and the (full) RAN sharing option discussed above. Basically, roaming provides a similar capability to RAN sharing where a subscriber of the home network can obtain services while roaming into a visited network. This can be viewed as a form of sharing where the visited network shares the use of its RAN with the home network for each subscriber roaming into it. However, there is a distinction to be considered between the

[2] These options are added here for completeness of the classification. More details on these two approaches have been covered in the previous subsections.

Table 5.5 Network sharing solutions between MNOs.

Sharing solution	Type of sharing	Characteristics
Site sharing	Passive	Operators collocate their own equipment. Very simple, it does not require operational coordination. Support equipment may or may not be shared
Tower sharing	Passive	Operators share the same mast or rooftop. Operators have own antennas, huge CAPEX reduction and environment benefits
Spectrum sharing	Active	Parts of the spectrum are leased by one operator to another. Improves spectrum efficiency and fights spectrum scarcity
Antenna sharing	Active	Antenna and all related connections are shared. Passive site elements are shared too
Base station sharing	Active	Operators maintain control over logical eNBs. Can operate on different frequencies, fully independent
(Full) RAN sharing	Active	RAN resources are combined: antennas, cables, BS and transmission equipment. Separate logical networks and spectrum
Core network element sharing	Active	Barely supported. Benefits are not as clearly defined as those for sharing the access network
Transmission network sharing (backhaul, backbone)	Active	Transport is shared, for example, fibre access and backbone networks

two cases. When roaming, the subscriber uses the visited network when outside of the home network geographic coverage and within the visited network geographic coverage. Instead, in a RAN sharing arrangement, all of the participants (hosting RAN provider and one or more participating operators) provide the same geographic coverage through the hosting RAN [62].

Apart from the network roaming and MVNO options, network sharing solutions today mostly focus on the access network, which is the most expensive part of an MNO's network. The benefits for sharing core network elements are not as clearly identified as those for sharing the access network. While it is conceivable that there may be some cost reductions in operations and maintenance, the scale and practicality of these remain more uncertain. Sharing solutions might need the involvement of a trusted third party, which installs and operates the shared infrastructure and acts as a broker among the sharing operators. Such a solution comes at the cost of losing some of the operational control of the network operators and reorganization of the involved departments. A detailed list of the passive/active RAN and core network sharing solutions, together with their characteristics, is provided in Table 5.5. While there may be significant commercial and practical hurdles to overcome, there are no fundamental reasons why multiple operators cannot share networks. Agreements may concern individual sites, a number of sites or particular regions. Passive sharing and RAN sharing do not require a fully merged network architecture, and there are examples of unilateral, bilateral (mutual access) or multilateral agreements. Further considerations on business drivers, regulation and technical and environmental issues of network sharing can be found in Ref. [61].

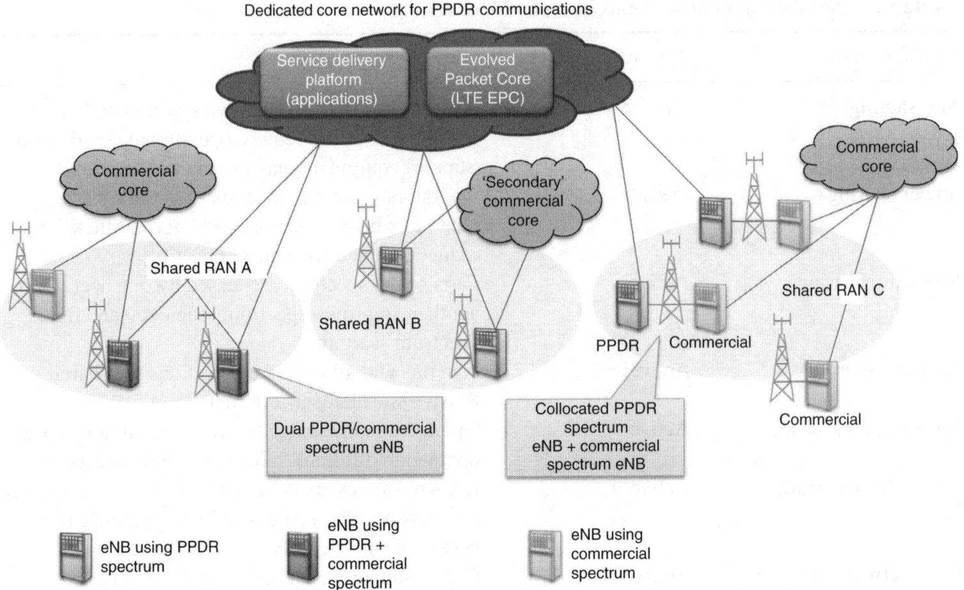

Figure 5.4 Different RAN sharing deployment options with MNOs.

While all of the previous options for sharing are technically feasible, the specifics and complexities arisen when determining the most appropriate form of sharing could vary largely depending on many factors (e.g. network roadmaps of the involved operators, business incentives, costs, level of control required by the involved partners, physical security at the shared radio sites, hardening requirements). Importantly, sharing raises critical issues about the governance of the shared resources (who manages and controls) that need to be resolved in the form of win–win solutions. In any sharing model, the PPDR operator or government responsible entity for PPDR communications would presumably need to enter into an SLA and/or roaming agreement with the commercial partners. That business negotiation is critical to the success of any potential RAN sharing solution. Therefore, solutions may differ per region/country. In the context of FirstNet in the United States, in addition to the use of passive RAN sharing that entails less complexity in terms of integration, the following range of solutions and associated technical issues have been identified [63]:

- **Greenfield RAN sharing**. This option is based on the deployment of 'dual-band' eNBs supporting both dedicated spectrum (i.e. Band 14 in the United States) and commercial spectrum as well as antenna infrastructure, backhaul and towers. This type of sharing is particularly appropriate for rural deployments where commercial networks are generally designed for coverage, similar to the needs of a PPDR network. This form of greenfield RAN sharing is depicted by 'shared RAN A' in Figure 5.4. Another realization of a greenfield RAN sharing is the case where the shared eNB sites only operate on the dedicated PPDR spectrum so that the MNO exclusively utilizes this spectrum on a secondary basis for its entire LTE service needs. This case is illustrated as the 'shared RAN B' shown in Figure 5.4.

- **Adding PPDR eNB to existing LTE RAN**. This option is based on the deployment of a new PPDR eNB alongside an existing MNO's site. This scenario, illustrated by 'shared RAN C' in Figure 5.4, leverages some or all existing infrastructure already deployed by the commercial carrier (e.g. antenna, tower, shelters, cabinets, backhaul). Nevertheless, it presents a variety of technical and service-related issues that deserve detailed consideration. For example, commercial operators generally design RAN for capacity in some areas, leading to much higher cell counts than might otherwise be required to provide service to PPDR users. This leaves the PPDR operator with two options: deploy a PPDR spectrum-specific eNB at each commercial site (i.e. a 1:1 overlay) or deploy fewer sites, enough to satisfy specific PPDR coverage and capacity needs. When deploying a 1:1 overlay, the addition of PPDR spectrum-specific eNB requires the insertion of additional radio components (diplexers) between a commercial operator's existing radio equipment and the antenna, a disruptive approach that may not be attractive to MNOs. Further, the risk of interference due to passive inter-modulation (cross-mixing of signals) created by the insertion of the new shared components needs evaluation at each affected site. If a 1:1 overlay is not used to reduce RAN costs, the existing commercial antennas likely cannot be shared due to the different vertical and horizontal adjustments required for the PPDR-enabled cell sites compared to commercial sites.

In addition to commercial mobile operators, the deployment of a dedicated network may be required to forge partnerships with a wider variety of entities that will be open to sharing infrastructure. As an example, in the case of the CO dedicated network model, sharing opportunities for essential (and expensive) components of the LTE infrastructure can be seized depending on the consortium participants and contracts, as well on the possibility for the user organizations to lease or share their own existing network elements to the mobile broadband network provider. In this regard, the APCO Broadband Committee has delivered a white paper [64] that explores business tools, methods and techniques that will fuel and support the leveraging of public and commercial assets for use in building and operating the network by FirstNet in the United States. The document focuses upon asset standardization and valuation methodologies that have the potential to greatly simplify the negotiation, cost modelling, assessment and eventual use of resources needed to establish the network readiness, deployment and ongoing sustainment. This document presents recommendations, insights and open questions for potential LTE PPDR network stakeholders, such as state and local agencies looking to provide in-kind contributions to the network.

5.5.4 *Network Sharing of Critical and Professional Networks*

This hybrid model involves players who have the need and possibility to deploy local dedicated, redundant and extremely dense networks over certain sensitive zones. This could be the case of critical infrastructures such as airports and other transportation facilities. In addition, locations such as shopping malls, stadiums, concert venues and the like could also be considered as long as the specifics of the facilities require the deployment of dedicated professional networks or installations (e.g. DAS for coverage due to complex building infrastructures). In this model, these dedicated communications systems deployed in specific areas or facilities will be leveraged or integrated as part of a global, interoperable PPDR communications system, as depicted in Figure 5.5.

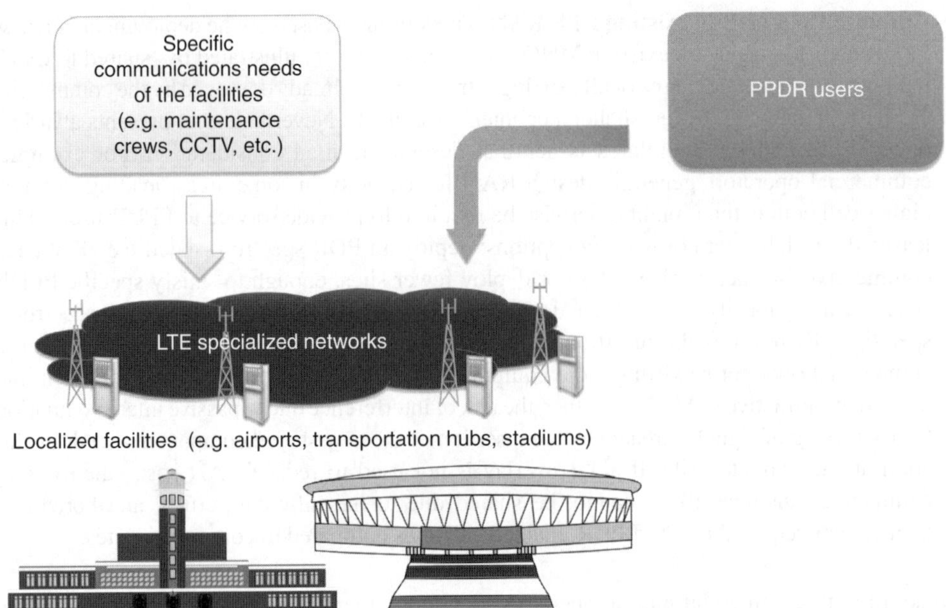

Figure 5.5 Sharing of specialist networks deployed in specific areas and facilities.

This model has been proposed by a major airport communications operator in France [54]. Airports are sensitive environments at the heart of which radio services are used to ensure the reliability of exchanges in real time, service performance and security of people. In both nominal mode and in crisis situations, airport operations require a secure and redundant radio network, a very high level of availability and a suitable coverage of buildings, technical galleries and runways. These different stakes are leading airport players to develop dedicated professional radio networks. In this context, [54] proposes a sharing service model by means of which the dedicated airport network could be integrated with the overall government solution for PPDR communications. The economic model proposed enables the absolute priority of government and airport PPDR uses over professional and potentially commercial uses. The model also enables much of the investment to be passed onto the specialist players. For its part, the government could keep the management of tactical networks and, more generally, the management of the overall service (networks, terminals, security applications). The interoperability and coordination of players are stressed as the main conditions for the success of this model. Commercial operators could also benefit from these dedicated networks, limiting the investment to be made to cover these zones and without competing with these specialist players.

5.6 Network Architecture Design and Implementation Aspects

This section describes some network architecture design and implementation aspects that are central for the realization of most of the previous described network delivery options for mobile broadband PPDR communications. To this end, the description of a reference architecture developed by ETSI that provides a high-level structure of a system intended to deliver mobile broadband mission-critical communications is firstly presented. On this basis, several constituent pieces are addressed in the subsequent sections, covering (i) interconnection options to get access to dedicated PPDR networks and SDPs from commercial networks, (ii) interworking solutions for the integration of broadband and narrowband legacy PMR services, (iii) the interconnection of deployable systems, (iv) the use of satellite communications and (v) the identification of the different types of connectivity services and frameworks for building the IP-based backbones and facilitating the interconnection of separate PPDR networks. Finally, and building on many of the previous constituent pieces, this section is concluded with the description of an MVNO-based delivery solution for mobile broadband PPDR communications, which is one of the approaches currently gaining more momentum as a viable short-term solution.

5.6.1 Reference Model for a Critical Communications System

A reference model for the complete architecture of a critical communications system (CCS) is specified by ETSI in Ref. [65]. A CCS is defined as the whole system that provides critical communications services to users in several professional markets (e.g. PPDR, railway, utilities). As such, the CCS covers the complete communications chain, including terminal, access network, core network, control rooms and applications. The reference model developed by ETSI establishes various interfaces and reference points that comprise the overall architecture together with a brief outline of some of the most important services that the architecture supports. The ETSI reference model for a CCS is depicted in Figure 5.6.

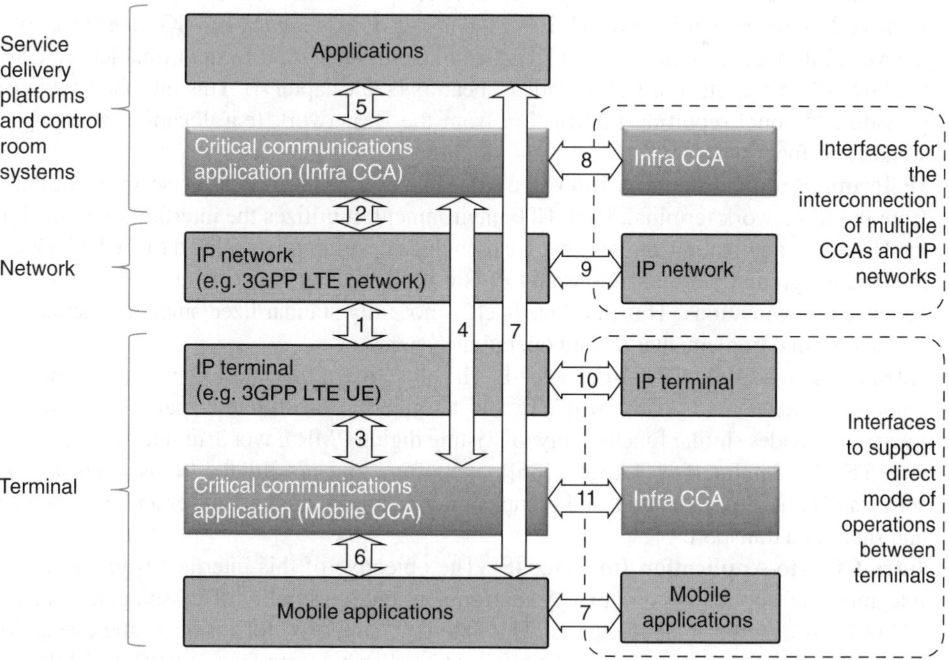

Figure 5.6 Reference model for critical communications systems specified by ETSI.

The central element of the architecture is the critical communications application (CCA), which can be seen as the SDP providing professional communications services to critical communication users. The CCA includes capabilities on the terminal side (Mobile CCA) and on the infrastructure side (Infra CCA). The CCA contains both application-related services (e.g. registration of users, affiliation to groups, call services, etc.) and control functionality (e.g. ability to set up bearers with the required characteristics to communicate with the terminal units and control the levels of priority of the various bearers in the access subsystem where this control is available). The CCA provides services to additional (third-party) applications, for example, to provide group addressed services or prioritized access services. These applications can reside in or be distributed across both terminals and infrastructure (e.g. control room applications).

One CCA may be connected to more than one broadband IP network. The broadband networks may be of the same type (e.g. multiple 3GPP LTE networks). The broadband networks may also be of mixed network types, such as a mixture of 3GPP LTE and Wi-Fi networks that provide service to the same CCA. Multiple CCAs may also share the same broadband IP access network. Therefore, there can be a many-to-many relationship of CCAs and broadband IP access networks. The reference model also considers the support of direct mode of operations (DMO) between terminals. The interfaces illustrated in Figure 5.6 are:

- **IP Core Network-to-IP Terminal Interface (1)**. This interface is specified according to the network protocols of the underlying IP network. If the underlying network is LTE, it consists of the 3GPP specified LTE UE-to-EPC interfaces.
- **Infra CCA-to-IP Core Network Interface (2)**. The objective of this interface is to allow interworking between an Infra CCA and IP core networks from different manufacturers. This interface is specified according to the network protocols of the underlying IP network. If the underlying network is an LTE EPC, it consists of existing Rx and SGi interfaces, plus the MB2 interface developed in the LTE Group Communication System Enablers (GCSE) to allow use and control of LTE broadcast bearers (see Chapter 4). This interface may also provide additional reporting information from the IP network (e.g. location, charging or some other function).
- **IP Terminal-to-Mobile CCA Interface (3)**. This interface relies on the services available from the IP network terminal. In an LTE environment, it utilizes the interfaces provided by the UE to any application and may evolve to include developments related to the LTE GCSE features for group communications and Proximity Services (ProSe) features for device-to-device communications. This interface itself is not fully standardized since it is dependent on the terminal implementation and operating system.
- **Infra CCA-to-Mobile CCA Interface (4)**. The objective of this interface is to allow interoperability between a CCA infrastructure and terminals from different manufacturers. This interface provides similar functionality to existing digital PMR Layer 3 air interface messages (e.g. TETRA Air Interface Layer 3 [66]), supporting, but not limited to, user registration, set-up and control of individual and group communications, media transfer and management and short data transport.
- **Infra CCA-to-Application Interface (5)**. The objective of this interface is to allow easy integration of applications in a CCA environment and portability of those applications to CCAs from different manufacturers. This interface supports, for instance, the dispatcher functionalities. This interface may be similar to a dispatch interface in existing PMR systems, extended to support multimedia capabilities.

- **Mobile CCA-to-Mobile Application Interface (6)**. The objective of this interface is to allow easy integration of mobile applications in a CCA environment and portability of those applications between terminals from different manufacturers.
- **Application-to-Application Interface (7)**. Some components of this interface may be defined by standards for specific applications that require generic formats to ensure interoperability between mobile applications and control room application (e.g. geo-location representation, video format, vocoders).
- **Inter-CCA Interface (8)**. The objective of this interface is to allow interoperability and interworking between CCSs. This interface supports interconnection of communications between users operating on different CCSs. This interface should also support mobility of users between different CCSs.
- **Core Network-to-Core Network Interface (9)**. This interface is determined by the underlying core IP network. If the underlying network is an LTE EPC, it makes use of 3GPP standard interfaces. This interface provides support and control of mobility and roaming of terminals between different core networks.
- **IP Terminal-to-IP Terminal Interface (10)**. This interface is determined by the terminal technology. If the terminals are LTE UEs, this interface will be a standard 3GPP interface, defined under the ProSe specifications in 3GPP Release 12. This interface supports direct communications between terminals as well as between terminals and relays (e.g. UE-to-UE Relay, UE-to-Network Relay).
- **Mobile CCA-to-Mobile CCA Interface (11)**. The objective of this interface is to provide control of direct CCA services between two or more terminals without any infrastructure path. If the terminals are LTE UEs, it relies on underlying services defined by 3GPP ProSe.

The CCS reference model also considers the support of interworking with legacy PMR networks through the definition of an interface between the CCS and existing legacy systems (interface 8b shown in Figure 5.7). This interface is intended to support communications interworking between users operating on a CCS and on a legacy PMR system. This interface is anticipated to support at least a minimum set of features for the interconnection of TETRA, TETRAPOL and P25 systems: individual calls, group calls and short data services. These interfaces can be based on existing Inter-System Interfaces (ISI) already specified for some PMR technologies (e.g. ETSI TETRA ISI, P25 ISSI). Some further details on these ISI are provided in Section 5.6.3.

Further details on the CCS reference model can be found in Ref. [65]. In addition to this reference model, ETSI is also standardizing the architecture and protocols for interface (4) between the Mobile CCA and Infra CCA with the aim to support a generic mission-critical service equivalent to the existing narrowband technologies over LTE [67].

5.6.2 *Interconnection to Commercial Networks*

The interconnection of commercial networks to dedicated PPDR SDPs is a central element for the implementation of the hybrid solutions described in Section 5.5. This section describes the different interconnection options from a network architecture perspective, illustrating the main involved network elements and interfaces. Additionally, some considerations on the

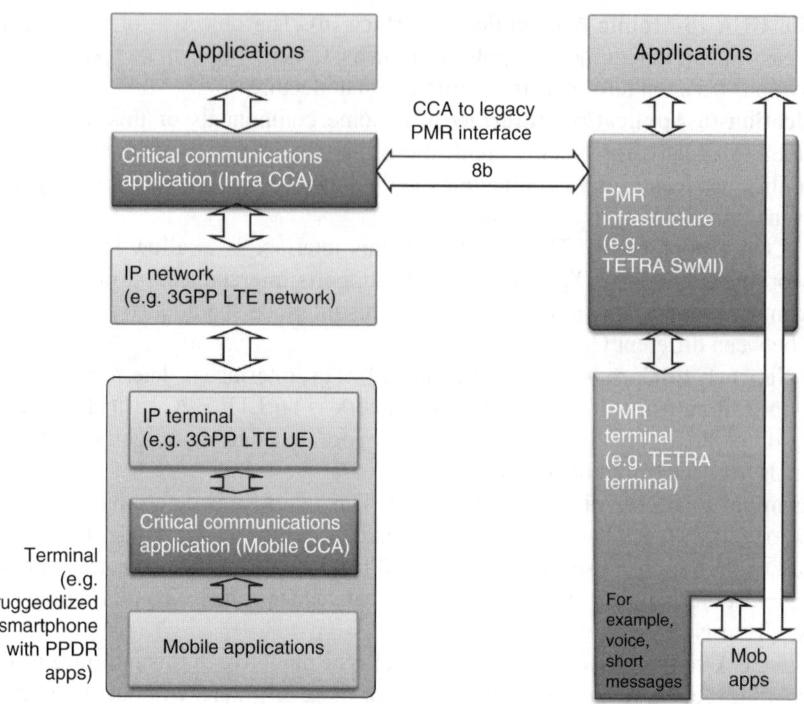

Figure 5.7 Interworking approach between ETSI CCS and legacy PMR systems.

suitability of the different interconnection approaches to support critical communications are given. The interconnection options can be categorized under three distinct approaches depending on the services that the MNO provides through the interconnection.

5.6.2.1 Interconnection Based on Private Connectivity Services

The approach based on private connectivity services can leverage the solution in use nowadays for MNOs to offer private intranet services to business and corporate customers. With this solution, a (dedicated/private) PPDR SDP is reachable to mobile terminals gaining IP connectivity through a commercial network without traversing the public Internet. The delivery of private connectivity services is commonly done through the assignment and configuration of a private Access Point Name (APN). The APN is the parameter used in LTE (as well as in UMTS/GPRS) to identify the external network to which the UE is gaining IP connectivity access (see Section 4.2.2 for more details on the LTE service model). By using a private APN, all 3G/4G traffic leaving a UE can be routed to an IP endpoint at the private/corporate network (instead of the public Internet). Access to the private network is restricted to users who have a subscription for the corresponding APN. Nowadays, private APNs are common offerings from most mobile operators. An alternative approach to a private APN is the use of a virtual private network (VPN) solution over the public Internet (i.e. secure communications between a VPN client installed in the mobile device and a VPN server at the

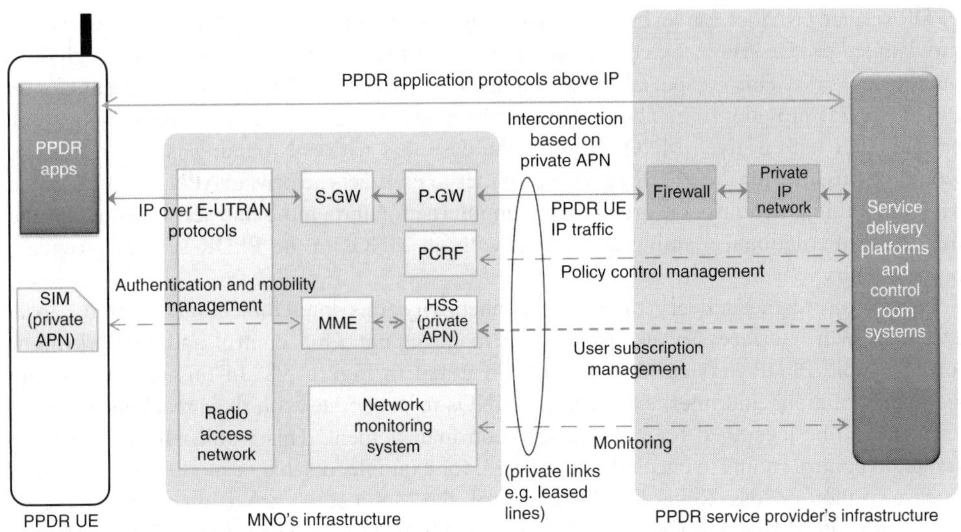

Figure 5.8 Interconnection solution based on private connectivity services (private APN).

company/organization network). Nevertheless, there are several benefits to using a private APN with mobile cellular devices [68]:

- External company/organization infrastructure is exposed only to the provisioned devices and not visible from the public Internet.
- All traffic from/to devices can be forced to traverse the corporate network, avoiding VPN deficiencies or uses in which the VPN allows split tunnelling (e.g. part of the traffic is not routed through the VPN) or do not offer an always-on VPN.
- The device itself is protected from attacks from other users on the cellular network as only other devices on the APN can route traffic to that device.
- Low-level malware such as rootkits that can bypass the VPN enforcement cannot bypass the APN and so will be easier to detect with corporate monitoring services.

In order to deploy a private APN, the interested organization should procure and provision an APN from a mobile operator and obtain SIM cards that are exclusively configured to use the private APN for their mobile data connection. Figure 5.8 shows a schematic of the network architecture based on the use of a private APN between an MNO's infrastructure and a dedicated/private infrastructure of a PPDR service provider hosting a PPDR SDP and other elements needed for the delivery of PPDR communications services. The interconnection between the commercial MNO and the PPDR operator is based on dedicated resources (e.g. leased lines), thus preventing risks related to security threats and/or traffic congestion. End-to-end security services (e.g. end-to-end encryption) remain within the responsibility of the PPDR organizations that control the PPDR applications in the terminals and SDPs. Nevertheless, authentication and mobility management are under the full responsibility of the MNO, which owns and manages the HSS and SIM cards. Therefore, PPDR organizations have to trust the MNO with regard to network access security. On the plus side of this approach, and given that the MNO is who undertakes authentication and mobility management, is that the

PPDR service provider can leverage any roaming agreement of the MNO and get access to the provisioned private APN service even when the PPDR terminals get connected through other mobile networks. This is especially relevant in the case of national roaming agreements for the PPDR service provider to be able to have multi-network access with the contract of a single private APN with a given MNO. Even in the case that national roaming is not in place, it would be possible for the PPDR service provider to contract the private APN service from an MNO in a foreign country or directly from an international roaming hub service provider that may hold international roaming agreement with all MNOs in the PPDR service provider's home country.

In addition to the delivery of private connectivity, this interconnection solution can be extended with interfaces for user/subscriber management, QoS control and network monitoring by the PPDR service provider, as illustrated in Figure 5.8. In this regard, existing technical solutions and interfaces used by MNOs to cooperate with their resellers and sales partners can be leveraged for user subscription management. This would allow PPDR personnel in control rooms to check the status and, for example, disable or modify a subscription in case of theft or loss. Regarding QoS control, the MNO could provision access to the Rx interface of its Policy and Charging Control (PCC) platform (see Section 4.2.3) so that the PPDR SDPs could use the PCC capabilities to send service-related information, including resource requirements, to be enforced in the IP connectivity services (e.g. activation of dedicated EPS bearers with specific QoS parameters indicated by PPDR application servers). However, this is not a common offer by MNO nowadays so that special arrangements would be needed between the MNO and the PPDR service provider to deploy such interface. Similarly, the ability to monitor the MNO's network for outages or other incidences is not a typical offering associated with private connectivity services. Therefore, a special arrangement with the MNO to get access to some information within its network monitoring system would also be required in this case.

5.6.2.2 Interconnection Based on Roaming Services

Another possible interconnection approach is based on the delivery of roaming services. In contrast to the private APN solution, in this case, the PPDR service provider deploys some LTE core network elements within its infrastructure, as illustrated in Figure 5.9. In particular, a P-GW and an HSS are under the control of PPDR service provider, along with the SIM cards used in the PPDR terminals. In this approach, standardized LTE roaming interfaces can be used for the interconnection of the two infrastructures (i.e. S6a and S8 interfaces, as described in Chapter 4). With this solution, authentication and mobility management are performed collaboratively, though the MNO does not hold the permanent secret keys within the HSS and SIM cards. This solution facilitates the deployment of the user/subscriber management and QoS control interfaces, which reside within the PPDR service provider infrastructure. Nevertheless, as in the case of private APN services, the ability to monitor the visited network is not a usual arrangement in roaming-based interconnections. Therefore, if this were a mandatory requirement demanded by the PPDR organizations, a special arrangement with the MNO would be necessary. Furthermore, the PPDR service provider has to administer its own national and international roaming agreements since the roaming agreements of the MNO cannot be leveraged.

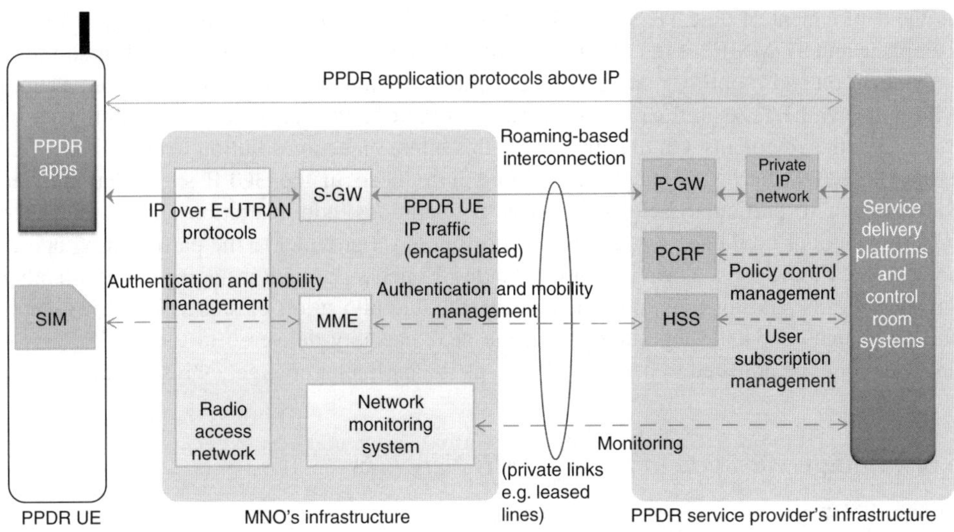

Figure 5.9 Interconnection solution based on roaming services.

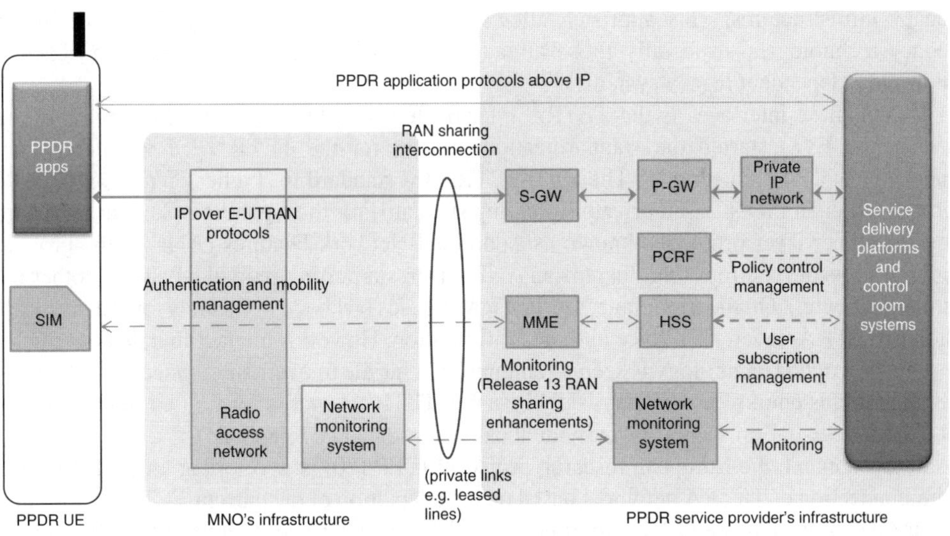

Figure 5.10 Interconnection solution based on RAN sharing services.

5.6.2.3 Interconnection Based on Active RAN Sharing Services

Finally, the third interconnection approach, illustrated in Figure 5.10, is based on the exploitation of RAN sharing services. It assumes that the PPDR service provider has deployed a full EPC and use the LTE RAN capacity deployed by one or several MNOs. Specifically, the interconnection solution depicted in Figure 5.10 corresponds to the MOCN solution supported in the 3GPP specifications, as described in Section 4.8. In this case, the interconnection is based on the deployment of the LTE S1 interface between the two infrastructures (i.e. S1-U for the user plane and S1-MME for the data plane, properly secured with NDS/IP). With this solution, authentication and mobility management are entirely handled by the PPDR service provider within its own EPC. As in the previous case, the PPDR service provider has to administer its own roaming agreements. On the other hand, this interconnection solution would allow leveraging the RAN sharing enhancements being introduced in the 3GPP specifications (see Section 4.8). In particular, selective access to operations, administration and maintenance (OAM) functions could be provided by the hosting RAN provider to the participating operators (i.e. the PPDR service provider among them) for network monitoring, together with other enhanced RAN sharing functions such as the flexible allocation of shared RAN resources and on-demand capacity negotiation (e.g. the peak capacity delivered to the PPDR service provider could be adjusted to better fit the needs if a particular emergency response).

5.6.3 Interconnection to Legacy PMR Networks

Current PMR systems are expected to continue being the central platform for mission-critical voice for many years. As illustrated in Figure 5.7, the interconnection approach for communications interworking under consideration in the ETSI reference model for CCSs is based on the deployment of an interface between the new PPDR SDPs and the core networks of the legacy infrastructures. This approach allows leveraging the existing interfaces within the legacy technologies, especially those already developed for the interconnection between multiple independent legacy systems.

One of these interfaces is the TETRA Inter-System Interface (TETRA ISI). Back in the 1990s, the ETSI started the standardization process for the TETRA ISI to interconnect independent TETRA networks. The ETSI TETRA ISI standard [69] relies on the QSIG/PSS1 protocol (i.e. an ISDN-based private signalling standard) for the signalling plane to exchange a set of upper layer protocols, known as Additional Network Features (ANFs), that provide different sets of functions (e.g. there is an ANF set for supporting individual calls, another for group calls, etc.). On the user plane, the TETRA ISI uses 64 kb/s E1 channels for the transmission of TETRA-coded user voice and data information. However, the fact that this is actually a circuit-switched technology is seen as an important hurdle to overcome. Indeed, the TETRA ISI standard is employed nowadays only by a few TETRA vendors for limited functionalities (i.e. basic registration scenarios, individual call, short data and telephone).

In this context, the European research project ISITEP [70] is developing a new ISI for the interconnection of TETRA networks based on the adaptation of the current TETRA ISI so that it can be deployed over an IP transport network. The new interface is referred to as IP ISI and, in addition to enable the interconnection of TETRA networks, is intended to be used also for the interconnection between TETRA and TETRAPOL networks and between TETRAPOL networks. While the ultimate goal of the ISITEP project by pursuing the development of the IP ISI is to facilitate the interconnection of the different national TETRA and TETRAPOL PPDR networks deployed across European countries, the consolidation of such IP ISI is expected to facilitate the integration of current narrowband PPDR networks with the forthcoming all-IP PPDR SDPs as well. The new IP ISI protocol is based on the Session Initiation Protocol (SIP), which is the current de facto standard for Voice over IP (VoIP) communications. The approach adopted in ISITEP is to allow that the already standardized ETSI ISI ANFs can be exchanged through SIP messages and use the Real-time Transport Protocol

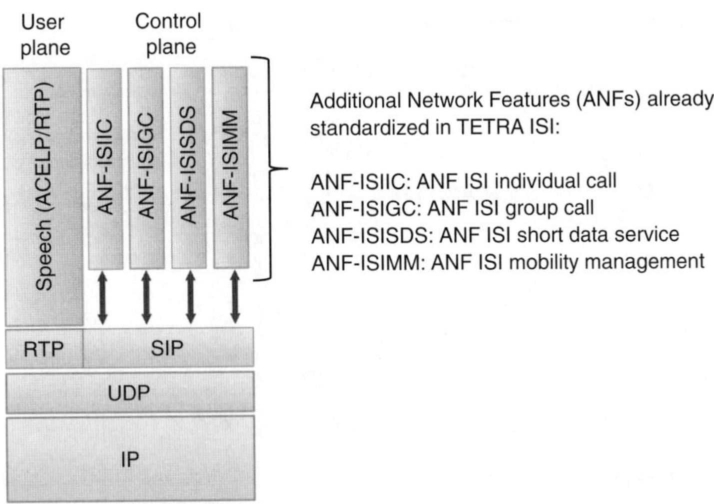

Figure 5.11 IP-based Inter-System Interface (IP ISI) for TETRA and TETRAPOL networks. From Ref. [71].

(RTP) for the voice traffic encoded with the corresponding codecs. A simplified view of the IP ISI protocol stack is depicted in Figure 5.11.

A similar approach is envisioned for P25 systems. The P25 standard has already defined an ISI known as Inter-RF Subsystem Interface (ISSI). It aims to connect disparate P25 networks, regardless of vendor. Its first commercial implementation dates back to 2010. P25 ISSI protocol is already based on IP, also relying on the IETF SIP for the control plane and the RTP to convey P25 voice frames [72]. Besides the connection between legacy P25 systems and new IP-based PPDR SDPs, another applicability of the P25 ISSI is found in the development of a standard for delivering P25 services directly to LTE terminals. This standard is being specified by the TIA Mobile and Personal Private Radio Standards Committee (TIA/TR-8) in the United States, and it is commonly referred to as P25 PTT over LTE (P25 PPToLTE). A preliminary architecture of this solution is depicted in Figure 5.12 [73]. It comprises a P25 PTToLTE server located at application level, within an SDP, reachable as an external network from a (commercial or dedicated) LTE network. The access point into the P25 network is via the standard ISSI protocol between an ISSI gateway (GW) in the P25 network and the P25 PTToLTE server. Then, a new protocol named Subscriber Client Interface (SCI) is used to connect the P25 PTToLTE server to a P25 PTToLTE client in the mobile terminal through the LTE network. The overall solution also incorporates the already standardized P25 Console Subsystem Interface (CSSI) protocol for the interconnection with control room dispatch consoles. Voice transmission relies on the standard P25 vocoders (voice coder/decoder) and allows up to AES 256-bit encryption services to be used at the LTE terminal. In this way, the P25 end-to-end encryption is preserved even when terminated in the LTE terminal.

5.6.4 Interconnection of Deployable Systems

As discussed in Section 5.3.1, deployable systems can be central for achieving cost-efficient network footprints. Furthermore, the use of deployables is instrumental for network restoration, network extension and remote incident response.

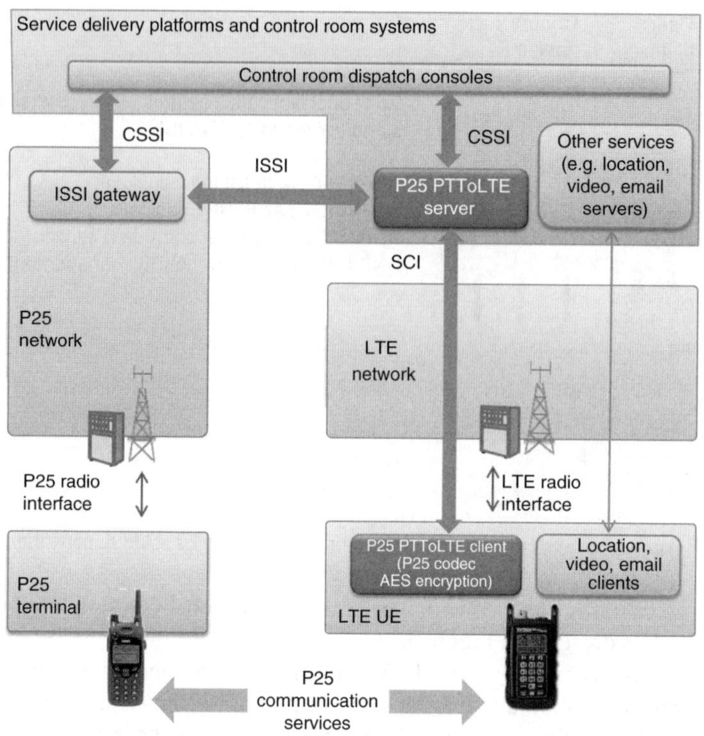

Figure 5.12 Interworking solution to extend P25 services to LTE terminals.

Deployable systems can be classified under two categories: cell on wheels (COWs) and system on wheels (SOWs). On the one hand, COWs typically include a BS (e.g. LTE eNB) along with one or more backhaul transports (such as microwave or satellite). COWs require connectivity to the core network where the LTE EPC and SDPs are located to be able to provide services to the mobile terminals under the coverage of the COWs. On the other hand, SOWs are fully functional systems that can act without backhaul connectivity. Indeed, SOWs are more complex systems (and consequently more expensive) than COWs. As a general approach, SOWs are more appropriate in rural environments and in disaster areas with heavy volumes of local traffic, where broadband backhaul connectivity is an issue. In turn, the use of COWs is envisaged in smaller incidents in urban and suburban environments, where better connectivity to the core network can be guaranteed. Both COWs and SOWs can adopt different form factors depending on their capabilities and design constrains (e.g. size, functionality, power supply, deployment time, environmental protection, etc.) and can be installed in trailers, mobile vehicles or transportable rack-mounted chassis.

Figure 5.13 depicts the main functional communications-related components integrated within COWs and SOWs together with their remote connectivity needs. As shown in the figure, SOWs include an eNB and a light implementation of the LTE EPC (i.e. HSS, P-/S-GW, MME, PCRF functions) to provide IP connectivity services to LTE devices. Additionally, SOWs include a local SDP to be able to offer services without the need to have wide area connectivity to remote SDPs. Optionally, SOWs could be supplemented with elements such as local Wi-Fi connectivity and PMR radio bridges to enable the communication with Wi-Fi-enabled devices

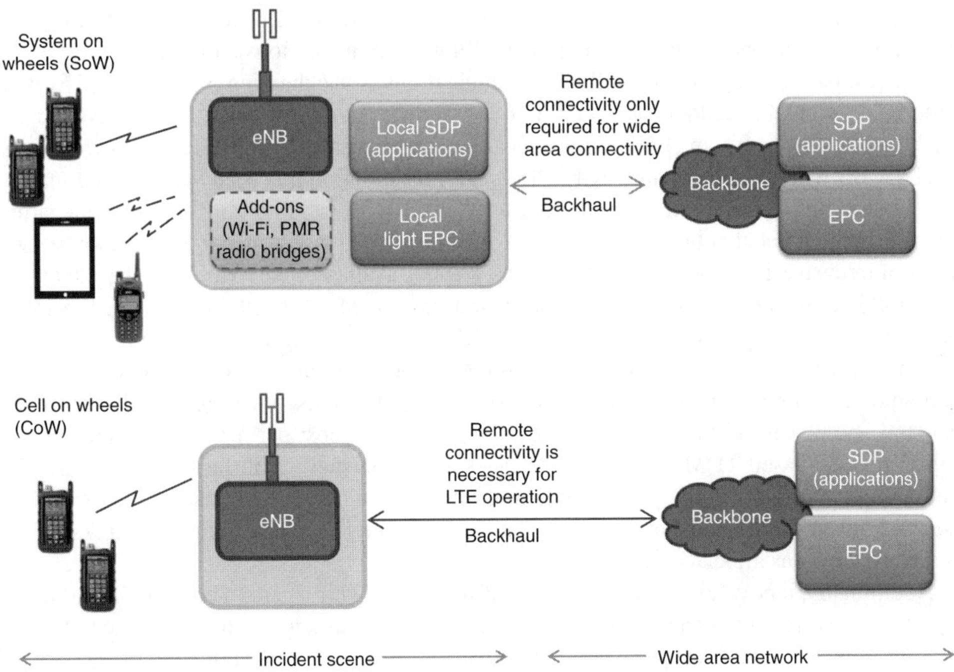

Figure 5.13 Components and remote connectivity requirements of SoWs and CoWs.

(e.g. tablets/laptops) and legacy PMR terminals (e.g. TETRA/P25 handheld radios) on the scene. In the case of COWs, the deployable typically consists of the eNB functionality alone. Therefore, wide area connectivity to the remote EPC and SDP is necessary for its operation.

Solutions for COWs and SOWs face important design challenges. One of these challenges is the 'smooth' integration and coordination of the radio settings of the deployed equipment with those used in neighbouring LTE sites (which may overlap in coverage) as well as other in other deployables that may operate in the local radio vicinity. This challenge is especially relevant if deployables are expected to use the same frequencies used in the WAN (e.g. use of a single PPDR dedicated 10+10 MHz allocation). In this regard, advanced inter-cell interference management capabilities of LTE, together with other features such as self-organization and handover management, are fundamental to support this scenario [74, 75]. Another important technical challenge is the integration of the different tiers of connectivity services and applications when SOWs are used. The target here is to achieve a consistent and interoperable operation of the local EPCs and SDPs within the deployables and the remote EPCs and SDPs so that all the components together work as a single wide area distributed system. In this context, distributed peer-to-multi-peer architectures for the application layer are being tested in FirstNet trial systems in the United States [76].

5.6.5 Satellite Backhauling and Direct Access

Satellite communications provide a unique and important method for PPDR to plan around the hazards of earth-based infrastructures that can be susceptible to all manners of natural and man-made catastrophes. Satellite communications can also be used in areas where existing terrestrial infrastructure is insufficient or overloaded. This turns satellite communications into an important component within the complete PPDR communications toolkit [77].

One primary application of satellite communications is backhauling of COWs and SOWs. Indeed, solutions for cellular backhauling over satellite have been around for many years to backhaul 2G BS traffic in some remote locations as well as used in PMR deployables (e.g. TETRA BS with satellite connectivity). However, its adoption has been very limited mainly due to the high operational expenditures associated with satellite connectivity, relegating satellite cellular backhauling to a sort of last-resort or worst-case transport solution only used when microwave links or wired alternatives are not available. Nevertheless, the advent of high-throughput satellite (HTS) technology and the use of IP-based TDMA satellite interfaces instead of single channel per carrier (SCPC) are changing the economics of satellite backhaul [78]. HTS operating in the Ka and Ku bands are able to exploit frequency reuse technologies along with multiple tightly focused spot beams, yielding greatly improved spectral and power efficiencies that are bringing down the cost per megabit per second. On top of this, IP-based TDMA satellite interfaces allow the sharing of the satellite bandwidth among many sites (i.e. capacity aggregation with trunking gains), allocating bandwidth on demand based on the real-time requirements at each site (in contrast to SCPC links where fixed capacity was allocated per site).

Additionally, innovation in the terrestrial infrastructure with regard to how satellite connectivity services are provisioned, customized and managed is anticipated to contribute to further driving down the costs. In particular, the advance towards a more efficient integration of satellite and terrestrial components can be fuelled by the use of technologies such as network functions virtualization (NFV) and software-defined networking (SDN). Enabling NFV into the satellite communications domain provides the operators with the tools and interfaces to establish end-to-end fully operable virtualized satellite networks over shared satellite communications platforms, which can be customized and offered to third-party operators/service providers (e.g. a PPDR operator could manage 'its' virtual satellite communications platform for the interconnection of its deployables). In turn, enabling SDN-based federated resource management between the terrestrial and satellite domains paves the way for a unified control plane that allow operators to jointly manage and optimize the operation of the overall communications chain encompassing the satellite and terrestrial equipment. In this respect, the European research project VITAL is working towards these objectives [79].

Figure 5.14 illustrates the applicability of satellite communications for the interconnection of COWs. The eNB is connected with the remote EPC through the satellite connection, which is used to support the S1 interface (see Chapter 4). The fact that the S1 interface is an IP-based interface enables the use of different bandwidth optimization techniques like header compression and performance-enhancing proxies (PEPs). Satellite terminals with transmit power less than 10W and antenna dishes of 75cm in diameter are common settings. State-of-the-art satellite interfaces are DVB-S2 and its upcoming evolution DVB-Sx for the forward link (from the hub/GW to the satellite terminal) and DVB-RCS2 for the return link. Typical achievable speeds can be up to 12Mb/s downlink and 3Mb/s uplink, being also possible to reach 50Mb/s downlink and 20Mb/s uplink in more demanding settings [78]. Figure 5.14 also shows that the satellite capacity can be shared among a number of COWs and SOWs as well as used in fixed sites, which is an example where the provisioned satellite connectivity can be used either for traffic overflow of congested sites or for improved resilience of some critical sites.

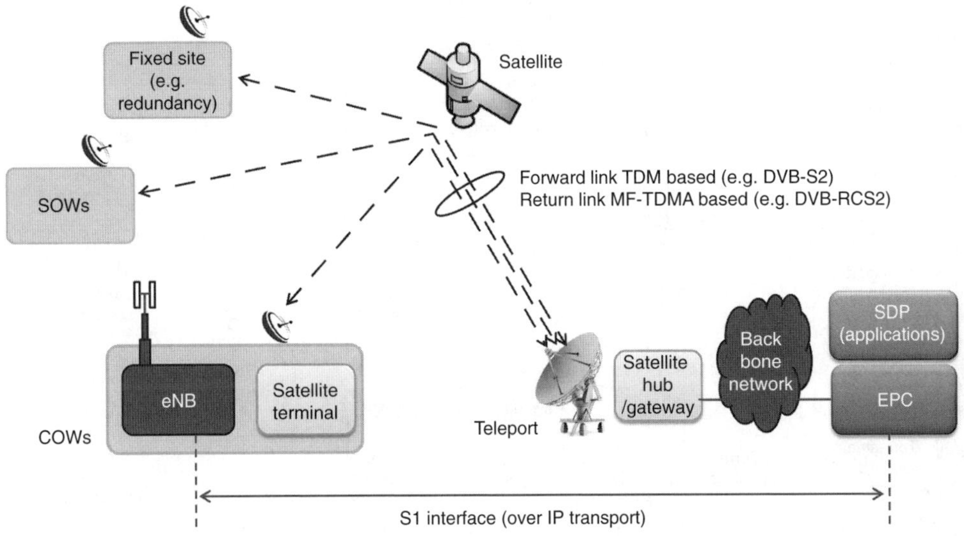

Figure 5.14 Deployment of S1 interface over a satellite connection.

A comprehensive generic architecture considering satellites as a transport solution in emergency communications has been developed by ETSI under the concept of Emergency Communication Cell over Satellite (ECCS) [80]. ECCS is a concept understood as a temporary emergency communication 'cell' supporting one or more terrestrial wireless standard(s) (e.g. cellular telephony, PSTN cordless, Internet access via Wi-Fi, PMR services), which are linked/backhauled to a permanent infrastructure by means of bidirectional satellite links. An ECCS system is intended to be a quasi-autonomous communications infrastructure in the field that enables communications between users inside and outside the disaster area using different sorts of communications devices. Indeed, the ECCS concept is not specific to communications between PPDR organizations but also applicable to solve the communications needs to the affected persons, victims or any other kind of involved people.

The ECCS architecture is shown in Figure 5.15. Despite the multitude of technical solutions that could be used to implement an ECCS system, the architecture is technology agnostic and mainly identifies the constituent logical blocks. The overall transmission path involves a number of domains, which compose an ECCS communications chain. The domains represent network elements involved in the ECCS communications chain, playing logically neighbouring functionalities in this chain or jointly enabling the provision of a given functionality (e.g. local access). A central role in the architecture is played by the ECCS servers placed on remote locations and the ECCS terminals on the field. These equipment together provide the backhauling of the different wireless services via satellite as well as, if necessary, the interconnectivity and interoperability between different services by means of GW or any other specific equipment (e.g. connectivity in the field between two PSTN cordless terminals can be achieved via a private branch exchange as part of the ECCS terminal). Indeed, different classes of ECCS terminals are likely to coexist due to different design constraints. Some examples are portable yet basic ECCS systems that can be easily packaged in an airborne-cabin-format suitcase, providing voice and data access via satellite; transportable and more powerful ECCS systems packaged in

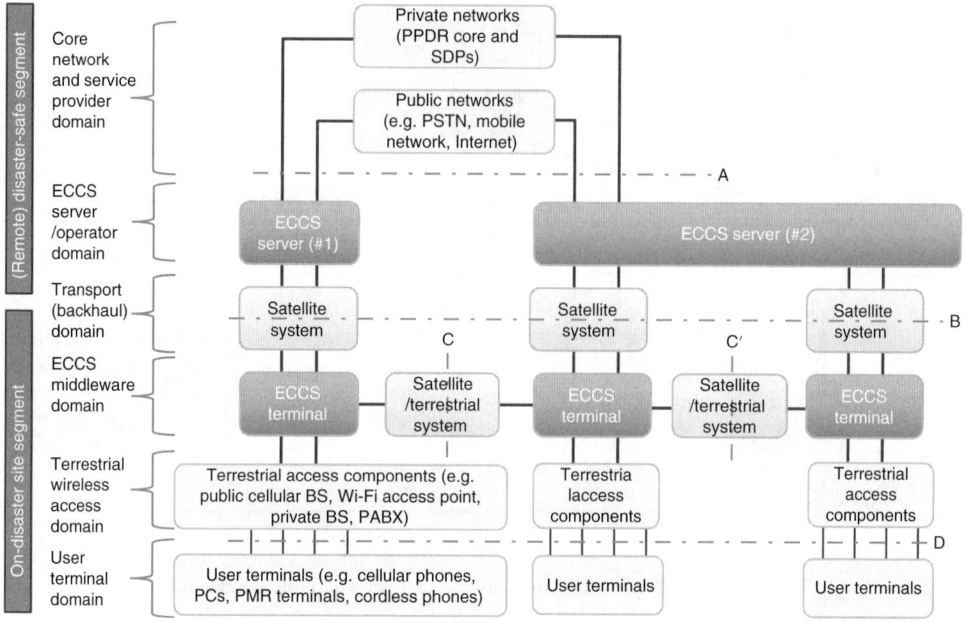

Figure 5.15 Generic architecture for Emergency Communication Cell over Satellite (ECCS).

an airborne-container format or multiple man-carried containers; and mobile ECCS systems with 'on-the-move' access. It is worth noting that the ECCS architecture assumes that an ECCS service provider operates the overall ECCS system and interacts with other actors (e.g. satellite capacity providers, PPDR operators, MNOs, etc.). The ECCS operator acts as a kind of 'concentrator' for a complete and tailored service provisioning, in terms of communications services, content and infrastructure, to the system users and should be their main/single direct interface. Figure 5.15 depicts the four main interfaces identified in the ECCS architecture:

1. **ECCS server–core networks** (A). These are the interfaces needed for the interconnection of the core networks of the different services being offered through the ECCS. For example, the ECCS server could provide access to a PPDR LTE EPC core and SDP, as shown in Figure 5.15.
2. **ECCS server–ECCS terminal** (B). These interfaces are related to the satellite transport connectivity. One or several satellite links, with the same or different satellite operator provider, could be established between ECCS servers and terminals.
3. **ECCS terminal–ECCS terminal** (C) and (C′). This interface enables direct connections between ECCS terminals, via either satellite or terrestrial links. While C′ interconnects ECCS terminals that share the same ECCS server (under the responsibility of the same ECCS operator or service provider), C interconnects ECCS terminals with different servers (and operators).
4. **ECCS terminal–user terminals** (terrestrial wireless) (D). This is the interface or set of interfaces used to provide on the field the terrestrial wireless services (e.g. Internet access via Wi-Fi, voice services via 2G/3G mobile interfaces, PPDR services via PMR interfaces/ LTE interfaces). These services can be considered as a small subnetwork that is connected via a satellite backhaul link to its core network.

Thus far in the previous discussion, the role of satellite communications for PPDR communications has only been addressed as a backhaul solution. However, mobile satellite service (MSS) systems can also form part of the toolkit of PPDR communications means. The fact that MSS systems provide very large coverage areas compared to terrestrial-based systems is a relevant feature that can be exploited in the context of PPDR communications. MSS systems currently in operation (e.g. Iridium, Inmarsat, Thuraya) are able to provide voice and data radiocommunications and access to the Internet. Further, these systems can facilitate access to public and private networks external to the MSS system. In terms of data transmission capabilities, current offerings provide data rates of up to 500 kb/s with flat antennas of the size of a laptop. These MSS terminals can even be carried by an individual. There are other MSS terminals that can be mounted on a ship, an airplane, a truck or an automobile and can deliver Communications-On-The-Move (COTM). An illustrative description of uses and examples of different MSS systems for emergency communications is provided in Report ITU-R M.2149-1 [81]. Recently, solutions that integrate the functionality of a satellite MSS terminal into smartphones have also been launched (e.g. Thuraya SatSleeve for a set of Samsung and iPhone models). The SatSleeve comes in the form of an adaptor that is plugged into the smartphone. Nevertheless, data capabilities of this solution are limited to short messaging, and data services with data speeds are up to 60 kb/s for downloads and up to 15 kb/s for uploads.

Finally, some consideration should also be given to MSS satellite networks that introduce a complementary ground component (CGC). The CGC consists of a core terrestrial network that is to be operated in an integrated fashion with a satellite segment of the network using the MSS frequency band. This network configuration combines the geographical coverage benefits of the satellite with the high throughput of terrestrial mobile communications networks. In Europe, the European Commission granted two European satellite operators the right to operate 2×30.0 MHz (2170–2200/1980–2010) of MSS spectrum (the so-called S-band) using a CGC network. The use of this spectrum asset was granted in 2009 for a period of 18 years and covers all EU member states. Indeed, an S-band satellite segment was launched in 2009 and could already support service launch for a portion of the EU member states. Additionally, the MSS spectrum band is set to become 3GPP compliant and hence can be deployed in Europe leveraging the economies of scale of the LTE ecosystem globally. Therefore, a satellite–terrestrial network on the S-band spectrum is an option to the PPDR sector to implement a dedicated network approach relatively quickly [82]. However, this option is not currently receiving widespread support across the PPDR and other critical communications users across, which is fundamental to move forward such a project [14]. Moreover, the current satellite system is limited in capacity and could not meet the entire demands of the European critical communications sector immediately. The development of the market on a pan-European basis would require a new high-performance satellite to be manufactured and launched.

5.6.6 Interconnection IP-Based Backbones

The interconnection of the multiple and diverse components of the overall PPDR communications system (e.g. radio sites, data centres hosting the EPC and SDPs, MNO's data centres with the commercial network equipment, PPDR deployables, emergency control centres and PSAPs, interconnection of regional/national PPDR networks, etc.) requires the deployment of a vast backbone network infrastructure. This backbone network or combination of networks is likely to combine different transmission technologies, from fibre to satellite links, and designed and operated with the sufficient redundancy and protection mechanisms to guarantee the high availability expected in this kind of critical infrastructures. This required interconnection infrastructure can be fully or partly owned by the government or public entities established for the delivery of PPDR communications, though it can also rely on the use of interconnection services and facilities provisioned by private carriers.

In this context, fibre has become the predominant technology for backbone networks, though some niche applications for microwave and satellite are in place [83]. Whereas in the 1990s backbone networks used dedicated technologies, such as Asynchronous Transfer Mode (ATM) and Synchronous Optical Networking (SONET) and Synchronous Digital Hierarchy (SDH), most of today's (Internet) backbone networks are built on the Ethernet suite of standards, which was originally designed for offices and data centres. The speeds of 1 and 10 Gb/s are now those most commonly used, with 100 Gb/s becoming increasingly available. Ethernet has become dominant because the high volumes used in data centres created a high-volume market that overshadowed the demands generated by the traditional telecom voice market. The Ethernet standard was expanded to support the requirements of carrier networks. SONET/SDH still remains central in backbone and interconnection for telephony services. In addition, modern fibre networks offer an additional level of flexibility through optical routing. Optical routing allows the network to route wavelengths irrespective of the content in the wavelengths.

Connectivity services offered nowadays by commercial carriers are typically categorized according to the OSI layer at which the interconnection service provider's systems interchange reachability information with customer sites:

- **L3 connectivity**. L3 is essentially synonymous of IP, so that L3 connectivity services are indeed IP connectivity services. This type of interconnection service is commonly referred to as IP virtual private network (IP VPN). State-of-the-art IP VPN services provide the performance, reliability and security of leased-line networks with the any-to-any scalability and flexibility of an IP network. They are usually based on the Multiprotocol Label Switching (MPLS) technology [84].
- **L2 connectivity**. VPNs provisioned using L2 technologies such as Frame Relay and ATM virtual circuits (VC) have been available for a long time [85], though over the past few years Ethernet VPNs have become more and more popular. Ethernet VPN, also sometimes referred to as Ethernet L2VPN, LAN VPN or Global LAN, represents the natural evolution for L2 interconnection technologies as well as for the delivery of dedicated capacity in the form of Ethernet private lines (EPL). As connectivity services provided by Ethernet VPN are below L3, the end users can retain exclusive control of their IP architecture, IP version, addressing scheme and routing tables. Carrier Ethernet is the marketing term being used by the Metro Ethernet Forum (MEF), which is the body leading the architectural, service and management technical specifications and oversees the certification programmes to promote interoperability and deployment of Carrier Ethernet worldwide [86]. In the case of Carrier Ethernet, the user-to-network interface is standard Ethernet (e.g. 10/100 Mb/s, 1 Gb/s). Indeed, in the so-called Ethernet VPN services, the carrier network behaves as an L2 Ethernet switch that might offers multipoint-to-multipoint connectivity between different customer sites.
- **L1 connectivity**. This provides a physical circuit between two interconnected sites. PDH/SDH is a suitable networking technology for delivering L1 connectivity through point-to-point digital circuits. This type of service is also commonly referred to as (traditional) leased

line, though the term leased line can also refer to other types of connectivity services based on the provision of private/dedicated capacity (e.g. renting private dark fibre could be an example of a (modern) leased line). Indeed, the main distinction of a leased line is that the line is rented from a third party rather than self-deployed. While private line services based on PDH/SDH technologies have been the predominant solution in the past, this type of service is also steadily being replaced by Ethernet (e.g. EPL) as telecommunications networks transitions towards all IP.

- **L0 connectivity**. With the increasing deployment of dense wavelength-division multiplexing (DWDM) equipment in the wide area backbone as well as in metropolitan area networks, ANSI and ITU-T initiated standards activity to define a common method for managing multiple wavelength systems – the Optical Transport Network (OTN). OTN has been designed to provide functionality of transport, multiplexing, switching, management, supervision and survivability of optical channels carrying a variety of client signals (e.g. Ethernet, SDH).

When it comes to interconnection solutions that may involve several carriers and service providers, the GSM Association (GSMA) has specified a managed IP network solution named IP exchange (IPX). IPX is a global, private, managed IP network solution that supports end-to-end QoS and the principle of cascading interconnect payments (i.e. all carriers in the interworking value chain are expected to receive a fair commercial return). While one of the motivations behind the specification of the IPX framework was the interconnection of mobile networks, the IPX service is not restricted to MNOs but intended to accommodate the connection to and between other communities such as fixed network operators (FNOs), Internet service providers (ISPs) and application service providers (ASPs). Collectively, all these types of potential IPX users are simply denoted as service providers within the IPX model.

IPX may be used to interconnect any IP service, either standardized (e.g. VoLTE services) or not. The IPX offers three standard modes of interconnection, which service providers are free to choose on a per-service basis: (1) bilateral transport only, where the IPX only provides transport at a guaranteed QoS between two service providers; (2) bilateral service transit, where the IPX provides QoS-based transport as well as service-aware functions; and (3) multilateral hub service, where the IPX provides QoS transport and service-aware functions to a number of interconnect partners via a single agreement between the service provider and IPX.

The IPX network model is illustrated in Figure 5.16. The IPX is formed from separate and competing IPX providers (or IPX carriers). Service providers are connected to their selected IPX provider(s) using a local tail interface (e.g. EPL). Service providers may be connected to more than one IPX provider. IPX providers connect to each other via peering interfaces. All parties involved in the transport of a service (up to the terminating service provider border GW/firewall) are bound by end-to-end SLAs. A common Domain Name Service (DNS) root database supports domain name resolution within the private network and E.164 NUmber Mapping (ENUM) capability to assist with the translation of telephone numbers to IP-based addressing schemes. This root database may be used by all IPX parties. IPX proxy elements within IPX provider networks can be used to support service awareness and interworking of specified IP services and make it possible to use cascading interconnect billing and a multilateral interconnect model (i.e. hub functionality within IPX proxies). Common rules regarding IP addressing, security, end-to-end QoS and other guidelines needed to ensure interoperability among service providers connected to the IPX backbone are described in the technical specification

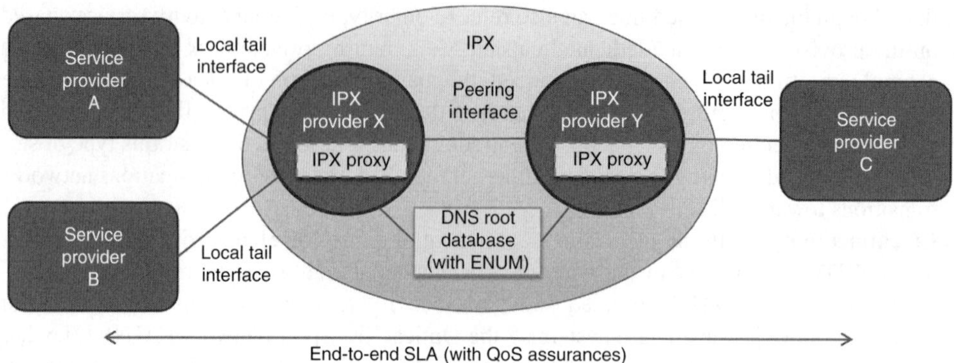

Figure 5.16 GSMA IPX model.

IR.34 [87]. Security guidelines established in IR.34 are complemented by another document, IR.77 [88], which concentrates on providing the detailed security requirements for IPX providers as well as service providers connecting to the IPX backbone.

In practice, a number of international carriers already provide IPX services (e.g. BT Wholesale, Telefonica International Wholesale Services, etc.) although some of them are in a limited form (e.g. support for bilateral transport only). In the context of PPDR communications, the use of a third-party IPX solution to achieve interconnectivity between regional networks was proposed in response to a FCC Notice of Inquiry (NoI) regarding the implementation options of the broadband PPDR network being built in the United States [89]. Currently, the IPX model is being considered, among other options, within the European research project ISITEP for the interconnection of national TETRA and TETRAPOL networks using the new IP ISI described in Section 5.6.3 [90].

5.6.7 Network Architecture for an MVNO-Based Solution

As discussed in Section 5.5.2, a mobile broadband PPDR communications delivery solution based on the MVNO model is the approach currently under consideration as a viable short-term solution in some European countries. Building on many of the network design and implementation aspects described in the previous subsections, this last subsection outlines the main features of an MVNO-based delivery solution developed within the EU research project HELP [37, 58]. A high-level view of the overall system architecture with the key network elements and interfaces is illustrated in Figure 5.17 and described in the following.

The system architecture defined in Project HELP considers a PPDR operator's core infrastructure that consists of IMS functions, application servers and EPC network components (i.e. HSS, PCRF and P-GW), all interconnected by means of a private IP network.

This core infrastructure is used to provide PPDR services to users in the field equipped with LTE-enabled PPDR terminals through a number of commercial LTE networks interconnected by means of standardized 3GPP interfaces (e.g. S8 for data transfer and S6a for signalling transfer, as depicted in Figure 5.17). National roaming agreements are assumed to be in place between the PPDR operator and multiple commercial MNOs. In addition to the use of the commercial capacity, it is considered that the PPDR operator has also deployed

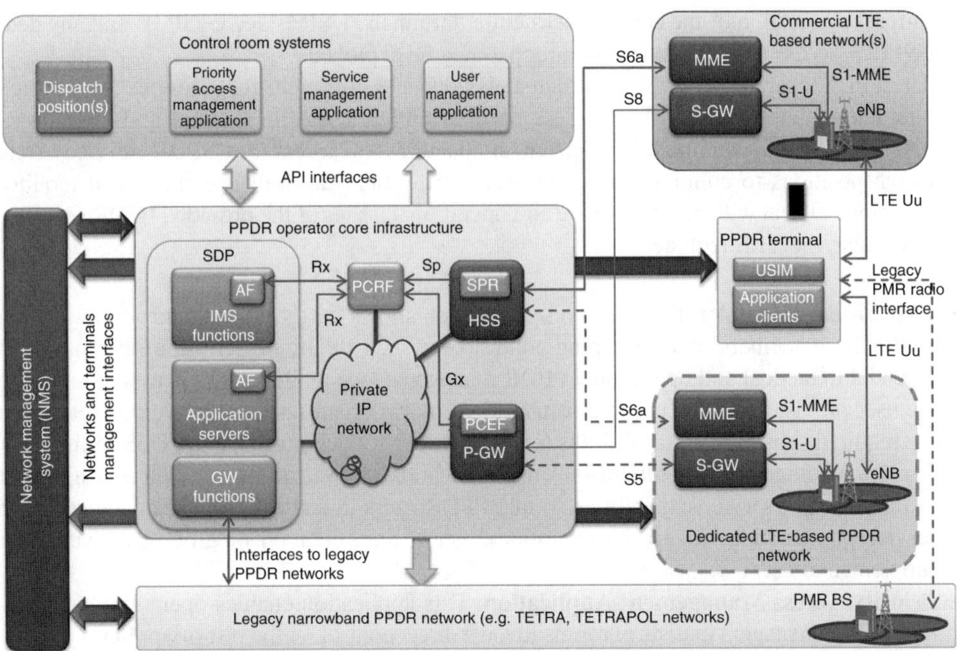

Figure 5.17 Project HELP system architecture for the delivery of PPDR communications.

dedicated LTE access capacity in some specific areas (represented as the dedicated LTE-based PPDR network in Figure 5.17). Hosting the P-GW within the core infrastructure provides a secure access to this critical infrastructure and allows the PPDR operator to fully manage the IP connectivity service (e.g. private APN and IP address allocation). Besides, mobility between multiple networks (commercial or dedicated) without service disruption is facilitated since the P-GW serves as a mobility anchor point for all PPDR traffic. On top of the IP connectivity service, commercial or PPDR-customized mobile VPN solutions would be used to add an additional security layer between the client and server application endpoints within the SDP.

The PPDR operator's core infrastructure also contains a GW for communications interworking with a legacy PMR network infrastructure. In the context of Project HELP, legacy TETRAPOL networks were under consideration, though the proposed solution approach is not specific to the TETRAPOL technology. The GW converts TETRAPOL protocols to SIP and RTP and vice versa, in order to keep the same group call service on both sides.

The PPDR operator controls the procurement and provisioning of the USIM cards used by PPDR subscribers. This approach avoids PPDR users ending up with a number of separate subscriptions to different commercial operators (i.e. handling multiple USIMs and using terminals with multi-SIM support) as well as provides independence from commercial MNOs through the ability to switch among them or support a number of them without changing PPDR users' USIM cards.

The network management system (NMS) illustrated in Figure 5.17 represents the collection of technical and operational management tools used by the PPDR operator to control and monitor the operation of the core infrastructure, terminals, USIMs, legacy PMR and any dedicated LTE access network infrastructure that may be deployed.

The PPDR operator's core infrastructure is connected through, for example, Application Programming Interfaces (APIs) to CRS used by PPDR users for tactical and operational management. CRS can include dispatch applications to access the PPDR services (e.g. dispatch positions to communicate with users in the field) as well as control and monitor applications to deal with administrative and operational issues of the provided PPDR services. In particular, the CRS include:

- **User Management Application**. This application allows for the control and administration of PPDR subscribers' data. Keeping control over such PPDR subscription information is essential to tactical and operational PPDR managers since it allows PPDR users to manage the user provisioning process (e.g. activation/removal of subscribers) as well as setting up the required subscribers' capabilities (e.g. subscriber service profiles with QoS settings).
- **Service Management Application**. This application allows for the dynamic management of the provided service capabilities so that PPDR users in control rooms can adjust PPDR service provisioning to specific operational needs (e.g. creation of groups, activation of supplementary services, etc.).
- **Priority Access Management Application**. This application enables operational and tactical PPDR managers in control rooms to have direct control on the priority policies applied to PPDR traffic. Policies may consider not just the relative priority of a particular user based on their agency affiliation but also the situational context of applications (e.g. PPDR users' role within an incident command structure, type of incident, those applications that are mission critical and must be prioritized, location of users, etc.). The support of prioritization is conceived as a realization of the MPS (described in Section 4.5.4), where the PCRF functionality is allocated within the PPDR operator's core infrastructure. In this way, PPDR managers are able to configure the information and rules used by the PCRF element for QoS control decision-making (e.g. choice of ARP and QCI values). As depicted in Figure 5.17, the proposed implementation also entails the deployment of other standardized 3GPP functional entities of the PCC subsystem (described in Section 4.2.3) as part of the PPDR core network infrastructure (i.e. Application Function (AF), Service Profile Repository (SPR) and Policy and Charging Enforcement Function (PCEF)).

References

[1] Bill Dean, Dave Kaun, Dave McCauley, Mike Milas, 'Options for Communications Governance and Cost Sharing', January 2009.

[2] Alphonso E. Hamilton, 'Why carrier-hosted P25 as a Service provides a roadmap for greater agency participation', Urgent Communications, January 2015.

[3] ITU-T Rec. Y.2011 (10/2004), 'Next Generation Networks – frameworks and functional architecture models. General principles and general reference model for Next Generation Networks', October 2004.

[4] NPTSC, 'Defining Public Safety Grade Systems and Facilities', May 2014.

[5] Claudio Lucente, 'Public Safety 700 MHz Mobile Broadband Communications Network; Operational Requirements', Document prepared for the Public Safety Canada/Interoperability Development Office and the Centre for Security Science, February 2012.

[6] Federal Communications Commission (FCC), 'Recommended Minimum Technical Requirements to Ensure Nationwide Interoperability for the Nationwide Public Safety Broadband Network', May 2012.

[7] ETSI TR 102 022-1 V1.1.1 (2012-08), 'User Requirement Specification; Mission Critical Broadband Communication Requirements', August 2012.
[8] National Public Safety Telecommunications Council (NPSTC), 'Public Safety Broadband High-Level Launch Requirements Statement of Requirements for FirstNet Consideration', 7 December 2012.
[9] National Public Safety Telecommunications Council (NPSTC), 'Public Safety Broadband Push-to-Talk over Long Term Evolution Requirements', 18 July 2013.
[10] Keynote presentation titled 'Getting Ready to Create FirstNet' at PSCR Conference, Westminster, Colorado given by Craig Farrill, Acting CTO and FirstNet Board Member, 5 June 2013. Reproduced in 4G Americas White Paper '4G Mobile Broadband Evolution: 3GPP Release 11 & Release 12 and Beyond'.
[11] ECC Report 199, 'User requirements and spectrum needs for future European broadband PPDR systems (Wide Area Networks)', May 2013.
[12] Ericsson, 'Key characteristics of a Public Safety LTE network', February 2014.
[13] Eric Wibbens, 'Public Safety Site Hardening: Site and System Considerations for Public Safety Grade Operations', White Paper by Harris Corporation, 2013.
[14] TETRA and Critical Communications Association (TCCA), 'A review of delivery options for delivering mission critical solutions', Version 1.0, December 2013.
[15] Drat ECC Report 218, 'Harmonised conditions and spectrum bands for the implementation of future European broadband PPDR systems', April 2014.
[16] Simon Forge, Robert Horvitz and Colin Blackman, 'Study on use of commercial mobile networks and equipment for "mission-critical" high-speed broadband communications in specific sectors', Final Report, December 2014.
[17] FM49(13) 071, 'On the Future Architecture of Mission Critical Mobile Broadband PPDR Networks', White Paper from German Federal Ministry of the Interior, Project Group on Public Safety Digital Radio; Federal Coordinating Office, Version 1.1, 19 November 2013.
[18] APT Report on 'PPDR Applications Using IMT-based Technologies and Networks', Report no. APT/AWG/REP-27, Edition April 2012.
[19] Ericsson White Paper, 'Public safety mobile broadband', Uen 305 23-3228, February 2014.
[20] R. Ferrús and O. Sallent, 'Extending the LTE/LTE-A Business Case: Mission- and Business-Critical Mobile Broadband Communications', Vehicular Technology Magazine, IEEE, vol.**9**, no.3, pp.47, 55, September 2014.
[21] Xu Chen, Dongning Guo and J. Grosspietsch, 'The public safety broadband network: a novel architecture with mobile base stations', Communications (ICC), 2013 IEEE International Conference on Communications (ICC) 20139–13 June 2013.
[22] 'FirstNet Board Meeting', 10 March 2014. Transcript available online at http://www.firstnet.gov/sites/default/files/FirstNet_board_committee_meetings_march_10_2014_transcript.pdf (accessed 29 March 2015).
[23] JerseyNet, Public Safety network of the State of New Jersey. Available online at http://jerseynet.state.nj.us/ (accessed 29 March 2015).
[24] Donny Jackson, 'PSCR testing extended-cell LTE for rural use', Urgent Communications, 18 December 2013.
[25] Christopher Redding and Camillo Gentile, 'Extended range cell testing', 2014 Public Safety Broadband Stakeholder Conference, 3–5 June 2014.
[26] Donny Jackson, 'Incentives needed to bolster in-building communications for first responders', Urgent Matters, October 2013.
[27] John Facella, 'Your Guide to In-Building Coverage: work is underway to revise the National Fire Protection Association (NFPA) standards to ensure all codes are streamlined and cohesive for public-safety officials', Mission Critical Communications Magazine, June 2014.
[28] Donny Jackson, 'Panel debates the challenges of providing indoor coverage to public safety', Urgent Communications, November 2014.
[29] TE Connectivity, 'TE public safety DAS as a discrete system', Application Note, 2014. Available online at www.te.com/WirelessSolutions (accessed 29 March 2015).
[30] Tracy Ford, 'Mission-critical implications of small cells', Mission Critical Communications Magazine, August 2014.
[31] Small Cell Forum. Available online at http://www.smallcellforum.org/ (accessed 29 March 2015).
[32] Donny Jackson, 'SouthernLINC Wireless announces plan to build LTE network', Urgent Communications, 12 September 2013.
[33] Donny Jackson, 'Public safety reconsiders who should use its broadband network', Urgent Communications, 15 May 2013.

[34] Donny Jackson, 'Webinar panel outlines the financial challenges facing FirstNet', Urgent Communications, 2 July 2013.

[35] Tammy Parker, 'Rivada sets sights on commercial market for its spectrum arbitrage platform', FierceWirelessTech, 26 January 2014.

[36] Bill Schrier, 'Band-14-every-cellular-device?', Urgent Communications, 12 June 2013.

[37] G. Baldini, R. Ferrús, O. Sallent, Paul Hirst, Serge Delmas and Rafał Pisz, 'The evolution of Public Safety Communications in Europe: the results from the FP7 HELP project', ETSI Reconfigurable Radio Systems Workshop, Sophia Antipolis, France, 12 December 2012.

[38] Rob Stafford, Todd Bohling and Tracy McElvaney, 'Priority, pre-emption, and quality of service', Public Safety Broadband Stakeholder Conference, June 2014.

[39] Kevin McGinnis, 'FirstNet Update', FirstNet Board Member, 13 November 2014.

[40] ITU-T Rec. E.106, 'International Emergency Preference Scheme (IEPS) for disaster relief operations', October 2003.

[41] Government Emergency Telecommunications Service (GETS). Available online at http://www.dhs.gov/government-emergency-telecommunications-service-gets.

[42] Wireless Priority Service (WPS). Available online at https://www.dhs.gov/wireless-priority-service-wps.

[43] Wireless Priority Service (WPS) in Canada. Available online at http://www.ic.gc.ca/eic/site/et-tdu.nsf/eng/h_wj00016.html (accessed 29 March 2015).

[44] UK Cabinet Office, 'Privileged Access Schemes: MTPAS, FTPAS and Airwave'. Available online at http://www.cabinetoffice.gov.uk/content/privileged-access-schemes-mtpas-ftpas-and-airwave (accessed 29 March 2015).

[45] Report ITS 22, 'Interworking Aspects Related to Priority Services in Swedish Public Communications Networks'. Available online at http://www.its.se/ITS/ss6363x/ss6363xx.htm (accessed 29 March 2015).

[46] PROSIMOS Project website. Available online at http://www.prosimos.eu/PROSIMOS/ (accessed 29 March 2015).

[47] Donny Jackson, 'Rivada Networks receives patent to enable public-safety preemption on commercial wireless networks', Urgent Communications, 10 September 2014.

[48] 3GPP TR 22.950, 'Priority service feasibility study'. Available online at http://www.3gpp.org/DynaReport/22950.htm (accessed 29 March 2015).

[49] 3GPP TR 22.952, 'Priority service guide'.

[50] 3GPP TS 22.067, 'Enhanced Multi-Level Precedence and Pre-emption service (eMLPP); Stage 1'.

[51] 3GPP TS 23.067, 'Enhanced Multi-Level Precedence and Pre-emption Service (eMLPP); Stage 2'.

[52] 3GPP TS 24.067, 'Enhanced Multi-Level Precedence and Pre-emption service (eMLPP); Stage 3'.

[53] Ryan Hallahan and Jon M. Peha, 'Policies for public safety use of commercial wireless networks', 38th Telecommunications Policy Research Conference, October 2010.

[54] Hub One and IDATE, 'Critical and professional 4G/LTE: towards mobile broadband', White Paper, 2014.

[55] European Union Agency for Network and Information Security (ENISA), 'National Roaming for Resilience: National roaming for mitigating mobile network outages', November 2013.

[56] Daniel Ericsson, 'The OTA Platform in the World of LTE', Giesecke & Devrient white paper, January 2011.

[57] Carlos Camarán and Diego De Miguel (Valoris), 'Mobile Virtual Network Operator (MVNO) basics: what is behind this mobile business trend', Telecom Practice, October 2008.

[58] R. Ferrús, O. Sallent, G. Baldini and L. Goratti, 'LTE: The Technology Driver for Future Public Safety Communications', Communications Magazine, IEEE, vol.**51**, no.10, pp.154, 161, October 2013.

[59] Christian Mouraux, 'ASTRID High Speed Mobile Data MVNO', PMR Summit, Barcelona, 18 September 2012.

[60] A. Khan, W. Kellerer, K. Kozu and M. Yabusaki, 'Network sharing in the next mobile network: TCO reduction, management flexibility, and operational independence', Communications Magazine, IEEE, vol.**49**, no.10, pp.134, 142, October 2011.

[61] GSMA White Paper, 'Mobile infrastructure sharing', 2012. Available online at http://www.gsma.com/publicpolicy/wp-content/uploads/2012/09/Mobile-Infrastructure-sharing.pdf (accessed 29 March 2015).

[62] 3GPP TR 22.852 V13.1.0 (2014-09), 'Study on Radio Access Network (RAN) sharing enhancements (Release 13)', September 2014.

[63] Alcatel-Lucent, Comments of Alcatel-Lucent to the 'National Telecommunications and Information Administration (NTIA)' in response to the NoI in the Matter of 'Development of the Nationwide Interoperable Public Safety Broadband Network', Docket no. 120928505-2505-01, November 2012.

[64] APCO International, 'Exploring Business Tools for Leveraging Assets', v10.0 final, July 2013.
[65] ETSI TR 103 269-1 V1.1.1, 'TETRA and Critical Communications Evolution (TCCE); Critical Communications Architecture; Part 1: Critical Communications Architecture Reference Model', July 2014.
[66] ETSI EN 300 392-1 V1.3.1, 'Terrestrial Trunked Radio (TETRA); Voice plus Data (V+D); Part 1: general network design', June 2005.
[67] Draft TS 103 269-2 V0.0.2,·'TETRA and Critical Communications Evolution (TCCE); Critical Communications Architecture; Part 2: Critical Communications application mobile to network interface architecture', December 2014.
[68] The UK Government's National Technical Authority for Information Assurance (CESG), 'End User Devices Security Guidance: Enterprise considerations', Updated 23 January 2014. Available online at https://www.gov.uk/government/publications/end-user-devices-security-guidance-enterprise-considerations (accessed 29 March 2015).
[69] ETSI EN 300 392-3-1, 'TETRA V+D ISI General Design', V1.3.1, August 2010.
[70] C. Becchetti, F. Frosali and E. Lezaack, 'Transnational Interoperability: A System Framework for Public Protection and Disaster Relief', Vehicular Technology Magazine, IEEE, vol.**8**, no.2, pp.46, 54, June 2013.
[71] ISITEP Deliverable D2.4.1, R. Ferrús (Editor), 'System subsystem design description (SSDD) candidate release', September 2014. Available online at http://isitep.eu/ (accessed 29 March 2015).
[72] Sandra Wendelken, 'New P25 ISSI Features Set to Boost Demand', Radio Resource Magazine, 2 July 2014. Available online at http://mccmag.com/onlyonline.cfm?OnlyOnlineID=466 (accessed 29 March 2015).
[73] W. Roy McClellan III and Michael Doerk, 'Standards for P25 over LTE', within book 'P25 What's Next for the Global Standard?', Mission Critical Communications, Education Series, 2013.
[74] D. Lopez-Perez, I. Guvenc, G. de la Roche, M. Kountouris, T.Q.S. Quek and Jie Zhang, 'Enhanced Intercell Interference Coordination Challenges in Heterogeneous Networks', Wireless Communications, IEEE, vol.**18**, no.3, pp.22, 30, June 2011.
[75] Ying Loong Lee, Teong Chee Chuah, J. Loo and A. Vinel, 'Recent Advances in Radio Resource Management for Heterogeneous LTE/LTE-A Networks', Communications Surveys & Tutorials, IEEE, vol.**16**, no.4, pp.2142, 2180, Fourth Quarter 2014.
[76] Joe Boucher and Mike Wengrovitz, 'Embracing FirstNet Collaboration challenges', Mutualink Inc. White Paper, version 1.1, November 2014.
[77] Satellite Industry Association (SIA), 'SIA First Responder's Guide to Satellite Communications', 2007. Available online at http://transition.fcc.gov/pshs/docs-basic/SIA_FirstRespondersGuide07.pdf (accessed 29 March 2015).
[78] iDirect, 'Extending 3G and 4G Coverage to Remote and Rural Areas Solving the Backhaul Conundrum', White Paper, November 2013.
[79] Horizon 2020 European research project VITAL. Available online at www.ict-vital.eu (accessed 29 March 2015).
[80] ETSI TR 103 166 v1.1.1, 'Satellite Earth Stations and Systems (SES); Satellite Emergency Communications (SatEC); Emergency Communication Cell over Satellite (ECCS)', September 2011.
[81] Report ITU-R M.2149-1 (10/2011), 'Use and examples of mobile-satellite service systems for relief operation in the event of natural disasters and similar emergencies', October 2011.
[82] Solaris Mobile, Luxembourg, 'Input on the 2 GHz MSS band and proposals for ECC Report B', FM 49 Radio Spectrum for BB PPDR, Brussels, Belgium, 21–22 March 2013.
[83] OECD, 'International cables, gateways, backhaul and international exchange points', OECD Digital Economy Papers, No. 232, OECD Publishing, 2014. Available online at 10.1787/5jz8m9jf3wkl-en (accessed 29 March 2015).
[84] IETF RFC 4364, 'BGP/MPLS IP Virtual Private Networks (VPNs)', February 2006.
[85] P. Knight and C. Lewis, 'Layer 2 and 3 Virtual Private Networks: Taxonomy, Technology, and Standardization Efforts', Communications Magazine, IEEE, vol.**42**, no.6, pp.124, 131, June 2004.
[86] Metro Ethernet Forum (MEF). Available online at http://metroethernetforum.org/ (accessed 29 March 2015).
[87] GSM Association, 'Guidelines for IPX Provider networks (Previously Inter-Service Provider IP Backbone Guidelines)', Official document IR.34, Version 9.1, May 2013. Available online at http://www.gsma.com/newsroom/wp-content/uploads/2013/05/IR.34-v9.1.pdf (accessed 29 March 2015).

[88] GSM Association, 'Inter-Operator IP Backbone Security Requirements for Service Providers and Inter-operator IP backbone Providers', Official document IR.77, Version 2.0, October 2007. Available online at http://www.gsma.com/technicalprojects/wp-content/uploads/2012/05/ir77.pdf (accessed 29 March 2015).
[89] Syniverse Technologies, Response to the Federal Communications Commission (FCC), 'Service Rules for the 698–746, 747–762 and 777–792 MHz Bands; Implementing a Nationwide, Broadband, Interoperable Public Safety Network in the 700 MHz Band; Amendment of Part 90 of the Commission's Rules', April 2011.
[90] ISITEP Deliverable D2.4.3, R. Ferrús (Editor), 'Network architecture candidate release', September 2014. Available online at http://isitep.eu/ (accessed 29 March 2015).

6

Radio Spectrum for PPDR Communications

6.1 Spectrum Management: Regulatory Framework and Models

Spectrum management is the process of regulating the use of radio frequencies to promote efficient use and gain a net social benefit [1]. An efficient spectrum management has to address three main interrelated problems. First, the correct amount of spectrum needs to be allocated to certain uses or classes of uses. Second, it needs to assign usage rights to certain users or groups of users. Third, it needs to adjust established policies as technology and markets evolve over time.

The radio spectrum is considered in most countries as an exclusive property of the state (i.e. the radio spectrum is a national resource, much like water, land, gas and minerals). As such, there are different radio spectrum items that need to be nationally regulated (e.g. frequency allocation for various radio services, assignment of licence and radio frequencies to transmitting stations, type approval of equipment, fee collection, etc.). To this end, national regulatory authorities (NRAs) are commonly established within sovereign countries as the competent legal regulatory bodies for spectrum management and regulation. However, due to the very nature of radio spectrum (radio waves do not respect administrative borders), international agreements for regulating the use of radio frequencies are necessary to, among others, coordinate wireless communications with neighbours and other administrations. These international agreements typically combine both multilateral and bilateral dimensions. Therefore, effective spectrum management requires regulation at the national, regional and global levels.

The principles of regulating access to radio spectrum have remained essentially the same during the history of radio [2]. Spectrum blocks are first allocated, through international agreement, to services broadly defined (e.g. mobile service, broadcasting service, satellite service, radio astronomy service, etc.). This process is called *allocation*. Then, the next step is to assign frequencies and grant authorizations by NRAs to specific users or classes of users. In

Mobile Broadband Communications for Public Safety: The Road Ahead Through LTE Technology, First Edition.
Ramon Ferrús and Oriol Sallent.

the end, authorizing the use of the spectrum is a national prerogative, subject to international obligations and, in some countries such as European Union (EU) member states, also subject to EU community law.

The remaining of this section provides an overview of the current key regulatory and legal instruments that govern the use of spectrum across the global, regional and national levels. The discussion on regional and national levels is primarily focused on the European context. The section is concluded with a discussion on the models and evolution of spectrum management practices in the quest for achieving a more efficient and flexible use of spectrum resources.

6.1.1 Global-Level Regulatory Framework

The international global regulatory framework is provided by the International Telecommunication Union (ITU), a specialized United Nations (UN) agency. Within the ITU, the ITU Radiocommunication Sector (ITU-R) plays the central role in the global management of the radio-frequency spectrum. ITU-R seeks to ensure the rational, equitable, efficient and economical use of the radio-frequency spectrum by all radiocommunication services. The main priorities of the ITU's regulation of radio spectrum are:

- To protect against harmful interference
- To allocate radio services to the various radio-frequency bands in the radio spectrum (including globally harmonized allocations for systems used in international air and sea travel), taking into account sharing and compatibility studies
- To promote the effective use of the spectrum

To do so, the most important ITU legal instrument is the Radio Regulations (RR) treaty [3], which is ratified by the ITU member states. The ITU RR is the global agreement on how the airwaves are defined, allocated and used without harmful interference between the various wireless services around the world. ITU RR are binding for the states, not for individuals or operators. The compliance with ITU RR therefore presupposes that each state will take the measures required (legislation, regulations, clauses in licences and authorizations) to implement domestically those obligations to other spectrum users (operators, administrations, individuals, etc.). As such, the ITU RR provide an international framework for effective spectrum management that is primarily structured by the need for global harmonization in various domains (satellite communication, maritime, civil aviation, scientific research, etc.), coexistence capability between different types of radio communication networks and physical properties of frequency bands. It has major implications for the industry in terms of economies of scale and therefore for the design of radio products.

The regulatory and policy functions of ITU-R are carried out through World and Regional Radiocommunication Conferences and Radiocommunication Assemblies. World Radiocommunication Conferences (WRCs) are regularly held every 4–5 years, and the decisions adopted are incorporated in the ITU RR. Ahead of the celebration of each WRC, there are many ITU-R study groups and working parties [4], with representatives providing support, research and background for ITU RR. Contributors to this work include representatives from state's regulatory authorities as well as from regional spectrum management organizations that represent the common interests of a group of states. The last WRC was held in January–February 2012 (WRC-2012), and the next one is planned for November 2015 (WRC-2015).

The ITU RR (Article 5 of the RR) contains the so-called Table of Frequency Allocations (TFA) that allocates frequency bands for the purpose of their use by one or more terrestrial or space radiocommunication services or the radio astronomy service under specified conditions. A radiocommunication service is defined as the transmission, emission and/or reception of radio waves for specific telecommunication purposes. Terrestrial services and space services can themselves be subdivided in several different types of services (fixed, mobile, broadcasting, etc.). Each frequency band can be allocated with one or more radio services (typically two to four). Frequency bands are allocated to radiocommunication services on a primary or secondary basis. Stations of a secondary service shall not cause harmful interference to stations of primary services and cannot claim protection against harmful interference from stations of a primary service. In order to recognize certain regional differences that may be necessary in the frequency allocations, the ITU has split the world in three geographical regions (1, 2 and 3), so that the TFA is particularized for each region.

In addition to the frequency allocations to radio services, the ITU RR also contain administrative provisions on the registration (and protection) of specific frequency assignments as well as on the use of radio frequencies. The principle underpinning most of the provisions of ITU RR stipulates that any new assignment (i.e. any new authorization to operate a radio station) must be made in such a way as to avoid causing harmful interference to services rendered by stations using frequencies assigned in accordance with the agreed TFA and the other provisions of the ITU RR, the characteristics of which are recorded in the Master International Frequency Register (MIFR). In particular, a new assignment can only be recorded in the MIFR after completion of a procedure aimed at ensuring that it will not cause harmful interference to assignments made in accordance with the RR and previously recorded systems.

6.1.2 Regional-Level Regulatory Framework

Regional spectrum management organizations also play a major role in the management of the radio spectrum resource. There are six main regional organizations in the world:

1. Inter-American Telecommunications Commission (CITEL)
2. European Conference of Postal and Telecommunications Administrations (CEPT)
3. Asia-Pacific Telecommunity (APT)
4. African Telecommunications Union (ATU)
5. Arab Spectrum Management Group (ASMG)
6. Regional Commonwealth in the Field of Communications (RCC)

These organizations seek the harmonization and coordination of spectrum use in their regions. These organizations also represent within global forums, such as ITU-R WRCs, the interests of their member countries and their NRAs along with telecommunications providers and the regional industry. In particular, regional spectrum organizations have a WRC preparatory function: administrations in each region submit draft proposals to the regional spectrum organizations, the regional organization adopts common proposals before the WRC in accordance with their own procedures, and the regional proposals are submitted to the WRC on behalf of all of their members.

At the European level, the Electronic Communications Committee (ECC) within the CEPT brings together 48 countries to develop common policies and regulations in electronic communications and related applications for almost the entire geographical area of Europe. The prime objective of the ECC is to develop harmonized European regulations for the use of radio frequencies. Permanent negotiation on conditions of use of spectrum is critical over Europe as it enables adapting spectrum use conditions to industry requirements and national situations. The CEPT ECC takes an active role at the international level, preparing common European proposals to represent European interests in the ITU and other international organizations. In order to achieve these objectives, the CEPT ECC endorsed the principle of adopting a harmonized European Table of Frequency Allocations and Applications (known as European Common Allocation (ECA) and published within the European Communications Office Frequency Information System (ECO EFIS)) to establish a strategic framework for the utilization of the radio spectrum in Europe [5]. The CEPT ECC delivers ECC decisions (measures that NRAs are strongly urged to follow), ECC recommendations (measures that NRAs are encouraged to apply) and ECC reports (result of studies conducted by the ECC) [6]. The implementation of ECC decisions and ECC recommendations by CEPT member NRAs is made on a voluntary basis since CEPT deliverables are not obligatory legislative documents. However, they are normally implemented by many CEPT administrations [6].

Legal certainty on the availability of the identified spectrum for a given usage and under specified conditions is actually accomplished among the European countries that form part of the EU. The European Commission (EC), the executive body of the EU, is responsible to undertake the EU radio spectrum policy aimed to coordinate spectrum management approaches across the EU. This radio spectrum policy was launched by means of the Radio Spectrum Decision (676/2002/EC) [7] in year 2002. Two complementary bodies were set up following the Radio Spectrum Decision to facilitate consultation and to develop and support radio spectrum policy across the EU member states:

1. The Radio Spectrum Policy Group (RSPG), a group of high-level representatives that advises on broad policy in the area. Comprising the EC and the spectrum authorities of the EU member states, the RSPG provides the EC with advice on high-level policy matters in relation to spectrum. Representatives of the European Economic Area (EEA) countries, the European Parliament and the regional and international bodies may attend as observers. Before being transmitted to the commission, the RSPG's expert opinions are submitted to public consultations of all spectrum users, both commercial and non-commercial, as well as any other interested stakeholders. Therefore, the RSPG constitutes a unique platform for member states, the EU and all relevant stakeholders to discuss and coordinate regulation of radio spectrum.
2. The Radio Spectrum Committee (RSC), which assists the EC in developing specific regulatory implementation measures and can send EC mandates[1] to the CEPT ECC. In this

[1] An EC mandate is a request from the EC via its Radio Spectrum Committee (RSC) to the CEPT to conduct technical studies in order to develop technical implementing measures at EU level. The ability of the RSC to mandate the CEPT was established under the Radio Spectrum Decision. The RSC and CEPT work closely together, and the CEPT reports or results of CEPT studies initiated by the mandates feed into and form the technical basis for RSC decisions. These in turn are the basis for EC decisions or proposals that set the framework for the EU radio spectrum policy environment.

way, EC can request CEPT to provide frequency allocations in support of EU policies, and CEPT outputs can be then codified into EC decisions, which are legally binding. An MoU between the EC and CEPT ECC defines the basis of their cooperation. In addition, an important element bridging the EC and the ECC regulatory frameworks is the ECO EFIS managed by ECC and that the EC decided to be 'the European Common Spectrum Information Portal' (EU member states are obliged to publish their frequency information in EFIS).

Therefore, harmonized conditions can be established on a European-wide basis either through an EC decision[2] (which is mandatory for EU member states to implement) or by implementing the aforementioned CEPT ECC decisions or recommendations.[3] Deviations from EC harmonization measures are possible but necessitates that the EU member states request derogation to be granted by the EC, expectedly for a limited duration. In addition, as a result of a recent update of the regulatory framework for electronic communications services (ECS), the EC is now allowed to submit legislative proposals to the European Parliament and Council for establishing multi-annual Radio Spectrum Policy Programmes (RSPP) [8]. The first RSPP, approved in 2012, created a comprehensive roadmap contributing to the internal market for wireless technologies and services, particularly in line with the Europe 2020 initiative and the Digital Agenda for Europe. The RSPP sets general principles and calls for concrete actions to meet the objectives of EU policies. Among the specific actions set out by the RSPP is making available sufficient harmonized spectrum for the development of the internal market for wireless safety services and civil protection. In the wider field of telecommunications regulation, the EU has also established a harmonized regulatory framework for rights of use in the context of electronic communications networks and services (ECN&S) [9]. The ECN&S regulation also establishes some provisions in relation with the management of radio frequencies for ECS. Among them, regulation mandates that member states shall ensure that spectrum allocation used for ECS and issuing general authorizations or individual rights of use of such radio frequencies by competent national authorities are based on objective, transparent, non-discriminatory and proportionate criteria.

Complementing the work of CEPT ECC and EC, the European Telecommunications Standards Institute (ETSI), an industry-led organization, also plays an important role in the European regulatory environment for radio equipment and spectrum. ETSI is officially recognized by the EU as a European Standards Organization. ETSI produces globally applicable standards for information and communications technologies (ICT), including fixed, mobile, radio, converged, broadcast and Internet technologies. In the context of spectrum regulation, among other things, ETSI has the faculty to develop the so-called system reference documents (SRDoc) that provide technical, legal and economic background on new radio systems under standardization. SRDoc are used to inform the ECC so that harmonization measures or other kind of actions might be started to support the new radio system or application.

[2] Legislation in force (decisions and directives), as well as other official documents (communications, recommendations, council conclusions), issued by the EC can be found at http://ec.europa.eu/digital-agenda/en/radio-spectrum-policy-document-archive.

[3] A library of decisions, recommendations and reports approved by the ECC can be found in the ECO Documentation Database available at http://www.erodocdb.dk/default.aspx. The database also provides main EC decisions related to spectrum regulation.

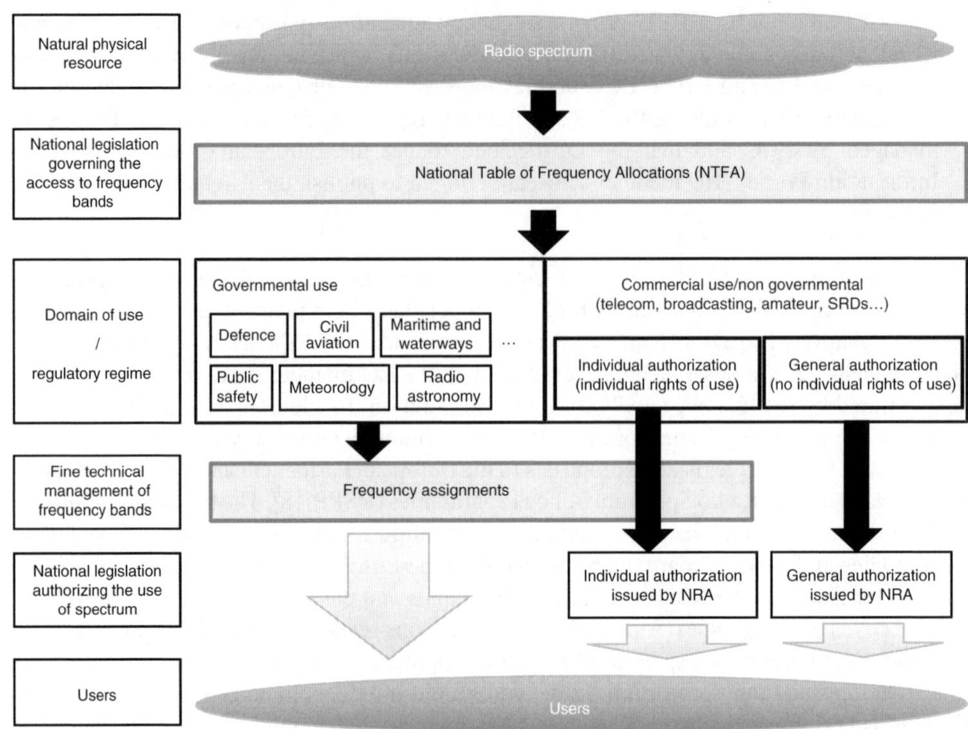

Figure 6.1 Baseline structure of national legislation on the use of the radio spectrum. From Ref. [10].

6.1.3 National-Level Regulatory Framework

At the national level, radio spectrum is managed by a designated competent government department or agency, which is commonly referred to as the NRA on spectrum matters. Examples of NRA are the Federal Communications Commission (FCC) in the United States, Ofcom in the United Kingdom and the Secretary of State for Telecommunications and Information Society in Spain.

Within the framework of action left by the observance of the regional and global agreements, NRAs must consider many factors, such as each country's socio-economic benefit. Therefore, NRAs will typically consult with all interested parties and seek to release as much spectrum as possible to allow the country to benefit from global economies of scale, interoperability, interference minimization (including with neighbour countries), international roaming and alignment with regional and global agreements. Many trade-offs are likely to be faced by NRA decisions when assigning spectrum. An example would be the conjugation of the direct economic benefits arisen from the assignment of frequencies to commercial mobile communications compared with the less direct economic purposes that nevertheless benefit society in many indirect ways, such as the designation of spectrum for PPDR communications.

Figure 6.1 illustrates the general structure of the national legislation that is commonly in place in most countries to regulate the use of radio spectrum [10]. As depicted in the figure, the elaboration of a National Table of Frequency Allocations (NTFA) constitutes the first step and the basis of radio-frequency spectrum management, being the main instrument of the national legislation to govern the access to frequency bands. An NTFA primarily specifies the radio services authorized by an individual administration along the frequency bands (referred to as 'allocations') and the entities that might have access to these allocations (e.g. government use, private use). An NFTA can also include specific frequency assignments to individual users (e.g. reserved bands for the Ministry of Defence) or installations at particular locations (e.g. protection of specific facilities from potential radio interference). Moreover, an NTFA typically shows the internationally agreed spectrum allocations of the ITU RR.

In a second step, specific national frequency assignments allow the fine management of frequency bands in accordance with the rules set in NTFAs, particularly in bands shared by different type of users as well as to ensure the proper coexistence of the spectrum users in adjacent bands (e.g. channel bandwidths, guard bands, transmission power settings).

Finally, users have to be authorized to be able to transmit on specific frequencies (i.e. users are granted rights of use of the radio spectrum, which is property of the state). This is in accordance with Article 18 of the ITU RR that stipulates: '*no transmitting station may be established or operated by a private person or by any enterprise without a license issued in an appropriate form and in conformity with the provisions of these Regulations by or on behalf of the government of the country to which the station in question is subject*'. The term 'licence' is understood here in its broad acceptance, basically establishing that the use of spectrum must be explicitly permitted.

The way that authorization is performed differs between the two possible domains of use: 'governmental' and 'non-governmental'.

'Governmental' use, or assimilated, covers various sectors (defence, civil aviation, maritime and waterways, public safety, meteorology, science, etc.). In this case, the rights of use are commonly limited to the rights described in the NTFA, and no additional legal acts are conducted to explicitly grant rights of use [11]. Access to spectrum resources by governmental users should be subject to regular review by the NRA.

In the case of 'non-governmental' uses of the spectrum, a public legal act issued by the NRA is necessary for the purpose of delivering spectrum usage rights to private entities or citizens. Under the EU regulatory framework, this legal act is referred to as 'authorization' (EU Authorization Directive [12]). On this basis, a legal differentiation is established between 'individual authorization' and 'general authorization' to reflect the obligation or not for the user to be granted individual rights of use before transmitting. For instance, radio applications that do not need individual frequency planning and coordination could be exempted from individual authorization and should be subject to a general authorization.

The different mechanisms in use nowadays to grant (individual or general) authorizations for the use of spectrum can be classified under three main approaches, referred to as 'licencing regimes'. These three main approaches are individual licencing (also denoted as 'traditional licencing'), light licencing and licence exempt (also referred to as 'unlicenced'). This classification, shown in Table 6.1, is established in CEPT ECC Report 132 [13] as a result of a review of the various terminologies used across the CEPT countries. The distinction between the three licencing regimes uses as a first-level differentiation factor the legal basis given under the EU regulatory framework for ECN&S (i.e. individual or general authorization). In this regard, while there is a clear association between 'individual licence' and 'individual authorization' as well as between 'licence exempt' and 'general authorization', a 'light-licencing' regime can be based on either an individual or a general authorization. In either of these two options, a light-licencing regime is characterized by the existence of an obligation for the spectrum user to register or notify its use to the NRA. This obligation is of administrative nature and necessitates, as a prerequisite for use, that the user contacts the NRA. Such

Table 6.1 Licensing regimes for the authorization of spectrum rights of use.

Individual licence(sometimes also referred to as 'traditional licencing')	Light licencing		Licence exempt(also referred to as 'unlicenced')
Based on **individual authorization** (individual rights of use are granted) Facilitates individual frequency planning and coordination Traditional procedure for issuing licences	Based on **individual authorization** (individual rights of use are granted) With limitations in the number of users Facilitates individual frequency planning and coordination Simplified procedure compared to traditional procedure for issuing licences	Based on **general authorization** (no individual rights of use are granted) Obligation for registration and/or notification No limitations in the number of users nor need for coordination Intended for radio applications that do not require individual frequency planning or coordination	Based on **general authorization** (no individual rights of use are granted) No obligation for registration nor notification Intended for radio applications that do not require individual frequency planning or coordination

provision remain in the field of 'general authorization' as long as they are only meant to allow controlling the deployment and use of the application so as to avoid harmful interferences on radio services, but not to restrict it. Conversely, this provision falls under the 'individual authorization' umbrella if associated with possible limitation of the number of users and specific requirements for coordination prior to use.

It is worth noting that the licences that are issued as a result of a 'light-licencing' regime based on 'individual authorizations' are not different from 'individual licences' from a regulatory perspective. In both cases, the licences should contain technical conditions that are necessary and sufficient to avoid harmful interference to other systems and users. However, the process of issuing licences could vary significantly between the light-licencing and traditional licencing schemes. While in traditional licencing relatively complex administrative processes such as beauty contests or auctions could take place, licences under light-licencing regimes may basically involve the use of dedicated IT systems for automatic frequency planning and licence assignment. Also of note is that, as captured in Table 6.1, a licence-exempt regime is a general authorization regulatory regime where radio equipment operates under a well-defined set of regulator-imposed rules (e.g. constraints on maximum transmit power, spectral mask, duty cycle, etc.) and where no provision for registration and/or notification is required to the users of these devices.

6.1.4 Spectrum Management Models

Spectrum management should be designed and carried out with the main goal of ensuring the efficient use of spectrum. Efficient use implies that, given the state of technology, spectrum is channelled to its most productive uses. Importantly, efficient use has to be properly conjugated with the achievement of other non-market objectives such as national security, safety and equal access goals. As spectrum uses change over time, an efficient spectrum management regime also needs to minimize the transaction costs associated with these adjustments.

Historically, access to and use of radio spectrum has been highly regulated in order to prevent interference among users of adjacent frequencies or from neighbouring geographic areas, particularly for reasons of defence and security [2]. Fuelled by technological advances and the consolidation of liberalized regimes in the telecommunications market, there have been in the last years significant innovations in the theory of spectrum management along with gradual changes in practice of spectrum management and regulation. This gradual change follows a growing consensus that past and current regulatory practices originally intended to promote the public interest have in fact delayed, in some cases, the introduction and growth of a variety of beneficial technologies and services or increased the cost of the same through an artificial scarcity. In addition to these delays, the demand for spectrum has grown significantly, highlighting the need for efficient use of all available spectrum in order to avoid scarcity. Those factors are making policymakers and regulators worldwide focus on new spectrum regulation principles with an increasing emphasis on striking the best possible balance between the certainty required to ensure stable roll-out of services and flexibility (or light-handed regulation) leading to improvements in cost, services and the use of innovative technologies. In particular, a pioneering role in the introduction of many innovative spectrum management approaches should be credited to national authorities for radio-frequency management such as the FCC in the United States and Ofcom in the United Kingdom.

In practical terms, three main spectrum management models can be distinguished nowadays:

1. **Command and control model**. Under this model, individual users are granted rights to use spectrum on an administrative basis in the form of individual licences to private users or just spectrum assignments for governmental use. The usage is often set to be exclusive: each band is dedicated to a single user, thus maintaining interference-free communication. The command and control management model dates back to the initial days of wireless communications, when the technologies employed required interference-free mediums for achieving acceptable quality. Thus, it is often argued that the exclusive nature of the command and control approach is an artefact of outdated technologies. However, an apparent advantage of this model is that services related to public interest could be sustained. This is especially true with regard to government usage of spectrum (e.g. military, PPDR, transportation) as well as for licencing of spectrum used for maritime and aeronautical services or even for services such as over-the-air television, which may not be as attractive as other potential commercial uses in terms of profitability, but they can be retained by the administration as beneficial for the society. Even beyond such uses, however, some governments retain a belief that regulators are best suited to determine which operators should be granted licences, and in some cases, there may be insufficient demand for spectrum to warrant competitive bidding.
2. **Market-based model**. In this model, individual licences with exclusive spectrum rights of use in certain bands are acquired by market mechanisms such as auctions. Auctions remedy some of the flaws of administrative model, allowing the placing of a market value on spectrum. In addition, secondary markets for spectrum and spectrum licences can be established based on spectrum trading regulation under which both the ownership and use of the spectrum rights of use can change in the course of a licencee's operation. Transfer

control of spectrum licences may be subject to government or regulatory approval. This is a major step beyond the simple auctioning of licences without granting any real flexibility of use. It does, however, require the full specification of what rights associated with spectrum use can be traded and utilized. The trading of spectrum rights combined with flexible usage conditions could substantially benefit economic growth. Typical users of this model are commercial terrestrial operators and satellite operators. The terms of *property rights* or *flexible rights of use* are also often used to describe the market-based approach.[4]

3. **Collective use of spectrum (CUS) model**. The CUS model is actually an umbrella term to designate all spectrum management approaches allowing more than one user to occupy the same range of frequencies at the same time [15]. Examples of licencing regimes that fit within the CUS model are licence-exempt regimes (in this case, the CUS model is typically referred to as 'spectrum commons') as well as light-licencing regimes that allow either a restricted or unrestricted number of users to share a common band. Typical users of bands operated under the CUS model are individuals, though commercial telecommunications providers also rely on the use of these bands to wirelessly access to their networks or for traffic offloading. Hence, typical uses of these bands are Wi-Fi as well as many other low-power devices (e.g. remotes, garage openers, sensors). The proliferation of products and services based on Wi-Fi and other short-range radio technologies is a clear exponent of the benefits and economic value that can be brought by using a CUS model to manage some amount of spectrum.

Governments have generally been cautious in determining which approach to use. The result, across the world, is a combination of spectrum management regimes that mainly incorporate legacy command and control regimes for government services, auctions and bidding for many commercial licences and licence-exempt uses for low-power devices. The three spectrum management models have unique advantages and disadvantages, and no single approach is superior on all counts. The optimal spectrum policy should determine the right mix of these methods rather than adopting one model [16].

6.2 Internationally Harmonized Frequency Ranges for PPDR Communications

The harmonization of spectrum for PPDR communications at the international level is recognized to offer many benefits. These include economic benefits (economies of scale in the manufacturing of equipment and more competitive market for equipment procurement), the facilitation in the development of compatible networks and services and the promotion of international interoperability of equipment for those agencies that require cross-border cooperation with other PPDR agencies and organizations (e.g. increased effective response to disaster relief).

Considering the above-mentioned benefits along with the growing telecommunication needs of PPDR agencies already anticipated in the late 1990s, the WRC-2000 resolved to

[4] The concepts of 'property rights' and 'flexible use' go hand in glove and often are used almost interchangeably. Technically, however, the property rights model originated to describe the concept of auctioning spectrum rights for original distribution, a concept in which the 'buyer' gains exclusive access to a block of spectrum (for a defined period of time) via a market transaction. Later policy innovations added the flexibility element, allowing licencees to determine how to use the spectrum licenced to them. The two concepts, however, are not identical, and in fact, many licenses have been auctioned to companies without granting them any real flexibility of use [14].

Table 6.2 Harmonized frequency bands/ranges established in ITU Resolution 646 (WRC-2003).

ITU-R region	Frequency bands/ranges
Region 1	380–470 MHz as the frequency range within which the band 380–385/390–395 MHz is a preferred core harmonized band for permanent public protection activities within certain agreed countries of Region 1
Region 2	746–806 MHz, 806–869 MHz and 4940–4990 MHz (Venezuela has identified the band 380–400 MHz for public protection and disaster relief applications)
Region 3	406.1–430 MHz, 440–470 MHz, 806–824/851–869 MHz, 4940–4990 MHz and 5850–5925 MHz (some countries in Region 3 have also identified the bands 380–400 and 746–806 MHz for PPDR applications)

invite ITU-R to study the identification of frequency bands that could be used on a global/regional basis by administrations intending to implement future advanced solutions for PPDR (ITU-R Resolution 645 [17]). In that context, the Agenda Item 1.3 (AI 1.3) was included in WRC-2003 to decide on which bands and to make regulatory provisions as necessary. Preparatory studies were carried out within ITU-R working groups that led to the delivery of Report ITU-R M-2033 [18], which constitutes a key reference in the context of PPDR communications. Report ITU-R M.2033 identified objectives, applications, general requirements, spectrum requirements and solutions to satisfy the operational needs of PPDR organizations. The report was notably based on the general assumption of a technology-neutral approach.

As a result of this previous work, ITU Resolution 646 was approved within AI 1.3 in WRC-2003. This resolution *strongly recommends* administrations to use regionally harmonized bands for PPDR to the maximum possible extent. The identified frequency bands/ranges, or parts thereof, to be considered by national administrations when undertaking their national planning are reproduced in Table 6.2. In the context of this resolution, the term 'frequency range' means a range of frequencies over which a radio equipment is envisaged to be capable of operating but limited to specific frequency band(s) according to national conditions and requirements (i.e. not all frequencies within an identified common frequency range are expected to be available within each country). Therefore, a solution based on regional frequency ranges enables administrations to benefit from harmonization while continuing to meet national planning requirements, which is important because the amount of spectrum needed for PPDR communications may differ significantly across countries. In any case, it's worth noting that ITU Resolution 646 is just a recommendation, and as such, it does not preclude the use of the identified bands/frequencies by any other application within the services to which these bands/frequencies are allocated.

It is worth noting that the focus in WRC-2003 AI 1.3 was to identify harmonized bands for mission-critical (narrowband) voice and low-rate data services for PPDR agencies, though it recognized the trend towards the evolution of wideband and broadband data applications (broadband applications such as video were thought to be relevant only for hot spot coverage). From 2003 till date, the ITU has been continuously working on preparing reports and recommendations on PPDR communications. Indeed, AI 1.3 has been kept linked to the PPDR spectrum issue across the subsequent conferences WRC-2007 and WRC-2012, with new documents released but with no changes introduced to date to the frequency ranges initially identified in ITU Resolution 646. The main relevant resolutions and recommendations produced on this matter since WRC-2003 are captured in Table 6.3.

Table 6.3 Key reference documents from ITU-R since WRC-2003 with regard to PPDR spectrum harmonization.

Document	Brief description
Report ITU-R M.2033, 'Radiocommunication objectives and requirements for public protection and disaster relief' (2003)	Radiocommunication objectives and requirements for public protection and disaster relief. This report was developed in preparation for WRC-03 Agenda Item 1.3. The document defines radiocommunication objectives and requirements for the implementation of future advanced PPDR solutions. The document also establishes reference terminology to precisely define and categorize public safety communications
ITU-R Resolution 646, 'Public protection and disaster relief' (approved at WRC-2003 and revised in WRC-2012)	This resolution encourages administrations to consider a set of frequency bands/ranges when undertaking their national planning for the purposes of achieving regionally harmonized frequency bands/ranges for advanced PPDR solutions There is a review by WRC.12, though no changes in frequency ranges were introduced
ITU-R Resolution 647, 'Spectrum management guidelines for emergency and disaster relief radiocommunication' (approved at WRC-2007 and revised in WRC-2012)	This resolution encourages administrations to consider global and/or regional frequency bands/ranges for emergency and disaster relief when undertaking their national planning and to communicate this information to the Radiocommunication Bureau of the ITU. A database system has been established and is maintained by the Radiocommunication Bureau
ITU-R Resolution 648, 'Studies to support broadband public protection and disaster relief' (WRC-2012)	This resolution invites ITU-R and administrations to study technical and operational issues relating to broadband PPDR and its further development and to develop recommendations, as required, on technical requirements for PPDR services and applications, the evolution of broadband PPDR through advances in technology and the needs of developing countries The resolution also resolves to invite WRC-15 to consider these studies on broadband PPDR and take appropriate action with regard to revision of Resolution 646 (Rev.WRC-12)
Recommendation ITU-R M.2015, 'Frequency arrangements for public protection and disaster relief radiocommunication systems in UHF bands in accordance with Resolution 646 (Rev.WRC-12)' (2012)	This recommendation provides guidance on frequency arrangements for public protection and disaster relief radiocommunications in certain regions in some of the bands below 1 GHz identified in Resolution 646 (Rev.WRC-12). Currently, the recommendation addresses arrangements in the ranges 380–470 MHz in certain countries in Region 1, 746–806 and 806–869 MHz in Region 2 and 806–824/851–869 MHz in some countries in Region 3 in accordance with resolutions ITU-R 53 and ITU-R 55 and WRC Resolutions 644 (Rev. WRC-07), 646 (Rev.WRC-12) and 647 (WRC-07)
Recommendation ITU-R M.1637, 'Global cross-border circulation of radiocommunication equipment in emergency and disaster relief situations' (2003)	This recommendation addresses issues to be considered in order to facilitate the global circulation of radiocommunications equipment to be used in emergency and disaster relief situations

Table 6.3 *(continued)*

Document	Brief description
ITU-R Recommendation M.1826 'Harmonized frequency channel plan for broadband public protection and disaster relief operations at 4940–4990 MHz in Regions 2 and 3' (2007)	This recommendation addresses harmonized frequency channel plans in the band 4940–4990 MHz for broadband public protection and disaster relief radiocommunications in Regions 2 and 3. The recommendation provides two frequency channelling plans for national administrations to consider when allocating spectrum for use by users who are directly involved with PPDR
ITU-R Recommendation M.2009, 'Radio interface standards for use by public protection and disaster relief operations in some parts of the UHF band in accordance with Resolution 646 (WRC-03)' (2012)	This recommendation identifies radio interface standards applicable for PPDR operations in some parts of the UHF band. The broadband standards included in this recommendation are capable of supporting users at broadband data rates, taking into account the ITU-R definitions of 'wireless access' and 'broadband wireless access' found in Recommendation ITU-R F.1399. This recommendation addresses the standards themselves and does not deal with the frequency arrangements for PPDR systems, for which a separate recommendation exists: Recommendation ITU-R M.2015

However, a key milestone is expected to be accomplished in WRC-2015 with regard to the identification of additional spectrum for mobile broadband PPDR use. In particular, WRC-2012 agreed to review and revise Resolution 646 for broadband PPDR under AI 1.3 in WRC-2015 so as to account for the new PPDR scenarios offered by the evolution of broadband technologies. To that end, Resolution 648 was approved in WRC-2012, inviting ITU-R to study technical and operational issues related to broadband PPDR and its further development and to develop recommendations, as required, on technical requirements for PPDR services and applications, the evolution of broadband PPDR through advances in technology and the needs of developing countries. Therefore, WRC-2015 AI 1.3 is anticipated to be instrumental in establishing new harmonized frequency bands for PPDR mobile broadband applications and the interoperability and economies of scale that will result. This is particularly relevant for Europe, which is consolidating a common position supporting the identification of new harmonized frequency band(s) (in addition to the 380–470-MHz band) for mobile broadband PPDR in Region 1, most likely within the range 694–862 MHz, while recognizing that it will be a national decision which band(s) is(are) selected for mobile broadband PPDR in each country.

The preparatory work for WRC-2015 AI 1.3 at CEPT level [19] states that, in order to establish a family of cross-border functioning broadband PPDR networks, it is not required to designate identical bands for this purpose, but rather choose the suitable bands out of the (eventual) harmonized frequency range(s) and adopt a common technology. Hence, according to CEPT's preparatory work, it may be more relevant to refer to 'harmonized conditions' rather than 'harmonized frequencies'. For broadband PPDR communications, harmonized conditions can be established if (i) a tuning range can be identified and (ii) a technical standard such as LTE can be harmonized. On this basis, there is a proposal to replace the term 'frequency range' used in Resolution 646 (WRC-2012) with the term 'tuning range', defined as a range of frequencies in which radio equipment is envisaged to be capable of operating, which may be limited to specific parts of the relevant frequency band(s) according to national circumstances and requirements. In addition, when the broadband PPDR networks use a common technical standard (such as LTE), it is argued that the tuning range could also be specified as one or several band classes. Using these definitions of harmonized conditions and tuning ranges, it will be possible to offer full flexibility for administrations to decide on their dedicated PPDR spectrum to meet national needs and demands chosen within the tuning range. The technology will then provide full roaming and will be open for interoperability even if the PPDR spectrum is not strictly harmonized.

6.3 Spectrum Needs for Mobile Broadband PPDR Communications

The need for spectrum suitable for the support of emerging broadband PPDR applications has been recognized for many years. This section first describes the different uses of spectrum needed across the different components of a PPDR communications delivery solution. Then, the methodologies commonly used for the computation of spectrum needs are explained. Finally, a number of estimates carried out by different organizations worldwide that quantify spectrum needs for mobile broadband PPDR applications are presented.

6.3.1 Spectrum Components

As outlined in Chapter 3 and further developed in Chapter 5, a multilayered communications approach is envisioned for future PPDR communications systems where complementary frequency bands and technologies can be used to deploy wide area (e.g. nationwide) coverage, with sufficient capacity to cater for routine, day-to-day communication requirements, together with the ability to extend coverage and capacity on an ad hoc basis to cope with challenging radio environments (e.g. tunnels, basements) or localized high-capacity demands. On this basis, spectrum needs related to different system components can be distinguished:

- **Spectrum for terrestrial wide area networks** (WAN). This is the spectrum to be used in the radio sites of the cellular network that delivers the PPDR communications services for day-to-day operations and most emergency scenarios. In this regard, considering the current situation in which voice communications are still supported over narrowband PMR technologies, spectrum estimations for WAN can distinguish between voice and broadband data communications.
- **Spectrum for localized, ad hoc deployments**. Radio equipment for large events and large disaster situations can be brought to the local area as required. This equipment may or may not be linked with the existing PPDR network infrastructure. Anyway, specific spectrum may be required for this use.
- **Spectrum for backhauling**. On the basis that additional local capacity can be deployed as required at the incident, it will be necessary to get data traffic to and from the locality, for example, to provide access to the Internet or other remote data sources and to maintain communications with the agency headquarters. Therefore, this may require the establishment of temporary fixed links using UHF or microwave frequencies or satellite links that may require some amount of spectrum to be available for such purpose. This is especially critical in the cases of unplanned events or major incidents, where there may not be any local alternative infrastructure available for backhauling.

- **Spectrum for direct mode operation** (DMO). DMO is currently an important means of communicating voice and narrowband data services when being out of network coverage. DMO operations could use spectrum designated for the permanent terrestrial WAN or need additional separate frequencies. In the case of supporting DMO also for broadband data services, the need of separate frequencies from the WAN would be dependent on the technical implementation of DMO into the future broadband PPDR solutions.
- **Spectrum for air–ground–air** (AGA) operations. In addition to the need for terrestrial operation, there may also be the need for AGA operation. AGA in this context means emergency services communication to or from low-flying (typically a few hundred meters over the ground level) airborne objects. These usually involve a video stream being relayed from a camera mounted on a helicopter or unmanned aerial vehicle (UAV) to a monitoring station on the ground. While this application could be regarded as a point-to-point link, it is difficult to deploy very directional antennas because of the aircraft movement. This turns into an increased risk of interference over a fairly wide area. In order to protect the land mobile infrastructure and to ease cross-border operation, there is a need to identify sub-bands or specific channels for airborne communications. As with the DMO case, AGA could use spectrum designated for the permanent terrestrial WAN or need separate frequencies.

6.3.2 Methodologies for the Computation of Spectrum Needs

In general terms, the methodologies used to determine the amount of spectrum required are based on (i) the characterization of the traffic demand in a given area or per incident, (ii) the estimation of the number of cell sites covering the area of interest and of any other radio equipment brought into (e.g. ad hoc network) and (iii) the characterization of the underlying wireless technologies mainly in terms of achievable spectrum efficiencies. This generic approach is applicable to the different spectrum components identified in the previous subsection, though specific assumptions and considerations have to be introduced for each of them. In the case of the estimation of spectrum needs for terrestrial WAN, the following aspects are considered:

- **Traffic demand**. This accounts for the amount of traffic generated by PPDR users when performing day-to-day tasks as well as when involved in different types of incident scenarios. It is typically represented in Erlangs[5] for voice services and in kb/s or Mb/s for data services. The demand can be computed from estimations of the number of PPDR users within local/regional/national areas (e.g. based on PPDR population density analysis) and from the characterization of the services that are going to be used. Demand can also be characterized per incident (i.e. incident-based approach characterization), based on the number of effectives needed within an incident response. Demand can be specified per service or as aggregated values for those set of services delivered through the same type of communications resources.
- **Number of cell sites covering the area**. This information is relevant to determine what fraction of the demand can be served by the radio resources available in a single cell site. The cell size directly impacts on the spectrum estimate. There could be incidents spanning

[5] An Erlang is a unit of telecommunications traffic measurement. Strictly speaking, an Erlang represents the continuous use of one voice path. In practice, it is used to describe the total traffic volume of 1 h.

over several cells as well as cells serving the traffic of several incidents. In this latter case, considerations on the location of the incidents within the cell coverage are also relevant for the spectrum computation (e.g. whether the incident place is close or not to the cell edge).

- **Characteristics of the technology providing the service**. Radio link spectral efficiency is a key performance indicator of radio technologies. The spectral efficiency is given in bit/s/Hz, thus measuring the amount of bits per second that can be transferred per unit of radio-frequency spectrum (Hz). In modern wireless technologies, spectral efficiency is not fixed but dynamically adapted to the radio link conditions (e.g. the achievable spectral efficiency varies between terminals located close to the cell sites and terminals operating at the cell edge). In addition to the characterization of the achievable spectral efficiencies from a system-wide perspective, it is also important to characterize the capacity of the system as to frequency reuse in neighbouring sites/cells. Systems currently used for narrowband PPDR voice services require frequency reuse patterns as high as 12–21 (i.e. the amount of spectrum needed in the network equals to the amount of spectrum needed in a single site multiplied by the reuse factor). In contrast, LTE technology can support much lower reuse factors, being even possible to reach full frequency reuse at the expenses of some capacity and interference trade-offs. In the end, taking into consideration all the factors that reduce the capacity over the air interface (e.g. guard bands, co-channel and adjacent channel interference, channels assigned to other purposes within the band), an overall system spectral efficiency can be quantified in bit/s/Hz/cell.

With all of the above elements, a coarse-grained estimation amount of spectrum can be straightforwardly derived. For example, in the case of data services, a common approach is just to compute the amount of spectrum (in Hz) needed in a single site/cell by dividing the estimated demand (in bit/s) by the achievable spectral efficiency (either average efficiency or efficiency at the cell edge). Once the spectrum needed in a single cell is estimated, the total amount of spectrum can be obtained from multiplying the single cell estimate by the applicable frequency reuse factor. A more fine-grained estimation, if necessary, can be pursued by considering more specifics on the network deployment and traffic distribution as well as other factors such as the possibility to use multicast/broadcast delivery for some applications or the introduction of weighting factors to capture the correlation between environments in coincident busy hours. These fine-grained computations may require the use of computational tools such as the ones used in the commercial domain for radio network planning and optimization. In any case, it is worth highlighting that the outcomes of any spectrum computation are usually very sensitive to the considered inputs and assumptions, so that spectrum estimations can differ considerably if no reference models are established.

A generic methodology for PPDR spectrum calculation based on the aforementioned principles is reported in ITU-R Report M.2033 [18]. The model describes how to address the characterization of four fundamental variables that determine the amount of spectrum required (i.e. demand for a given area, number of sites/cells covering the area, spectral efficiency of the technology providing the service and amount the technology is able to reuse frequencies). The model provides a spectrum calculation for each type of PPDR service, considering both voice calls and data services. The basic equation employed in the model is as follows:

$$\text{Spectrum (MHz)} = \frac{\text{Traffic per cell (Mb/s/cell)}}{\text{Net system capacity (Mb/s/MHz/cell)}}$$

This fundamental equation is applied for the various service categories in both the uplink and downlink paths. In calculating the total demand per cell, the ITU-R model factors in the total seconds of usage per user, the average users per cell and the bit rate for the application in question. Protocol overheads (e.g. channel coding) and service-grade parameters are considered in the computation of the traffic per cell (e.g. introduction of multiplicative factors to account for blocking probabilities). On the other hand, the net system capacity is directly a computation of the system spectral efficiency where factors such as frequency reuse, guard bands and the use of some resources for signalling channels are considered.

Despite some differences in the computation of the model's inputs and in the applicability of the previously mentioned formula per incident and per cell or accounting for the overall system efficiency, the ITU-R spectrum model calculator has been considered as the baseline approach in several other studies that provide estimations for broadband PPDR spectrum (e.g. CEPT ECC Report 199 [20], US NPSTC [21], Canada DRDC CSS [22]). As an example to illustrate the applicability of this type of methodology, the computation of spectrum needs for day-to-day operations scenarios is briefly explained in the following as addressed in ECC Report 199. The proposed methodology used for such scenarios consists of the following five steps:

Step 1: Define the incidents (scenarios).
Step 2: Estimate the total traffic requirement per incident including background traffic.
Step 3: Calculate the link budgets and cell size.
Step 4: Estimate the number of incidents that should be taken into account simultaneously per cell.
Step 5: Estimate the total spectrum requirement based on assumptions on the number of incidents per cell, location of incidents within a cell and spectrum efficiency per incident.

With regard to the computation of the traffic demand, this methodology follows an incident-based approach where traffic is summed over several separate incidents and background traffic is then added. The outputs of each step are summarized in Table 6.4.

In addition to the earlier incident-based methodology, CEPT Report 199 also provides spectrum estimations based on a slightly different methodology that does not treat traffic separately from several incidents. This alternative methodology is based on the use of the Law Enforcement Working Party (LEWP)/ETSI Matrix described in Chapter 2. In particular, individual incidents and background traffic are combined together into the peak load in the busy hour in normal conditions and in emergency conditions. This is considered through a detailed characterization of the data requirements per application addressed within the LEWP/ETSI Matrix. Then, for each application, it is decided whether the application is 'incident-centric' or a background application spread out across the cell. This distinction is made to be able to consider either average or incident-specific spectrum efficiencies to be accounted for each application. In some way, this approach can be thought as a bottom-up approach where the estimated traffic from each individual user application and the number of users of that application are considered in creating the load. This is in contrast to the approach followed in the incident-based methodology to account for the background traffic load, which is completely separate from the incidents' traffic load and estimated mainly based on the characteristics of the likely applications used in routine activities.

Table 6.4 Example of the computation of spectrum needs for day-to-day operations scenarios as addressed in ECC Report 199.

Methodology	Description
Steps 1 and 2	Characterization of the traffic demand in the following type of incidents[a] and for background traffic: • Road incident: 1300 kb/s • Traffic stop police operation: 1300 kb/s • Background traffic: 1500 kb/s
Step 3	LTE technology has been chosen as the reference technology. Based on parameters from 3GPP performance specifications for LTE Release 10 and on the selection of the reference modulation for the communication link (i.e. modulation and coding scheme that would be available at cell edge), the following cell ranges are estimated for different frequency bands to achieve an uplink spectrum efficiency of 0.31 bit/s/Hz under different areas (Urban, Rural, Open): • Band 420 MHz: 1.9 km (urban), 3.3 km (suburban) and 10.4 km (open) • Band 750 MHz: 1.4 km (urban), 2.6 km (suburban) and 8.8 km (open)
Step 4	The number of incident per cell is computed taking into account the population within a cell and the number of PPDR incidents per population. The population in a cell is given by the size of the cell multiplied by the density of population. The number of incidents in the cell is then given by the size of population multiplied by the rate of incidents per population. Multipliers applied to allow for uneven distribution by time of day and geographic location. Based on statistics for the number of incidents occurring simultaneously across Germany at busiest times estimated from publicly available data, the following estimations are considered: • For a 420-MHz cell, up to four incidents per cell sector should be taken into account • For a 750-MHz cell, up to three incidents per cell sector should be taken into account It is considered that number of incidents does not vary across urban, suburban and open cells due to differences in population density that counterbalances the differences in cell sizes
Step 5	The estimate of the total spectrum takes care of the location of incidents within the cell. In particular, it is assumed that one incident is located at cell edge, and the rest of traffic is evenly distributed on the cell coverage. Focusing on the uplink case, spectrum efficiency is considered to be 0.31 bit/s/Hz at cell edge, and two values are considered for the average spectrum efficiency across the whole cell: 0.64 (pessimistic) and 1.49 (optimistic). Based on these considerations together with the total traffic requirements from Steps 1 and 2, uplink spectrum requirements result in the following ranges: • For a 420-MHz cell, 8.0–12.5 MHz • For a 750-MHz cell, 7.1–10.7 MHz

[a] Additional details on the estimates of the throughput requirements for the mobile broadband data applications in use in these scenarios are provided in Chapter 2.

Methodologies such as the ones described previously can also be applied to some extent to the computation of the other spectrum components identified in Section 6.3.1. As a particular case, the basic methodology proposed in Ref. [18] has also been used to estimate the minimum bandwidth requirement for broadband disaster relief (BBDR) applications in localized areas using spectrum in the 5-GHz band [26]. However, the specifics of the localized area as well as the diversity of uses for that spectrum may prevent the ability to firmly determine specific spectrum needs. For example, in the case of spectrum for localized and ad hoc deployments, the applications can range from point-to-point, line-of-sight, broadband links with very high spectral efficiency to local area networking achieving much lower spectral efficiencies (e.g. Institute of Electrical and Electronics Engineers (IEEE) 802.11 type deployments with non-light-of-sight connectivity). The net spectral efficiency will then depend on the individual mix of applications. In addition, frequency reuse may also be highly variable. For example, if a region uses a given spectrum band for airborne operations on multiple assets, nearly the entire band can be exhausted by this application because of the difficulty in reusing frequencies.

6.3.3 Spectrum Estimates

Numerous studies have substantiated the spectrum needs for mobile broadband PPDR applications in different countries and regions across the world [20–26]. The main outcomes from a number of studies are summarized in Table 6.5.

There are many other similar studies addressed to compute the spectrum requirements. A study released in June 2013 considered eight Asian countries, namely, Australia, China, Indonesia, Malaysia, New Zealand, Singapore, South Korea and Thailand. The study supports that a minimum of 10 MHz for broadband PPDR is required on the basis of the opportunity cost argument [27]. In a separate study, China estimated between 30 and 40 MHz of spectrum required for broadband. The United Arab Emirates' telecommunications regulatory authority (TRA) also conducted a PPDR spectrum study that concluded that PPDR use could in theory be supported in as little as 2 × 5 MHz of spectrum and that an allowance of 2 × 10 MHz would enable reasonable future growth [28].

Summing up, despite the differences across some of these estimates, reserving a minimum of 2 × 10 MHz for mobile broadband PPDR is becoming the prevailing option, though not excluding additional country allocations to meet specific needs.

6.4 Existing Spectrum Assignments for PPDR and Candidate Bands for Mobile Broadband

While national administrations do consider the harmonized frequency bands/ranges, or parts thereof, established in ITU Resolution 646 (WRC-2003) for the assignment of spectrum for PPDR communications within their jurisdictions, not all PPDR spectrum assignments fit within the ITU harmonized ranges. Indeed, due to the specific country needs and particular organization of the PPDR sector, the total amount of spectrum assigned to PPDR and the adopted frequency ranges differs significantly between countries. A description of the current PPDR spectrum availability is provided in the following, focusing on existing assignments as well as on the candidate bands under consideration in some regions for the delivery of mobile broadband PPDR communications.

Table 6.5 Studies addressing the assessment of spectrum needs for PPDR communications.

Study	Spectrum requirements
CEPT ECC Report 199, 'User Requirements and Spectrum Needs for Future European Broadband PPDR Systems (Wide Area Networks)', May 2013	The amount of spectrum for future European broadband PPDR wide area networks (WAN) is estimated in the range of 2×10 MHz The report also concludes that there could be additional spectrum requirements on a national basis to cater for DMO, AGA, ad hoc networks and voice communications over the WAN
ETSI SRDoc TR 102 628, 'Additional Spectrum Requirements for Future Public Safety and Security (PSS) Wireless Communication Systems in the UHF Frequency Range'	The ETSI report establishes the following requirements: • 2×3MHz for narrowband PPDR • 2×3MHz for wideband PPDR • 2×10 MHz for broadband PPDR The report proposes the spectrum to be allocated below 1 GHz, narrowband/wideband within the tuning range of existing allocations and a separate band for broadband A revision of the document is underway in ETSI to include more detailed calculations to justify the requirements and to update the document to reflect the changes in the external environment since publication of the initial version of the SRDoc in 2010
WIK Consulting and Aegis Systems, 'PPDR Spectrum Harmonization in Germany, Europe and Globally', December 2010	Minimum spectrum requirements below 1 GHz for Germany are estimated to be 15 MHz (uplink) and 10 MHz (downlink) The spectrum already identified for public safety use in the 5150–5250-MHz band, augmented if possible with spectrum from the largely unused 1452–1479.5-MHz band (currently intended for T-DAB use), should be adequate to address capacity 'hot spots' arising from major events or incidents in Germany A minimum of 15 MHz (unpaired) somewhere between 1 and 5 GHz is estimated to be required on a harmonized European basis to support AGA video links, with a further Germany-specific 7.5 MHz potentially required. Coordination with the military could be considered Wireless backhaul requirements for the WAN can be met from existing microwave fixed link bands, possibly augmented by satellite in remote areas
NPSTC, 'Public Safety Communications Report: "Public Safety Communications Assessment 2012–2022, Technology, Operations, and Spectrum Roadmap"', Final Report, June 5, 2012	Estimates are provided for four disaster scenarios (a hurricane, a chemical plant explosion, a major wild land fire and a toxic gas leak)[a] The lowest demand is estimated at 6 MHz (downlink) and 7.5 MHz (uplink), representing a total of 15 MHz in case of paired allocations The highest demand is estimated at 8.9 MHz (downlink) and 13.8 MHz (uplink), representing a total of 26.7 MHz in case of paired allocations The report states that 20-MHz allocation (of 10 MHz (uplink) and 10 MHz (downlink)) for the highest demanding incident would be nearly 4 MHz insufficient on the uplink

Table 6.5 (*continued*)

Study	Spectrum requirements
Defence R&D Canada – Centre for Security Science (DRDC CSS), '700 MHz Spectrum Requirements for Canadian Public Safety Interoperable Mobile Broadband Data Communications', February 2011	The key conclusions that are derived from this study are: • 10 + 10 MHz is an insufficient bandwidth to support the needs of public safety in the 10–15-year horizon • Improvements in spectral efficiency will likely outpace public safety's demand for data, and as a consequence, the requirement for bandwidth should begin to attenuate beyond 10 years, which is the point when penetration of LTE devices in the public safety community is expected to saturate • Despite the rapid pace of technical innovation, the ability to meet the needs of public safety with 10 + 10 MHz of spectrum in a distant future, that is, beyond 15 years, is not evident, but it is likely that 10 + 10 MHz will not be sufficient at that time either

[a] Additional details on these scenarios are provided in Chapter 2.

6.4.1 European Region

At the European level, the band 380–385/390–395 MHz is in widespread use for permanent narrowband[6] PPDR systems (mostly TETRA and TETRAPOL national and regional networks). Indeed, this 5 + 5-MHz block is the only harmonized band at the European level for narrowband systems, as established in ECC Decision (08)05 [29] and reflected in ITU Resolution 646 as the preferred core harmonized band within the tuning range of 380–470 MHz. Certain channels within this band have also been identified at the European level for DMO and AGA purposes to protect the land mobile infrastructure and to ease cross-border operation (provisions established in decisions ERC/DEC/(01)19 and ECC/DEC/(06)05).

ECC Decision (08)05 also establishes that sufficient amount of spectrum shall be made available for wideband[7] digital PPDR radio applications within the available parts of the frequency range 380–470 MHz. However, the reality is that these frequencies are being heavily used by non-PPDR applications (e.g. civil PMR systems) in most European countries, preventing the designation of additional spectrum for the use of wideband PPDR systems [23].

In addition to the frequency designations for narrowband and wideband PPDR systems, ECC Recommendation (08)04 [30] establishes that administrations should make available at least 50 MHz of spectrum for digital BBDR radio applications. In particular, two possible frequency bands are identified: 5150–5250 MHz (preferred option) and 4940–4990 MHz. Such amount of spectrum is expected to allow PPDR agencies to implement on-scene broadband wireless networks (e.g. 'hot spot' access points in localized areas, temporary incident command centres erected at an incident scene) as well as deploy temporary point-to-point radio links. Nevertheless, the implementation of this ECC recommendation so far is quite

[6] Narrowband systems use channel spacing up to 25 kHz.

[7] Wideband systems can use channel spacing of 25 kHz or more.

Table 6.6 Main bands available across European countries for PPDR communications.

Frequency band (MHz)	Available bandwidth (MHz)	Comments
68–87.5 146–174	— —	Many European countries have national frequency designations for PPDR in the VHF frequency range. However, these are not harmonized allocations across Europe
380–385/390–395	10	Extensively used for the operation of current PPDR narrowband wide area networks (TETRA and TETRAPOL). This is indeed the only actual European harmonized band
385–390/395–399.9 410–430 450–470	10 20 20	These bands form part of the tuning range identified in ECC Decision (08)05 for potential PPDR use, including wideband services (e.g. TEDS), across Europe. However, in many countries, these frequencies are mostly used for non-PPDR applications (e.g. civil PMR systems)
5150–5250(alternatively 4940–4990)	50	Designated for local and temporary use through an ECC recommendation. Not actually implemented in most countries

limited across European countries (according to Ref. [31], only one third of CEPT countries have actually implemented it).

It is worth noting that many European countries also have national frequency designations for PPDR in the VHF frequency range, which are not harmonized throughout Europe. Moreover, some countries have already reserved spectrum specifically for AGA use (e.g. the United Kingdom has aggregated approximately 42 MHz between 3.1 and 3.4 GHz for digital video from airborne vehicles; Germany has allocated spectrum for that purpose in the 2.3-GHz band). However, at the moment, there is no European harmonized frequency band to support this type of applications.

A summary of the main bands with PPDR spectrum designations across European countries is given in Table 6.6.

It is remarkable that no spectrum designations exist as of today in Europe for the deployment of WAN for mobile broadband PPDR services. Recognizing the need to provide the PPDR community with mobile broadband services and the benefits associated with European-wide harmonization of spectrum use for PPDR, the identification of new suitable spectrum bands for mobile broadband PPDR is on the agenda of the European institutions. Importantly, the RSPP approved in 2012 [8] adopted the following commitment in its Article 8.3:

> The [European] Commission shall, in cooperation with the Member States, seek to ensure that sufficient spectrum is made available under harmonised conditions to support the development of safety services and the free circulation of related devices as well as the development of innovative interoperable solutions for public safety and protection, civil protection and disaster relief.

This commitment arrived after a series of efforts and initiatives fostered at different levels across the EU. In fact, back in 2009, the Justice and Home Affairs (JHA) Council of the EU approved a recommendation (Recommendation 10141/09, commonly referred to as the COMIX recommendation [33]) on improving radio communication between operational units in border areas. The COMIX recommendation concluded that law enforcement and public safety radio communication systems will need to support and to be able to exchange high-speed mobile data information beyond the capabilities of current networks, and a common standard operating in a harmonized frequency band will make this possible. Consequently, the COMIX recommendation suggested that CEPT/ECC should be tasked to study the possibility of obtaining sufficient additional frequency allocation below 1 GHz for the development of future law enforcement and public safety networks. Accordingly, the LEWP supporting the JHA Council activities established a Radio Communication Expert Group (RCEG) to work on relevant radio and spectrum matters, which asked CEPT/ECC to take into account the PPDR needs for a mission-critical broadband solution and for this purpose to designate harmonized frequencies [34].

It is worth noting that the current approach towards the harmonization of the broadband PPDR sector in Europe is not targeting a common frequency designation across Europe. Instead, a number of deployment options and spectrum bands for the delivery of broadband PPDR services in Europe are intended to be recognized [32]. In this way, national administrations could opt for the most suitable option or combination of options according to its particular national circumstances. At the end, this approach is believed to provide for the minimum harmonization needed to facilitate adequate interoperability and contribute to maximizing the benefits from the economies of scale for the delivered PPDR solutions.

The gross of the work towards the specification of this harmonized mobile PPDR communications framework in Europe is currently being conducted within the Frequency Management Project Team 49 (FM PT 49) at the CEPT/ECC level. This work is being addressed in cooperation with ETSI and other key organizations such as the RCEG/LEWP. It is expected that FM PT 49 work will eventually culminate in a CEPT/ECC deliverable by 2016 defining the harmonized framework for PPDR communications in Europe. More details on the expected deliverables and the roadmap established by FM PT 49 are given in Section 3.4.2.

Concerning the spectrum bands under consideration for this harmonized European PPDR framework, two candidate spectrum ranges have been identified within FM PT 49: 400 MHz (410–430 and 450–470 MHz) and 700 MHz (IMT-band, 694–790 MHz). The selection of the most suitable spectrum band for mobile broadband PPDR needs to assess and balance many different aspects such as:

- The potential availability of the band(s) for broadband PPDR and the associated timeframe
- The cost of the band's refarming
- The available bandwidth and contiguity of the band
- The radio propagation conditions
- The risks and character of likely interference
- The potential for the development of business ecosystem in this band(s)
- The potential to achieve harmonization with band plans in other regions to benefit from economies of scale

Some main considerations in this regard concerning the two identified candidate spectrum ranges are given in Table 6.7. As previously mentioned, it is important to stress that the provisioning of broadband PPDR in Europe is not intended to be restricted to the use of 400- or

Table 6.7 Candidate bands considered in the harmonization of mobile broadband PPDR solutions across European countries.

Frequency band	Comments
410–430 and 450–470 MHz	This band shows very good propagation characteristics, potentially reducing the number of sites needed to provide the necessary coverage (rural areas) This band would also allow benefiting from infrastructure from narrowband PPDR networks deployed in the 380–400-MHz band and facilitate a progressive implementation of LTE systems (e.g. in gradual steps of 1.4, 3, 5 MHz) which may end up in reusing at longer term the 380–385-MHz band (and duplex) for high data rates services by carrier aggregation of different sub-bands However, this band is not considered to be a stand-alone solution for the full 2×10-MHz need for BB PPDR due to its limited availability in some European countries as well as the high cost that re-planning of this band might have (in addition to current uses, this band is also being considered as a suitable band to support M2M communications such as smart metering and smart grid in some European countries). For this reason, the deployment of a PPDR broadband solution in this band is mainly regarded as a complement to another solution (e.g. a 700-MHz solution or a roaming agreement with a commercial LTE operator) Mobile broadband PPDR in this band could be realized as a 2×5-MHz LTE solution in the frequency range 410–430 and/or 450–470 MHz. Compatibility studies are being addressed at CEPT based on LTE technology (3GPP Release 12) with channel bandwidths of 1.4, 3 and 5 MHz. Results of these studies are expected to form part of ECC Report 218, to be released by mid-2015. Indeed, 3GPP has already added support for the use of LTE in the band 452.5–457.5 MHz paired with 462.5–467.5 MHz, which is specified as Band 31 and commonly known as LTE 450 MHz. Furthermore, the 450 MHz Alliance, an industry-based organization representing interests of 450 MHz spectrum stakeholders, is also pushing for the development of the LTE 450 MHz ecosystem. In this regard, first LTE 450 network deployments in the commercial domain have already been announced in 2014 [35]
694–790 MHz	This band is supported as the main candidate spectrum option for mobile broadband PPDR across Europe. Many countries see an opportunity for a national decision within this range for BB PPDR services, either as a dedicated network, a commercial solution or a combined (hybrid) solution The 700-MHz band in Europe is expected to be dedicated to wireless broadband by 2020 (±2 years) [38]. The band is currently occupied by terrestrial TV broadcasters and wireless microphones. This band repurposing is known as the second digital dividend (the first was the repurposing of the 800-MHz band) The harmonized technical conditions for the use of the 700-MHz band (694–790 MHz) for wireless broadband in the EU have been already detailed in CEPT Report 53, in response to a European Commission Mandate. The CEPT Report 53 sets out the channelling arrangement based on a paired 2×30-MHz scheme (703–733/758–788 MHz) and a flexible approach to accommodate up to four blocks of 5 MHz for supplementary downlink (SDL) in the 738–758-MHz part of the duplex gap. This channelling arrangement of the 700-MHz band is well aligned with the Asia-Pacific Telecommunity (APT) 700 band plan, which is adopted by most countries across the Asia-Pacific and Latin America regions. The APT700 segmentation is based on two overlapping duplexers of 30 + 30 MHz, in which the lower duplexer band fits perfectly in the European scheme. Hence, a possible decision by national administrations could be the use of one or a number of blocks within this 2×30-MHz pairing for PPDR purposes

Table 6.7 *(continued)*

Frequency band	Comments
	In addition, the channelling structure under consideration in CEPT also provides special arrangements that could be used as national options for PPDR usage (and/or other possible applications such as PMSE and M2M), as illustrated in Figure 6.2. One of these special arrangements for PPDR is to use the guard bands (i.e. 698–703- and 788–791-MHz blocks, paired with some spectrum within the duplex gap of the paired 2×30-MHz block in 733–758 MHz). In particular: • A 2×5-MHz FDD channelization on the frequency bands 698–703 MHz (UL) and 753–758 MHz (DL) (with a conventional duplex) • A 2×3-MHz FDD channelization on the frequency bands 733–736 MHz (UL) and 788–791 MHz (DL) (with a conventional duplex) Another special arrangement is to use the 25-MHz duplex gap of the 2×30-MHz paired scheme to support a 10+10-MHz allocation for PPDR between 733 and 758, with an internal gap of (only) 5 MHz, even though this option faces important technical difficulties associated with the duplexer design for a 5-MHz duplex gap if the terminal has to cover the whole frequency range [36, 37] On the plus side of using special arrangements for PPDR is that it could increase the chances for PPDR allocations without directly competing for spectrum with commercial operators. On the downside, a dedicated sub-band would create a niche market for PPDR, with possible negative effects on economies of scale and interoperability. In addition, PPDR adjacent to the 700-MHz band leaves less protection for broadcast networks. Compatibility and sharing studies for broadband PPDR systems considering these special arrangement options are currently being addressed in CEPT ECC. Results of these studies are expected to form part of ECC Report 218

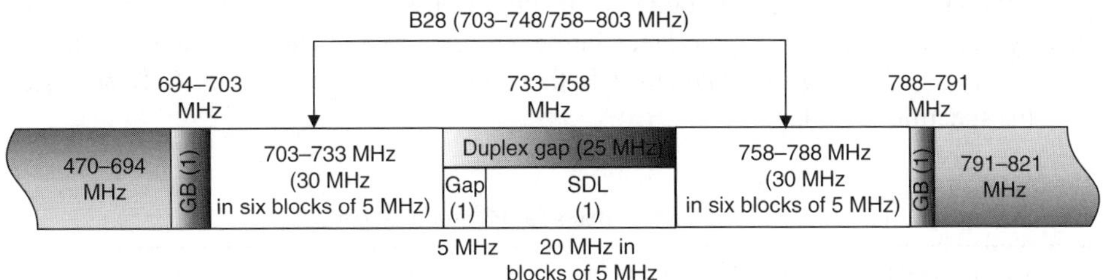

Figure 6.2 The channelling arrangement for the 700-MHz band in Europe 25. (1) indicates that the usage of the guard bands and of the duplex gap of the paired band plan (733–758 MHz) may also be considered at national level for PPDR use and/or other possible applications (e.g. PMSE, M2M).

700-MHz spectrum. Instead, it is envisioned that PPDR capability can be enhanced by allowing mobile broadband PPDR services to be delivered across different bands (both below and above 1 GHz) and networks (both dedicated and commercial). Therefore, it is central that manufacturers produce multiple band integrated chipsets, including the designated PPDR ranges, using a common technology for PPDR user terminals ideally on a global or regional basis. In fact, leading chipmakers in the mobile industry have nowadays solutions that support access to over 12 spectrum bands, many of them under 1 GHz [39].

In addition to the harmonization of spectrum suitable for the deployment of wide area mobile broadband PPDR networks, CEPT is also conducting studies that may end up with the allocation of additional harmonized spectrum for other PPDR-specific uses [40]. In particular, ECC is developing CEPT Report 52 in response to a recent EC mandate to undertake studies on the harmonized technical conditions for the 1900–1920 and 2010–2025 MHz frequency bands (unpaired terrestrial 2-GHz bands) in the EU. The aim of the report is to assess and identify alternative uses of the underused unpaired terrestrial 2-GHz bands other than for the provision of mobile broadband services through terrestrial cellular networks, as well as the development of relevant least restrictive technical conditions for spectrum use. Potential harmonized uses of this band consider BB PPDR ad hoc communications (e.g. local video links) including, for some countries, PPDR broadband AGA applications (e.g. video links between terrestrial units and helicopters). In addition, spectrum at the 5-GHz frequency range (i.e. 4940–4990 MHz) identified for BBDR radio applications in ECC/REC/(08)04 may be subject to further harmonization activities within CEPT.

6.4.2 North America

In the United States, PPDR agencies can licence spectrum in seven separate frequency bands where the FCC has allocated spectrum for public safety use over the years [21]. Those bands, together with indications on the amount of spectrum available in each and its uses, are listed in Table 6.8. As shown in the table, there are several bands in use for narrowband voice and low-speed data systems. Each voice band has unique propagation characteristics, and each band is good or bad for different types of systems. The 30–50-MHz band is primarily used in some statewide systems to provide mobile coverage of highways. The VHF band is a good band for rural areas, while the 450- and 700/800-MHz bands are used in urban and suburban areas where good portable coverage is needed. The 700/800-MHz bands are best suited for trunking systems and increasingly are being used for large regional and statewide systems to provide improved communications and interoperability across multiple agencies and jurisdictions. Although not shown in the table, some PPDR entities in the United States currently utilize systems in some geographic areas in the 470–512-MHz band, known as the T-Band. However, the US Congress mandated in 2012 that public safety agencies with T-Band systems vacate the spectrum by 2021.

Table 6.8 Spectrum available in the United States for PS communications.

Frequency band (MHz)	Available bandwidth (MHz, approximate)	Comments
25–50	6.3	Used for narrowband services
150–174	3.6	Used for narrowband services
220–222	0.1	Used for narrowband services
450–470	3.7	Used for narrowband services
809–815/854–860	3.5	Used for narrowband services
806–809/851–854	6	Used for narrowband services
758–768/788–798	20	Wide area broadband
768–769/798–799	2	Guard
769–775/799–805	12	Used for narrowband services
4940–4990	50	Short-range broadband and point-to-point links

With regard to mobile broadband PPDR, the frequencies that have been allocated in the United States are those in the range 758–768/788–798 MHz. Indeed, this is the spectrum assigned to FirstNet for the deployment of the nationwide public safety network. It is also worth noting that the Spectrum Act passed in 2012 also allows a flexible use of broadband in the 700-MHz narrowband spectrum (769–775 and 799–805 MHz), but any move to do so would first need to consider the potential for interference between broadband and narrowband systems.

In addition, the 4.9-GHz band is available in the United States for short-range broadband data and point-to-point data links. The FCC reallocated this band (4.94–4.99 GHz, 50 MHz of spectrum) from federal government use in 2002 and adopted rules designating the band for public safety services in 2003. The band can be used for any terrestrial-based radio transmission including data, voice and video. All multipoint and temporary (<1 year) point-to-point links are primary users of the band. Permanent point-to-point links are secondary users and require separate site licences. The FCC has implemented a geographic licencing scheme for mobile applications. A licence grants a PPDR agency authorization to use all 50 MHz of spectrum within its legal jurisdiction whether that jurisdiction is a state, town, city or county. Licencees must share the spectrum and coordinate frequency use. However, the overall operation can be made very reliable because the general public is prohibited from using the band. This band is currently under review by the FCC to promote greater use of the band, including opening up a part of the band for commercial applications and finding ways to complement the nationwide interoperable LTE public safety broadband network currently in development [41].

The United States was the first country to allocate broadband PPDR spectrum in the 700-MHz band. Initially, 5 + 5-MHz spectrum was allocated, though early in 2012 the US Congress passed a law consolidating an additional 5 + 5 MHz (the so-called D-Block) that sat contiguous to the initial allocation. The channelling arrangement for the 700-MHz band in United States is illustrated in Figure 6.3. Canada is following a similar allocation.

For the deployment of LTE equipment in the US band plan, the 3GPP has designated four operating bands: 12, 13, 14 and 17 (Band 17 is a subset of Band 12). The spectrum already assigned for mobile broadband falls within Band 14 (758–768 and 788–798).

6.4.3 *Asia-Pacific and Latin America*

In the Asia-Pacific region, the 806–824-MHz segment is widely used for the provision of narrowband communications in support of PPDR applications. This allocation is harmonized through ITU Resolution 646.

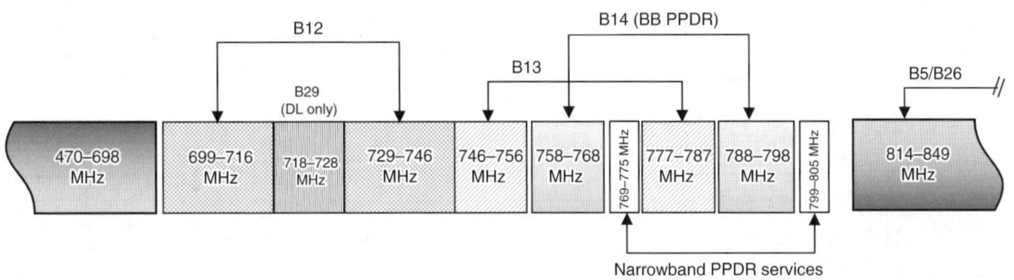

Figure 6.3 The channelling arrangement for the 700-MHz band in United States and PPDR allocations within this band.

The 800-MHz band is also considered by some countries (e.g. Australia, Singapore) for the designation of new spectrum for mobile broadband PPDR, while other national administrations consider the 700-MHz band (South Korea, Mexico, United Arab Emirates) [42]. In the case of Australia, the NRA (ACMA) is undertaking a number of initiatives to improve spectrum provisions for public safety. The most important are:

- Making provision for 10 MHz of spectrum from the 800-MHz band for the specific purpose of realizing a nationally interoperable a PPDR mobile broadband cellular 4G data capability. This band supports 4G (LTE) systems, and as such, it is considered to be 'beach front' spectrum by carriers and PPDR agencies alike. The actual frequencies to be provided within the 800-MHz band will be determined later in the context of the ACMA's review of the 803–960-MHz band.
- Enabling 50 MHz of spectrum from the 4.9-GHz band for PPDR agencies. This spectrum is recognized internationally as a PPDR band in Regions 2 and 3, capable of extremely high-capacity, short-range, deployable data and video communications (including supplementary capacity for a WAN in locations of very high demand).
- Implementing critical reforms in the 400-MHz band, where spectrum has been identified for the exclusive use of government, primarily to support national security, law enforcement and emergency services, is ongoing.

The channelling arrangements for the 700 and 800 MHz are illustrated in Figure 6.4, showing some of the considered allocations for mobile broadband PPDR.

The segmentation of the 700 MHz follows the harmonized APT band plan, commonly known as APT700 band plan. This band plan has been standardized by the 3GPP into two operating bands: Band 28 for FDD operation and Band 44 for TDD operation. The FDD option (Band 28) has attracted the most support from the industry so far. Indeed, many countries have already allocated, committed to or recommend allocating APT700 FDD (Band 28) spectrum for LTE deployments. It is worth noting that the APT700 band plan is not compatible with the 700-MHz arrangement adopted in the United States, since both rely on different channel bandwidths and channel locations within the 700-MHz band.

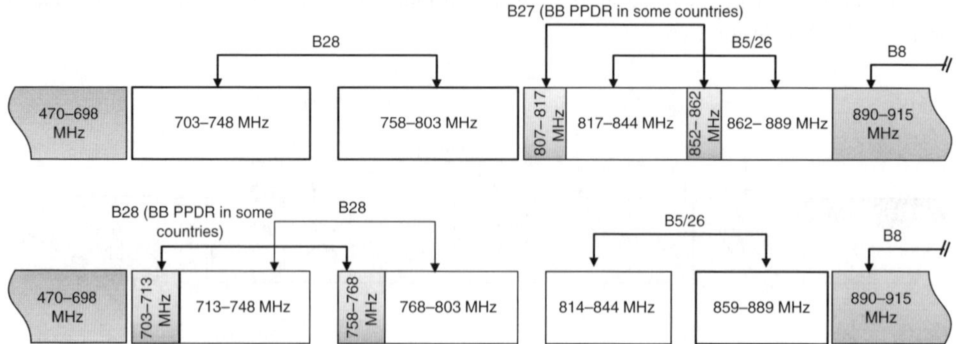

Figure 6.4 Channelling arrangements for the 700 and 800 MHz in the Asia-Pacific region and main PPDR designation within these bands.

In a further boost to the dominance of the 700-MHz band for PPDR LTE systems, it was recently announced that Brazilian public safety, national defence and critical infrastructure services will be allocated spectrum in this band. Significantly, in a departure from the common South American practice of adopting the US spectrum planning arrangements, the announcement stated that the APT 700-MHz band (Band 28) would be used [43]. In a related effort, the CITEL recommended to its member states across North, Central and South America that PPDR broadband spectrum allocations should be made along the 700-MHz band, considering either the APT 700-MHz band (Band 28) or the US band (Band 14).

6.5 Spectrum Sharing for PPDR Communications

The case of assigning a minimum amount of dedicated spectrum to support the delivery of wide area mobile broadband PPDR services is gaining momentum in many countries.[8] As discussed in the previous section, some national administrations have already designated spectrum blocks of 10 + 10 or 5 + 5 MHz in the 700- or 800-MHz bands for mobile broadband PPDR use. This amount of dedicated spectrum is estimated to be enough to satisfy PPDR needs for mission-critical communications in most operational scenarios, as concluded in some of the studies presented in Table 6.5. However, it is also recognized in some of those studies that this amount of spectrum might fall short to satisfy the capacity requirements in a major incident. Indeed, estimations by ECC PT49 [20] and NPSTC [21] show some worst case scenarios where more than 10 + 10 MHz of spectrum would be required. Nevertheless, it is also acknowledged that it would be highly inefficient to dimension PPDR spectrum provisions around what might be a once-in-a-generation event, likely resulting in a very low utilization of a significant portion of the dedicated spectrum during most of the time.

In this context, the introduction of spectrum sharing approaches between PPDR and other services in some specific bands could be instrumental to have the required degree of flexibility that would allow PPDR to use substantially more spectrum at times of stress while ensuring the utilization of this spectrum for other uses at times of day-to-day PPDR activity. Spectrum sharing refers here to the application of technical methods and operational procedures to permit multiple users to coexist in the same region of spectrum [44]. The use of spectrum sharing for PPDR use constitutes a plausible approach to complement a dedicated assignment with additional spectrum that could be required to handle an exceptional demand or just for the deployment of particular PPDR applications in a more efficient manner at specific times and locations. Likewise, the introduction of spectrum sharing could become a facilitator for administrations to designate a larger amount of spectrum for PPDR use (e.g. 20 + 20 MHz instead of 10 + 10 MHz) under the basis that (a portion of) this spectrum can be effectively shared with others when not actually in use by PPDR applications [45].

The consideration of a spectrum sharing framework for PPDR needs to evaluate two main central requirements: availability (i.e. guarantees that the sufficient amount of spectrum will be available when and where it is needed) and responsiveness (i.e. how fast the spectrum is ready for PPDR use since the need arises). Depending on the level of fulfilment of these

[8] There are only few exceptions such as the United Kingdom that favours a model based on procuring commercial services off a mobile network operator to provide broadband PPDR services, without reserving any specific spectrum for PPDR use.

requirements, a particular spectrum sharing solution can be considered suitable to support mission-critical applications or just intended to provision additional capacity for non-mission-critical applications.

The possibility to rely on a higher spectrum amount on a temporary basis with respect to that used in day-to-day conditions is captured in ITU-R Resolution 646, 'Public Protection and Disaster Relief', from WRC-2012, where it is recognized that 'the amount of spectrum needed for public protection on a daily basis can differ significantly between countries, that certain amounts of spectrum are already in use in various countries for narrow-band applications, and that in response to a disaster, access to additional spectrum on a temporary basis may be required'. Furthermore, there are several ongoing initiatives at regulatory and standardization level that envision the applicability of spectrum sharing techniques for PPDR communications to some extent:

- Standardization mandate (M/512) by the EC to the European standardization bodies to identify an approach and a number of issues where standardization should enable the development and use of Reconfigurable Radio Systems (RRS) technologies in Europe [46]. This EC mandate was issued in November 2012 and work is currently underway. In particular, Objective C of this mandate is targeted to explore potential areas of synergy among commercial, civil security and military applications. These include architectures and interfaces for dynamic use of spectrum resources among these three domains for disaster relief.
- Communication by the EC (COM(2012) 478) on the promotion of the shared use of radio spectrum resources in Europe [47]. This EC communication proposes the development of two tools to provide more spectrum access opportunities and to incentivize greater and more efficient use of existing spectrum resources: (1) an EU approach to identify beneficial sharing opportunities (BSOs) in harmonized or non-harmonized bands and (2) regulatory tools establishing the so-called shared spectrum access rights (SSARs) to authorize licenced sharing possibilities with guaranteed levels of protection against interference. On this context, the communication identifies as a potential use case: 'Incumbent rights holders could benefit from the mutual reassurance of an appropriate sharing contract by proposing BSOs, e.g. public entities could offer access to spectrum capacities to commercial operators in return for co-funding of network infrastructures for broadband public protection and disaster relief (PPDR) applications'.
- New Citizens Broadband Radio Service (CBRS) in the 3.5-GHz band under development by the FCC in the United States. In December 2012, the FCC issued a notice proposing to allocate 3550–3650-MHz band for small cells via a shared access scheme [48]. This band is currently used for US naval radar operations. The FCC notice proposes a three-tiered shared access system enforced by a spectrum access system (SAS) and the use of geo-location-based opportunistic access technology, in which PPDR applications would be afforded quality-assured access to a portion of the 3.5-GHz band in certain designated locations.

The next subsection develops some fundamental concepts and provides a categorization of spectrum sharing models for PPDR communications together with a discussion on the suitability of each model. On this basis, the subsequent subsections further develop two possible solution frameworks for PPDR spectrum sharing: one based on applicability of the Licenced

Shared Access (LSA) regime and another exploiting secondary access to TV white spaces (TVWS) for PPDR use.

6.5.1 Spectrum Sharing Models

Spectrum sharing may be achieved by numerous methods such as coordinating time usage, geographic separation, frequency separation, directive antennas and so on. In the past, the employment of spectrum sharing mechanisms has typically been on a static, pre-planned basis. For example, the simplest means of spectrum sharing is the operation of systems in the same frequency band but in different geographical areas. Nevertheless, spectrum sharing can be more complex and allow sharing frequencies in the same geographical area. For instance, cognitive radio (CR) technologies (e.g. geo-location databases (GLDBs), spectrum sensing) can be utilized to adapt, in a dynamic and flexibly manner, the spectrum used by PPDR communications equipment considering the presence and activity of other users of the shared band.

A general classification (not specific to PPDR) of spectrum sharing models is based on two defining features [49]:

1. Whether the spectrum sharing arrangement comprises **primary–secondary sharing** or **sharing among equals**. In the former case, some systems have the right to operate as a primary spectrum user, and policy mandates that secondary devices are not allowed to cause harmful interference to a primary system.[9] In the latter case, all devices have equal rights, and typically there is more flexibility about how to behave in the presence of peers.
2. Whether sharing is based on **cooperation** or **coexistence**. In a model based on cooperation, systems or devices sharing the band must communicate and cooperate with each other to avoid mutual interference. With a coexistence model, devices try to avoid interference without explicit signalling (at most, devices sense each other's presence as interference and apply 'good-neighbour' sharing practices to use the common resource).

Based on these two above-mentioned features and on the types of licencing regimes and spectrum management models discussed in Section 6.1, three categories of spectrum sharing models for PPDR use can be distinguished [50]:

1. **Models based on the dynamic transfer or coordination of individual spectrum rights of use** between sharers so that, at a given time and location, there is only one user authorized to use the spectrum.
2. **Models based on primary–secondary sharing**, where there is a primary user that holds individual rights of use for a given spectrum band but multiple secondary users are allowed to access the spectrum in an opportunistic manner whenever the primary user is not affected. Two variants of this model are distinguished based on the use of either cooperation (through coordination mechanisms) or coexistence approaches between the primary and secondary users.

[9] The concepts of 'primary' and 'secondary' uses are actually defined under the ITU RR (see Section 6.1), which establishes that frequency bands are allocated to radiocommunication services on a primary or secondary basis. Stations of a secondary service shall not cause harmful interference to stations of primary services and cannot claim protection against harmful interference from stations of a primary service.

3. **Models based on the CUS of a shared band**, where multiple distinct users with equal rights operate in the same range of frequencies at the same time and in a particular geographic area under a well-defined set of sharing conditions. Two variants of this model are also distinguished based on the use of either cooperation (through coordination mechanisms) or coexistence approach between the sharers.

The feasibility of each sharing model for PPDR communications primarily depends on the type of users involved in the sharing framework, which may involve users across the commercial, military and PPDR domains. The adoption of a given sharing model may require changes to the organizational structures and relationships among these users. In some cases, these changes are just an extension of existing agreements (e.g. joint procedures for disaster management between military and public safety entities in large natural disasters). In other cases, new agreements (e.g. sharing rules or conditions among users) must be put in place. Moreover, the amount of changes required in the existing infrastructures managed by these different communities is another important aspect to consider. Any proposed sharing model should minimize the changes to the existing infrastructures.

The adoption of a sharing model is also dependent on the development of suitable technologies and regulatory frameworks. Different sharing models may require more or less complex modifications to existing standards and undertake different technical challenges. In some cases, the technical requirements for specific functions (e.g. spectrum sensing case of models built upon CR technologies) may be difficult to implement with existing technological capabilities (e.g. computing/processing power). Furthermore, international and national spectrum regulations must be modified to permit the deployment of some of the sharing models.

Based on earlier observations, Tables 6.9, 6.10 and 6.11 describe the principles, applicability and some examples for the three identified categories of spectrum sharing models. The tables also provide a discussion on the suitability of each model considering organizational and operational aspects of the involved users as well as relevant technical and regulatory initiatives that can contribute to pave the way towards their adoption.

6.5.2 Shared Use of Spectrum Based on LSA

LSA is a new spectrum regulatory approach that facilitates the introduction of new users in a frequency band while maintaining incumbent services that may exhibit low or localized utilization in the band. LSA fits under an individual licencing regime, so that *LSA licences* are to be granted to the new users. LSA aims to ensure a certain level of guarantee in terms of spectrum access and protection against harmful interference for both the incumbent(s) and LSA licencees, thus allowing them to provide a predictable quality of service (QoS) (i.e. each user would have exclusive individual access to a portion of spectrum at a given location and time). Therefore, LSA excludes concepts such as 'opportunistic spectrum access', 'secondary use' or 'secondary service' where the new user has no protection from primary user(s). Remarkably, LSA applies only when the incumbent user(s) and the LSA licencees are of different nature (e.g. governmental vs. commercial users) and operate different types of applications, being subject to different regulatory constraints. Hence, an LSA sharing framework is expected to have limited impact – likely no impact – on the market regulation policy objectives since incumbent and LSA licencees belong to two different vertical markets. From the incumbent's

Table 6.9 Models based on the dynamic transfer or coordination of individual spectrum rights of use.

Sharing principles	Individual rights of use are dynamically transferred or coordinated between users by means of, for example, spectrum leasing procedures or technical mechanisms that guarantee that there is only a single user authorized to use the spectrum at a given time and location. Prioritization and pre-emption principles can be considered in the transfer or coordination to give preferential access to some users
Applicability	Spectrum bands where access authorization relies on holding individual spectrum rights of use. This includes traditional licences as well as LSA licences[a]
Illustrative examples	*Temporary transfer of spectrum usage rights from non-PPDR to PPDR users.* PPDR users involved in an incident response or in a major planned event (e.g. Olympics games) could request, for example, military authorities or other private holders to lease part of its spectrum for PPDR use *Temporary transfer of the spectrum usage rights from PPDR to non-PPDR users.* In this case, when there is no emergency situation and part of spectrum designated for PPDR is not being used, spectrum usage rights could be leased to other users such as telecom operators. This leasing could be interruptible under strict guarantees when required by the PPDR spectrum licencee
Suitability considerations	Transfer of exclusive spectrum rights of use is already regulated in many countries [51–53]. Current spectrum transfer procedures can take some days, which is suitable for long-planned events (e.g. G20 summit or Olympic games). Operation at lower timescales needs further regulatory and technical developments (i.e. new technical capabilities are needed for executing the transfer in the order of 30 min/1 h, which is the timescale for the initial response to an emergency crisis) The advantage of this solution is that spectrum availability is guaranteed with exclusive spectrum usage rights for all the duration of the lease. Therefore, if a framework is defined and deployed to guarantee also the timeliness of the provision of the spectrum, this is a feasible model even for mission-critical PPDR applications. Its realization needs cognitive or (at least) tuneable radios that can be configured to operate in different spectral bands Initial deployment could be restricted to spectrum transfers among PPDR users, including the possibility to create spectrum pools contributed by multiple licencees for mutual use [54]. Extension to other governmental and/or commercial marketplace users could be addressed in a subsequent stage Newly allocated bands could explicitly be designated by regulatory authority for spectrum sharing through dynamic transfer of rights of use. Assignment of spectrum usage rights could be managed through a centralized mechanism in the form of spectrum coordination server or spectrum broker [55]. A new spectrum regulatory model such as LSA can facilitate this approach Dynamic transfer of exclusive rights of use can be applied between military domain and PPDR organizations, but it will require new regulatory frameworks and new procedural interfaces between the correspondent control centres [56]

[a] Licenced Shared Access (LSA) is a new spectrum regulatory approach that facilitates the introduction of new users in a frequency band while maintaining incumbent services that may exhibit low or localized utilization in the band. The new users hold LSA licences that grant individual spectrum usage rights subject to some utilization conditions that account for the existence of incumbent users. More details on LSA are covered in Section 6.5.2.

Table 6.10 Models based on primary–secondary sharing.

Sharing principles	Secondary access is allowed in an opportunistic manner in a band where there is a primary user that holds the spectrum rights of use. Secondary access is opportunistic (i.e. time and location availability of the spectrum are dependent on the actual activity of the primary user that can be dynamic) and in a non-interference basis (i.e. the primary user shall not be affected by secondary transmissions). Two possible variants are distinguished: 1 A primary–secondary coordination mechanism is used to allow the primary user to have some control on the secondary access (e.g. dynamically decide whether secondary access is allowed or not) 2 There is no primary–secondary coordination mechanisms so that primary users don't have control over secondary access (i.e. primary and secondary users coexist without explicit interactions)
Applicability	Spectrum bands where there is an (primary) owner of spectrum rights of use and the NRA decides to authorize secondary access (e.g. TV UHF bands in the United States and United Kingdom)
Illustrative examples	*Secondary access allowed to PPDR users in non-PPDR bands*. For example, communication devices such as ad hoc communications systems brought in the incident area by PPDR agencies could use, for example, a military band on a restricted geographical basis [57]. Coordination mechanism between military and PPDR agencies could be established *Secondary access is allowed in PPDR bands*. For example, a PS network operator may advertise that part of spectrum is not being used and so make this spectrum available for secondary access. Secondary users can be restricted to other PPDR applications or be open to non-PPDR services like critical infrastructure agencies (e.g. energy and other utilities) or even commercial use
Suitability considerations	Solutions for PPDR spectrum sharing may benefit from proposals and achievements within the TV white space domain, based at present on the usage of a geo-location database [58] If PPDR is primary use, it is a feasible model even for mission-critical PPDR applications. Indeed, secondary access to (primary) PPDR spectrum is not precluded by ETSI in Ref. [23] under a strict pre-emptive regime to ensure the performance of PPDR communications. Using PPDR spectrum for commercial use with preferential access given to PPDR in case of emergencies was also considered by FCC in an intent to promote the deployment of a joint-use network employing both PS and commercial spectrum (i.e. D-Block) [59] If PPDR is secondary use, it can provide an opportunistic additional capacity to alleviate congestion problems for mission-critical applications as well as facilitate the deployment of non-mission-critical applications. PPDR secondary access to (primary) military spectrum is a possible approach, considering that military organizations possess considerable regions of spectrum that may not be used in the location of the incident. A three-level sharing scheme, where military is the primary user, PPDR is a second-tier primary user and commercial networks are the secondary users, is discussed in Ref. [23] Sharing between military users and PPDR users can be done on the space dimension or the time dimension as described in Ref. [57]. In the space dimension, PPDR users can use the spectrum in an opportunistic way if the spectrum is used by the military only in specific areas (e.g. military compounds). In the time dimension, spectrum can be shared as in the case of radar [60]. For example, low-power systems could potentially share with radar if the radar sweep can be detected and the transmission of the device timed to avoid interference. This solution would require a close coordination or reliable technical solutions to ensure that the military radio communication services are not impacted by harmful interference. A potential challenge is the lack of network interfaces between military networks and PPDR networks due to security reasons Coordination models can offer more QoS guarantees at the cost of added complexity. Coordination can be addressed at the communications system level (e.g. beacon signals broadcasted by PPDR networks to enable/disable secondary access) or at the organizational and procedural level (e.g. extension of existing procedures for coordinated disaster management in the case of military–PPDR spectrum sharing) Coexistence models are necessary in cases that coordination fails or is not possible (e.g. geo-location databases not reachable). CR technology is particularly relevant in these cases

Table 6.11 Models based on a collective use of spectrum.

Sharing principles	Multiple users are authorized to use the band as a result of either a general authorization regime (e.g. licence-exempt band with no limitations in the number of users) or a light-licencing regime (e.g. limits on the number of authorizations might be in place) Two variants: 1 Coordination among authorized users/devices is required through a common management protocol in order to cope with mutual interference 2 No common management protocol is defined among authorized devices. Instead, coping with mutual interference is mainly pursued through the compliance of devices to the specific regulator-imposed rules (commonly denoted as 'spectrum etiquettes')
Applicability	Spectrum bands where collective usage rights are in place instead of individual (exclusive) usage rights
Illustrative examples	*Shared access in a spectrum band designated only for PPDR use*. For example, all registered and explicitly authorized PPDR agencies might use this band to set up fast deployable equipment (e.g. wireless access points, point-to-point links). Coordination for, for example, channel assignment could be carried out through a common protocol supported by all authorized devices. The development of such a common protocol is facilitated by the restriction of this band to PPDR applications *Shared access in a general purpose licence-exempt band such as the 2.4- or 5-GHz ISM bands*. The use of this band can bring additional capacity in the incident area for local area communications, yet no preferential access or coordination mechanisms will be available for PPDR users to control the interference from any other legitimate user of the band (e.g. personal devices or private/public wireless access networks)
Suitability considerations	Application-specific bands for PPDR communications are already available in the United States (4.9-GHz band) and in some European countries (broadband disaster relief [BBDR] band in the 5-GHz frequency range), especially to implement on-scene broadband wireless networks. Authorized users are responsible for interference prevention, mitigation and resolution coordination among them The establishment of frequency planning coordination processes would be also a plausible option to share these bands with other non-PPDR users such as utilities or transportation. This approach has been recently proposed by the National Public Safety Telecommunications Council (NPSTC) with regard to allowing access to 4.9-GHz spectrum to critical infrastructure industries, including the energy sector, on a shared, co-primary basis with public safety Coordination can be also based on technical mechanisms (e.g. use of technologies such as IEEE 802.11y that could allow frequency coordinators to have dynamic control over channel access in a PPDR shared band) Shared spectrum can be used for mission-critical PPDR applications if suitable coordination mechanisms such as those referred above are in place Coexistence approaches (e.g. using existing general purpose licence-exempt bands such as ISM bands at 2.4 and 5 GHz) cannot offer QoS guarantees for mission-critical PPDR. However, as proven by the massive adoption of Wi-Fi devices, the reality is that achieving some additional capacity with good perceived QoS in these bands is not so unlikely. This fact can be even more evident in the utilization of those bands in non-residential areas (e.g. crisis incident in rural areas)

perspective, LSA could be an alternative to spectrum refarming processes that might be too costly to be implemented, not possible in a reasonable timeframe or simply not desirable.

The LSA[10] framework was proposed by the EC's RSPG in 2011 [61]. The RSPG Opinion on LSA [62] approved in November 2013 provides the following definition:

> A regulatory approach aiming to facilitate the introduction of radiocommunication systems operated by a limited number of licensees under an individual licensing regime in a frequency band already assigned or expected to be assigned to one or more incumbent users. Under the Licensed Shared Access (LSA) approach, the additional users are authorised to use the spectrum (or part of the spectrum) in accordance with sharing rules included in their rights of use of spectrum, thereby allowing all the authorised users, including incumbents, to provide a certain Quality of Service (QoS).

It is worth noting that the LSA concept embraces the so-called Authorised Shared Access (ASA) concept, which was firstly proposed by an industry consortium (involving Qualcomm and Nokia) to provide shared access to IMT spectrum under a licensing regime in order to offer services with a certain quality of service [63]. In particular, it could be said that both LSA and ASA concepts are equivalent excepting the fact that ASA was proposed in the context of IMT bands, while the LSA concept has been defined with a broader scope of applicability, not limited to IMT bands. Therefore, the terms LSA and ASA terms are often used interchangeably, and even sometimes some documents use the notation 'LSA/ASA' model [64].

A report on LSA (ECC Report 205 [10]) was released by CEPT ECC in February 2014. ECC Report 205 establishes the scope and components of the LSA regulatory approach and provides detailed considerations towards the potential implementation of LSA in the 2300–2400-MHz ('2.3-GHz') frequency band in the EU to provide access to additional spectrum for mobile broadband services (referred to them as mobile/fixed communications networks (MFCN)). In this regard, technical harmonized conditions for this band to support wireless broadband ECS are being developed by CEPT in response to a mandate by the EC [65]. In particular, ECC Report 55 [66] and related ECC Decision 14(02) [67] set out the technical conditions to allow coexistence between wireless broadband applications in the 2.3-GHz band and to ensure coexistence with the services and applications above 2400 MHz (e.g. Wi-Fi networks). The ECC Report 55 also contains guidelines on the additional conditions to be applied, on a case-by-case basis, for coexistence between wireless broadband and services below 2.3 GHz. This band is rather different from others where similar initiatives have been applied, because of a range of important incumbent services in limited but specific parts of Europe. In this context, the ECC intends to develop further guidelines to aid administrations in developing an appropriate sharing framework for coexistence between communications networks and incumbent services and applications at the national level (e.g. see [68]).

In parallel, and in close cooperation with regulatory efforts, the ETSI technical committee on RRS is addressing the standardization of an LSA technical framework. In this regard, an SRDoc proposing the adoption of LSA usage in 2.3–2.4 GHz was issued in 2013 [69], and LSA system requirements were released in October 2014 [70]. Work is currently ongoing

[10] The basic principle of LSA can be found already in the Authorized Shared Access (ASA) concept proposed first by Qualcomm and Nokia [50] and intended to act as a regulatory enabler to making available, in a timely manner, harmonized spectrum for IMT/mobile broadband services.

towards the specification of a system architecture and high-level procedures (to be reported in ETSI TS 103 235 [71]).

A description of the regulatory and technical frameworks of the LSA model is provided in the next two subsections. Then, the use of a potential applicability of the LSA framework for PPDR communications is discussed.

6.5.2.1 Regulatory Framework

The central piece for the implementation of LSA is the establishment of the so-called sharing framework to define the spectrum that can be made available for alternative usage under an LSA model.

The sharing framework can be understood as a set of sharing rules or sharing conditions, which have to be established under the responsibility of the NRA. In particular, the sharing framework has:

- To materialize the change, if any, in the spectrum rights of the incumbent(s) (e.g. delimitate the geographical areas where incumbent spectrum rights are actually necessary)
- To define the spectrum, with corresponding technical and operational conditions, that can be made available for alternative usage under LSA while guaranteeing the protection of the incumbents' services

The definition of an effective sharing framework requires the involvement of all relevant stakeholders (illustrated in Figure 6.5):

- The Administration/NRA
- The incumbent(s) (most likely to be governmental bodies)
- The prospective LSA licencee(s) (e.g. mobile network operators (MNOs) in the case of the 2.3-GHz band intended to be open for mobile broadband services across Europe)

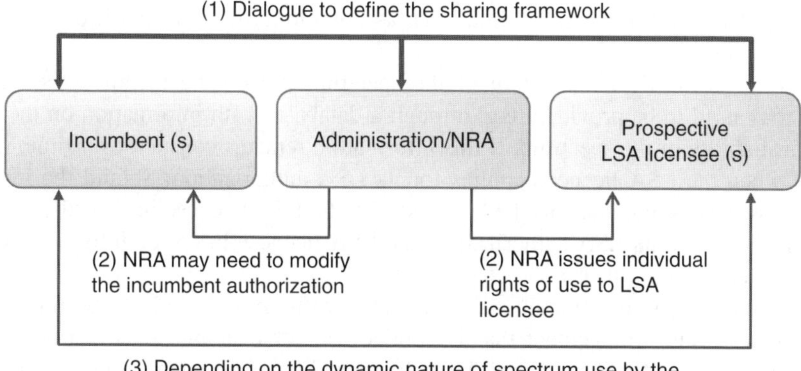

Figure 6.5 Actors and regulatory processes in the establishment of the LSA framework.

The Administration/NRA should identify the relevant parties to be involved in the development of the sharing framework. After this, a dialogue between Administration/NRA, incumbent(s) and prospective LSA licencees should be initiated, with the aim of determining the terms of the sharing framework:

- The incumbent should report on the conditions under which LSA can be facilitated. These should include its statistical current and future spectrum requirements in order to operate its services in the band. In particular, it may report frequency band, predefined time, frequency use by geographical area and statistical use of the band as well as other technical conditions such as pre-emption conditions, in case of urgency, where the incumbent may retrieve use of the spectrum.
- The prospective LSA licencees should provide some indication of the minimum duration of the sharing framework required to enable an adequate return on investment. It may also be useful for the LSA prospective licencees to report on the frequencies, locations and times where spectrum is most acutely required. These conditions are needed to ensure the proper spectrum usage by both the incumbent and the LSA licencee in adjacent time/space/frequency domain(s).
- The Administration/NRA should determine the relevant conditions in particular to ensure operations of the incumbent services to be protected. Based on these conditions, the Administration would set the sharing framework, which can be eventually referenced under the NTFA. The administration may also need to modify the incumbent authorization accordingly.

Then, in accordance with the established sharing framework, the Administration/NRA would set the authorization process with a view to delivering, in a fair, transparent and non-discriminatory manner, individual rights of use of spectrum to LSA licencees. LSA does not prejudge the modalities of the authorization process to be set by NRA taking into account national circumstances and market demand (e.g. traditional licencing, light licencing with individual authorization). Granting LSA individual rights of use can also be associated with a number of obligations, as commonly done in traditional (exclusive rights of use) licences.

Depending on the dynamic nature of the spectrum use by the incumbent(s), the LSA licencee may need to be provided (e.g. through a database) with information on the area(s)/time of availability of the spectrum. If this information remains constant over time, it can be provided when the LSA licencee applies for its LSA authorization. Should the incumbent needs to have access to (a part of) the band used by the LSA licencee in accordance with the conditions defined in its LSA authorization, the LSA licencee has to be informed by agreed means and has to modify its use.

The concept of 'sharing framework' associated with LSA should not be mixed with a conventional sharing arrangement that is applied for, for example, fixed services such as microwave links or PMR-like services. In such cases, there is no 'incumbent' having priority or exclusive spectrum access across a territory, and new systems are typically introduced on a first-come/first-served basis by applying appropriate coordination measures (e.g. geographic frequency separation measures).

6.5.2.2 Technical Framework

ETSI is standardizing a technical framework for the implementation of an 'LSA system' in the context of enabling access for MFCNs to the 2.3-GHz band. As such, an 'LSA system' comprises one or more incumbents, one or more MFCNs (LSA licencees) and the means to enable the coordination between the incumbents and the LSA licencees, such that the latter may deploy their networks without harmful interference.

The LSA system is designed considering that sharing may in general be dynamic (i.e. the requirements of the incumbent may be such that some portions of the spectrum are not permanently available to the LSA licencee in any given location). In this regard, the set of practical details for sharing a given LSA spectrum resource is referred to as 'sharing arrangement', which may be subject to change but should always remain consistent with the 'sharing framework' defined by the Administration/NRA. For instance, a particular spectrum sharing arrangement between an incumbent and an LSA licencee may include constraints on the potential variations of resource availability to, for example, facilitate the implementation and operability of the sharing system. Examples of such constraints are:

- Changes in the spectrum resource availability may only occur at preset times (e.g. periodic).
- There may be minimum allowed intervals between successive changes (in general or affecting a given area).
- The availability of spectrum resources may be preconfigured (only a finite set of possible combinations in space/frequency is allowed).
- Changes may not be allowed if they violate certain statistical criteria (e.g. overall availability of a certain resource in a given time frame).

On this basis, the proposed LSA system architecture at the time of writing (work is still in progress) is shown in Figure 6.6. It consists of two functional entities:

1. **LSA repository (LR)**. The LR supports the entry and storage of information describing incumbent's usage and protection requirements. It is able to propagate this information to

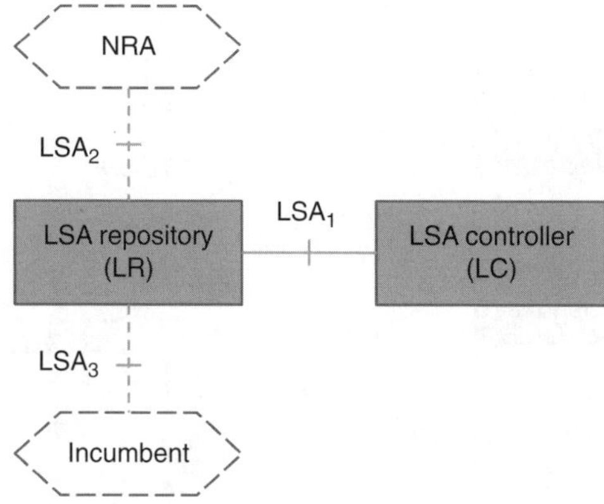

Figure 6.6 LSA Architecture Reference Model.

authorized LSA controllers (LC) and is also able to receive and store acknowledgement information received from LC. The LR also provides means for the NRA to monitor the operation of the LSA system and to provide the LSA system with information on the sharing framework and LSA licence details. The LR enforces the sharing framework and the licencing regime and may in addition realize any non-regulatory details of the sharing arrangement.

2. **LC**. The LC is located within the LSA licencee's domain and enables the LSA licencee to receive or request LSA spectrum resource availability information from the LR and to provide acknowledgment information to the LR. The LC interacts with the licencee's network in order to convey availability information and support the mapping of this information into appropriate radio transmitter configurations and receive the respective confirmations from the LSA licencee's network.

Three reference points are defined:

LSA1: Reference point between LR and LC
LSA2: Reference point for Administration/NRA interaction with the LR
LSA3: Reference point for Incumbent interaction with the LR

An illustrative scenario implementing this LSA system architecture is shown in Figure 6.7. The scenario considers the case of an MNO who holds two (traditional) licences to operate some amount of spectrum (referred to as licenced band/carrier A and licenced band/carrier B) over the entire coverage area where its network is deployed. In addition, it is considered that the MNO also holds an LSA licence to use some additional spectrum that is only available in a smaller restricted area due to the presence of incumbents in the rest of the network coverage (referred to as LSA licenced band/carrier C). In this situation, the MNO is responsible for ensuring that only the appropriate base stations (i.e. those out of the grey area depicted in

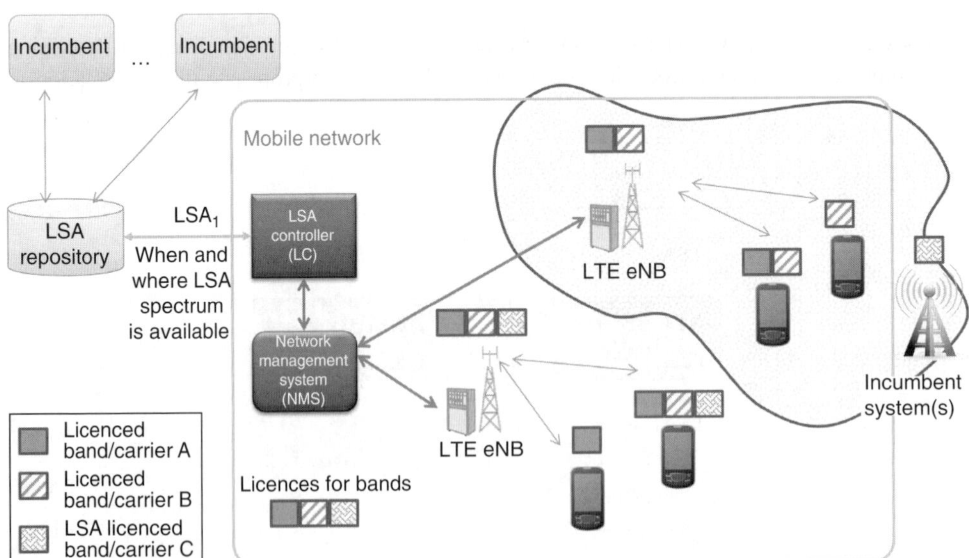

Figure 6.7 Illustrative scenario implementing the LSA architecture.

Figure 6.7) can actually operate in the LSA spectrum, either using this spectrum in a standalone manner or combining its use with the other licenced bands if carrier aggregation capabilities are in place. To this end, an LC would form part of, or interact with, the network management system (NMS) of the MNO. The LC will receive information on LSA spectrum resource availability over the LSA_1 interface from the LR that governs the access to this band. On this basis, the NMS then would translate the information on spectrum availability obtained by the LC into the appropriate radio resource management commands to the base stations in the operator's network. Hence, a user terminal located in the area where the LSA band is available can have access to either of the licenced bands and the LSA bandusi or to all of them if it has the appropriate carrier aggregation capabilities. A user terminal located in the area where the LSA band is not available can only use the licenced bands.

From MNO's point of view, LSA introduces a variability in the amount of spectrum that can be utilized for its normal operation, and in principle, the mechanisms required to utilize the LSA band are similar to the inter-RAT or inter-band load balancing means that already exist in the NMS and are widely deployed in today's cellular systems. Hence, using the LSA spectrum will require the following steps:

- At the appropriate time indicated by the LC, the MNO's NMS instructs the relevant base stations to enable transmission in the LSA band.
- If needed, reconfigurations and system information updates of the other networks operating in the underlying band are performed.
- Existing load balancing algorithms in the radio access network (RAN) will make use of the newly available resources and transfer devices to the new band as needed.
- Transferring the devices can be achieved using different techniques such as:
  - Reselection procedures: User terminals in idle mode that migrate in and out of the coverage area of the LSA frequency may reselect on such frequency.
  - Inter-frequency handover procedures: The RAN initiates handover procedures to transfer user terminals in connected mode from the underlying band towards the LSA band.
  - Carrier aggregation procedures: The RAN reconfigures appropriate user terminals (i.e. terminals supporting carrier aggregation between the underlying band and the LSA band) to start operating in a carrier aggregation mode.

Vacating the LSA spectrum will require the following steps:

- When the granted time period for the operation by LC in the LSA band expires or when due to emergency situations, the incumbent requires its spectrum back (e.g. public safety or other incumbents that require that stipulation), the existing load balancing algorithms in RAN will ensure that devices are transferred back to the underlying band.
- Transferring the devices back can be done via the same aforementioned reselection, inter-frequency handover and carrier aggregation procedures.

6.5.2.3 Applicability to PPDR Communications

Some considerations on the potential applicability of an LSA framework to support PPDR communications were addressed within the European research Project HELP [72]. In this project, a functional architecture in line with the LSA reference model described in the previous section was proposed for the dynamic coordination of some amount of spectrum between a PPDR operator and other users in a manner that predictable QoS was ensured for all sharers. Project HELP pointed out two candidate bands where the applicability of this model would deserve special consideration: the 700 MHz, where mobile broadband services are to become co-primary services in Europe after WRC-2015, and the 225–380-MHz band, used for military applications in NATO countries. In the first case, an LSA model could be adopted involving a PPDR network operator, who would have the role of the incumbent, and a commercial MNO, who would be granted the LSA licence. In the second case, the LSA model would involve the military authorities, who would act as the incumbent, and a PPDR network operator, now in the role of the LSA licencee.

Similarly, a study by a consultancy firm requested by the TCCA on the matter of the need for PPDR broadband spectrum in the bands below 1 GHz in Europe [45] argued that consideration should be given to the shared use of spectrum for PPDR and commercial LTE based on a sort of LSA model,[11] on the basis that the case for a minimum of 10 + 10 MHz of dedicated PPDR was already taken for granted. In particular, the study identified the following approaches for allocating and assigning spectrum so as to permit varying degrees of flexibility between PPDR use and commercial use by an MNO:

- **Inflexible**: A fixed (dedicated) assignment of bands (nominally 2 × 10 MHz) to broadband PPDR is conducted. No ability to use more spectrum for PPDR at times of stress is enabled. Unused spectrum at times of low PPDR activity is not permitted to be exploited by others.
- **Flexible**: A fixed assignment of bands (nominally at least 2 × 10 MHz) to broadband PPDR is conducted, though the PPDR operator has the option to sub-licence rights of use to any portion of the bands that is unused (based on time or geography) pursuant to a sort of LSA arrangement. This type of arrangement would permit unused PPDR spectrum to be exploited by commercial operators only at times or locations of low PPDR activity.
- **Highly flexible**: A fixed assignment of somewhat larger bands (such as 2 × 20 MHz) to broadband PPDR is conducted. The PPDR operator has the option to sub-licence rights of use for a substantial portion of the bands (e.g. 2 × 10 MHz) pursuant to a sort of LSA arrangement. However, and differently from the flexible case, the PPDR operator would only keep rights to reclaim the spectrum back only under limited circumstances (such as a declared emergency). Therefore, this arrangement permits unused spectrum to be exploited by commercial entities at times and locations of low PPDR activity, in addition to providing commercial entities the ability to use substantially more spectrum at times of stress.

More recently, a use case entitled 'Licensed Shared Access for Supplemental BB PPDR' has been included within ETSI TR 103 217 [73], which reports on a feasibility study[12] that explores the potential areas of synergies between commercial, civil security and military domains in the medium/long term (5–15 years) in response to the EC mandate for RRS (M/512). This use case proposes to apply LSA for providing supplemental data broadband connection to existing narrowband PPDR. In particular, the use case states that administrations may consider introducing BB PPDR application as primary user into a newly refarm frequency band but permit

[11] Work in Ref. [45] refers specifically to the Authorised Spectrum Access model, which is equivalent to the LSA model as explained at the beginning of Section 6.5.2.

[12] The work is expected to be concluded along 2015 and intended to provide the baseline for the definition of the standardization work programme in a subsequent phase of Objective C within the EC Mandate M/512.

spectrum resources to be shared based on LSA with commercial systems such as mobile broadband. The document recognizes that PPDR operations that are by nature largely unexpected would be an important drawback for commercial systems as it means uncertainty access to spectrum resources, additional risk and complexities associated with spectrum sharing and could make investment by commercial systems less certain compared to the case of exclusive spectrum. However, the document ascertains that this drawback should be balanced with the fact that PPDR operations are geographically localized and limited in time, which may result in a few percent of resource spectrum unavailability for commercial systems at the scale of a country with national coverage.

In the light of the previous text, the sharing case based on LSA between PPDR and commercial operators could actually represent an efficient approach worthy of further investigation to permit varying degrees of flexibility for PPDR BB spectrum use. Technology is not seen as an impediment for this kind of approach. Nonetheless, the definition of the 'sharing framework' becomes the central question to be answered, setting up crystal clear sharing rules along with the mission-critical levels that would trigger spectrum resource pre-emptions. Doing so, an LSA framework may provide adequate guarantee in terms of spectrum access at a regional or national scale to commercial LSA licencees in order to incentivize and secure investments in network and equipment.

6.5.3 Shared Use of Spectrum Based on Secondary Access to TVWS

A white space can be defined as '*a part of the spectrum, which is available for a radiocommunication application (service, system) at a given time in a given geographical area on a non-interfering/non-protected basis with regard to other services with a higher priority on a national basis*' [74]. The concept of allowing additional transmissions in white spaces is a technique to improve spectrum utilization as well as to unlock some spectrum for new uses, provided that the risk of harmful interference to the existing licenced users of the spectrum can be appropriately managed. Most efforts currently underway related to the exploitation of white spaces are focused on the parts of the VHF and UHF frequency bands used for TV broadcasting (i.e. TVWS). The existence of a significant amount of unused parts of spectrum within these bands has been reported in some studies (e.g. see [75]). However, the actual availability of white space spectrum differs significantly among countries (depending on the relevance that terrestrial TV broadcasting service plays in front of other TV delivery platforms such as cable TV or satellite TV) as well as among areas within the same country (e.g. urban and rural areas).

The favourable propagation of radio waves in the VHF/UHF bands and their ability to penetrate deep inside buildings make this frequency range quite valuable. Indeed, many and diverse potential use cases for the exploitation of unused TVWS have been identified such as broadband Internet access in rural areas, wide area machine-to-machine communications, wireless backhauling, rapid deployed networks, in-home networking, etc. [76].

In this context, consideration should be given to the use of TVWS by PPDR equipment to leverage on the long-reaching and highly penetrative signal capabilities of this band, especially in hard-to-reach areas (e.g. tunnels, building basements) as well as in low or very low populated areas where an important part of this spectrum might remain underutilized.

In the regulatory domain, some of the world's most influential regulators, including FCC in the United States, Ofcom in the United Kingdom and the European CEPT/ECC, Japanese

Ministry of Internal Affairs and Communications (MIC) and Singapore's Info-Communications Development Authority (IDA), are at the forefront of developing the rules for the use of TVWS [58]. In the standardization and industrial domains, the ETSI, IEEE, Internet Engineering Task Force (IETF) and ECMA International have also started several activities to shape future solutions for the so-called white space devices (WSD) or TV band devices (TVBD).

The regulatory and technical frameworks that set the playing field for the potential exploitation of TVWS by PPDR applications are described in the following two subsections. On this basis, a functional architecture of a technical solution for the control and use of TVWS in a LTE-based PPDR network is presented in the last subsection, together with the identification of some regulatory actions that would further contribute to increase the level of dependability that PPDR users could have on this kind of spectrum sharing solution.

6.5.3.1 Regulatory Framework

A first regulation on the use of TVWS was released by the FCC in the United States in 2010 [77] and went officially into effect in early 2011. Since then, some corrections to the rules have been introduced [78], and the operation of TVBDs for commercial use is now a reality in the United States in the TV channels frequencies across the VHF and UHF bands [79].

In Europe, leading efforts are taking place in the United Kingdom towards the allowance of WSDs in the spectrum range between 470 and 790 MHz. While this spectrum is mainly used for digital terrestrial television (DTT), other services such as programme making and special event (PMSE) systems (e.g. use of wireless microphones, talkback systems and in-ear monitors in concerts, sport events and others) have also to be accounted as incumbents to be protected. Regulatory requirements have been already drafted by Ofcom [80], and white space technology is currently being piloted in the United Kingdom [81].

Furthermore, possible harmonization measures for the use of WSDs in the band 470–790 MHz across Europe are also under consideration within CEPT/ECC, which established the Frequency Management Project Team 53 (FM PT 53) in September 2012 [82] to study the potential use of TVWS in Europe. Among others goals, FM PT 53 is intended to provide a master set of the overall requirements for CEPT countries and develop, if appropriate, harmonized regulatory measures to complement the related standardization activities in ETSI with the aim of enabling the development and deployment of WSD while ensuring protection to incumbent services.[13] Thus far, some technical studies have been performed within CEPT on the technical and operational requirements for the possible operation of (so-called) cognitive radio systems (CRS) in the frequency band 470–790 MHz (ECC Reports 159, 185 and 186 [74, 83, 84]). Based on these CEPT/ECC studies, ETSI produced a harmonized standard that went into effect in September 2014 [85]. The ETSI harmonized standard covers the essential requirements that must be met by WSDs to be conformant with the R&TTE Directive that regulates the placement of radio and telecommunications terminal equipment in the EU market.

[13] Besides terrestrial TV broadcasting service, other incumbent radio services/systems authorized for operation across CEPT countries within or next to this frequency band with a regulatory priority include programme making and special event (PMSE) systems including radio microphones in particular, radio astronomy service (RAS) in the 608–614-MHz band, aeronautical radio navigation service (ARNS) in the 645–790-MHz band and mobile service (MS) below 470 MHz and above 790 MHz.

Other countries that are moving forwards in the regulation of TVWS are Canada, Singapore and South Africa.

All these regulatory frameworks advocate for a licence-exempt regime for spectrum authorization where WSDs learn about the available WS channels that can be used at a given time and location from a GLDB. The main components of the regulatory framework enabling the operation of WSDs are depicted in Figure 6.8 and briefly described in the following.

The existing regulations consider both fixed and portable WSDs. Fixed WSDs are intended to be deployed at a fixed location with a fixed antenna height. Fixed WSDs typically must store their location, which can either be entered by a professional installer or self-determined using geo-location technologies. On the other hand, portable WSDs must be able to self-geo-locate whenever they move more than a specific distance (typically, it could be 50 m, which is related to the accuracy of the geo-location technology in use). The issues inherent with self-geo-location (e.g. latency, accuracy and time to fix) mean that the use cases for these devices tend to be more limited than those for fixed devices.

WSDs provide their location to a given GLDB, which returns information on the available frequencies and permitted transmission settings (e.g. a list of channels on which they may operate, maximum transmit power, etc.). This approach shifts the complexity of spectrum policy conformance out of the device and into the database. This approach also simplifies the adoption of policy changes, limiting updates to the GLDB, rather than numerous devices. It also opens the door for innovations in spectrum management that can incorporate a variety of parameters. In the current approaches, WSD queries typically provide device location, device information (e.g. type, serial number, certification ID, etc.), antenna height (for fixed) and additional identifying information (e.g. device owner). In the future, GLDB can include other

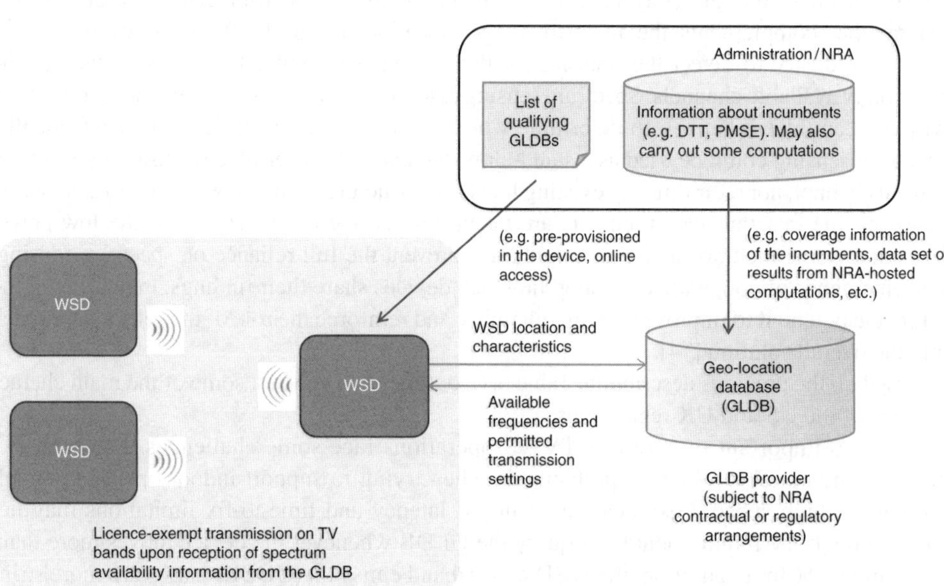

Figure 6.8 Framework for the operation of white space devices (WSDs).

parameters, such as user priority, signal type and power, spectrum supply and demand, payment or micro-auction bidding and more.

GLDBs are likely to be provided by third parties authorized or contracted by regulatory agencies. There is also the case that a regulatory agency decides to manage the GLDB on its own, much like an online licencing system. Regulatory agencies can opt to have single or multiple GLDB providers. Competition among multiple database providers can be beneficial to end users, as it is likely to drive innovation and give users greater choice. Regulatory agencies will collect the details of the incumbents' usage (e.g. DTT, PMSE systems). Incumbent information can range from very detailed planning information gathered from the incumbent user (e.g. detailed planning of the DTT broadcast network) to the information gathered from the registration processes used to protect entities that are eligible to receive interference protection in TV spectrum (e.g. online registration of licenced wireless microphones to operate in a given location). This information can be delivered to the GLDB providers for them to conduct the calculations to apply protection criteria to these systems (e.g. computation of exclusion zones for specific TV channels and sometimes for the adjacent channel as well). Alternatively, a regulatory agency may decide to carry out the bulk of the calculations in-house and deliver to the database very specific information datasets such as the maximum equivalent isotropic radiated power (EIRP) values that a WSD can use at all locations for a combination of parameters. These two approaches are not exclusive and can be combined (e.g. a regulatory agency must decide to keep all the computations for the DTT service and task the GLDB providers with the computations concerning the protection of PMSE equipment). The protection criteria of the methodologies for the protection calculations are commonly established through regulatory proceedings allowing feedback from all stakeholders.

It is worth mentioning that spectrum sensing can also be used to determine the availability of unused channels, though today's regulations mainly consider spectrum sensing as an optional feature that complements the mainstream geo-location approach. With spectrum sensing, WSDs would try to detect the presence of the protected incumbent services in each of the potentially available channels. Spectrum sensing essentially involves conducting a measurement within a candidate channel, to determine whether any protected service is present. Ideally, spectrum sensing could be used as a stand-alone technique to determine the usability or not of a given channel, not requiring any existing local infrastructure such as connection to a database. However, the fact that measurements are taken only at a given location and the low-power levels that some incumbent signals might have prevent the full reliance on spectrum sensing. The emergence of cooperative sensing, in which devices share their findings, may bring in the future the potential to improve sensing reliability and reinforce the role of the sensing approach into the overall solution [74].

Based on the previous description, Table 6.12 outlines and contrast some of the main characteristics of the US and UK regulations.

It is also important to note that TVWS operations face some challenges. For example, geo-locating portable WSDs is problematic when trying to support indoor operation or full mobility at high speed. Geo-location accuracy, latency and time-to-fix limitations may not easily support the requirement to re-query the GLDB whenever the device moves more than, for example, 50 m. In addition, the WSD out-of-band emission (OOBE) masks are more stringent than for most unlicenced devices in other bands in order to protect operation close to TV

Table 6.12 Main characteristics of the US and UK regulatory frameworks for the use of TV white spaces.

	FCC regulation [77–79]	Ofcom draft regulation [80]
Types of TVWS devices	Fixed devices. Operate at fixed positions and obtain channel availability from databases Portable/mobile devices. The location of these devices might change. Two different subtypes: 1 Mode I. Do not use an internal geo-location capability and do not access to databases. Channel availability is obtained from Mode II or fixed devices 2 Mode II. Use an internal geo-location capability and access to databases for channel availability	Two main, non-exclusive, dimensions to classify devices: 1 Fixed or mobile/portable devices, according to whether the location of the device is permanent or can change 2 Master or slave. A master device directly communicates with a database, while a slave device can only obtain operating parameters from a master device
Information provided to the GLDB	Unique device identifier Antenna geographical coordinates and their accuracy (50 m) Device type (fixed, Mode I, Mode II) Device's antenna height above ground level (meters) Contact information of the owner of the device and of the person responsible for the device's operation	Unique device identifier Antenna geographical coordinates and their accuracy (50 m) Technology identifier Fixed or portable/mobile nature Lower and upper frequency boundaries and maximum EIRP spectral densities of in-block emissions Indoor or outdoor nature (optional) Antennas' characteristics (optional)
Power limits	4 W EIRP for fixed devices 100-mW EIRP for personal/portable devices. This is further reduced to 40-mW EIRP in case of operation of these devices in channels adjacent to occupied TV bands channels 50-mW EIRP for sensing-only devices	Variable maximum transmit power per location. Power limits are specified in terms of maximum permitted EIRP spectral density (in units of dBm/0.2 MHz). The power limit is a function of the quality of the DTT coverage in the geographical area where the DTT receiver is located. Spatial resolution of the power values that is based on 100 × 100 m geographic pixels. The area of the United Kingdom is covered by over 20M pixels
	Additional limits on power spectral density (PSD) are defined considering a uniform distribution of transmit power limits over the channel frequencies. Adjacent channel emission limits are also specified	Additional limits are to be set out in terms of through minimum adjacent channel leakage ratio (ACLR) settings
Specification of authorized frequency	A list of permitted TV channels	A list of lower and upper frequency boundaries (not restricted to TV channel boundaries) with a resolution of 100 kHz. Maximum permitted EIRP spectral density for each frequency boundary pair is indicated

(*continued*)

Table 6.12 (*continued*)

	FCC regulation [77–79]	Ofcom draft regulation [80]
Authorization principle	Licence-exempt regime based on geo-location and database access for spectrum authorization Spectrum sensing is not mandatory but optional feature Sensing-only devices also permitted (these devices have additional approval requirements and lower power limits)	Licence-exempt regime based on geo-location and database access for spectrum authorization Spectrum sensing is not mandatory but optional feature No provisions for sensing-only devices
Authorization validity and rechecking	Database must be checked at least once per day. If channel list cannot be refreshed, it times out next day at 11:59 p.m. Mode II device needs to check its location at least once every 60 s, except in powered-down modes. Rechecking database must be done if location changes by more than 50 m with respect to previous consultation Mode I devices must either receive a special signal from the Mode II or fixed device that provided the list of available channels to verify that it is still in reception range of that device or contact a Mode II or fixed device at least once every 60 s to re-verify/re-establish channel availability	Time validity of parameters (in minutes) is to be provided by the database as part of the channel availability response The database can send instruction for a WSD to cease transmission in 60 s (i.e. the so-called 'kill-switch' feature)
Database awareness of the operation of WSDs	Fixed devices shall register their operation in the database No provisions for any kind of reporting back to databases from WSDs once channel availability information has been provided	Registration is not mandated Requirement to have an acknowledgement of receipt of information from WSDs on used channels and EIRPs

receivers. This OOBE limit has proven difficult to achieve at a cost point low enough to make WSDs further broadly attractive.

Moreover, regulatory uncertainty is also hindering the uptake of TVWS operations. For instance, in the United States, there is an ongoing incentive auction proceeding by the FCC in the 600-MHz band aimed to repack and repurpose some spectrum now used for TV stations and auction it off to mobile licenced services. This poses uncertainties on how much spectrum will be available for TVWS following that auction. Similarly, the future spectrum availability for TVWSD within the TV UHF band in Europe is also unclear [86, 87]. In this regard, co-primary allocations for mobile and broadcasting services between 694 and 790 MHz will be a fact for the EU after WRC-15, so that this band could be used for mobile services in some countries in the near future (around 2020, with some countries like Germany and France already planning the auctioning of this spectrum along 2015), importantly reducing the availability of TVWS. In addition to the 700-MHz band, the possible need and point in time for a further co-primary allocation for broadcasting and mobile services below 700 MHz, that is, 470–694 MHz, is also under discussion for the longer term (beyond 2030) in Europe [38]. However, no consensus has been reached in this case since the sustainability of the current European audiovisual model is highly dependent on the availability of this core spectrum.

In spite of these technological challenges and regulatory uncertainties, the capability of database-enabled devices to operate in vacant TV spectrum without causing interference is already proven (there are hundreds of registrations of database-controlled fixed TVBDs in the United States), paving also the way to apply this concept in other spectrum bands other than the TV bands.

6.5.3.2 Technical Framework

Standards related to the use of TVWS are being addressed in multiple bodies, including the ETSI, IEEE, IETF and ECMA International. A summary of these standardization activities is provided in Table 6.13.

Within the ETSI, standardization activities related to TVWS are mainly undertaken within the technical committee on RRS (ETSI TC RRS). In particular, TC RRS has developed the following elements:

- Use cases for operation in white space frequency bands (ETSI TR 102 907).
- System requirements for operation in UHF TV band white spaces (ETSI TS 102 946).

Table 6.13 Summary of standardization initiatives related to TVWS.

Organization	Standardization activities	Comments
ETSI	TS 103 143, EN 303 144	System architecture and protocols for the information exchange between different geo-location databases (GLDBs)
	TS 103 145, EN 303 387-1	System architecture and high-level procedures for coordinated and uncoordinated use of TV white spaces
	ETSI EN 301 598	Harmonized EN covering the essential requirements of article 3.2 of the R&TTE Directive
IEEE	802.11af	Specifying enabling technologies in TV white spaces for WLAN systems
	802.22	IEEE 802.22 is a standard specifying WRAN communication systems operating in TV bands
	802.15.4	Specifying enabling technologies in TV white space for low-rate WPAN systems
	802.19.1	Coexistence framework for dissimilar systems in 802 standards
	1900.X	Specifying enabling technologies in TV white space communications
ECMA	ECMA-392	Specifies a medium access control (MAC) sub-layer and a physical (PHY) layer for cognitive wireless networks operating in TVWS bands
IETF	PAWS	Extensible protocol to obtain available spectrum from a geospatial database by a device with geo-location capability

- A system architecture for the information exchange between different GLDBs. The architecture is specified in ETSI TS 103 143 [88]. A related European norm ETSI EN 303 144 [89] defines the parameters and procedures for information exchange between different GLDBs.
- A system architecture and high-level procedures of a system that can allow operation of WSDs based on information obtained from GLDBs, considering both uncoordinated use of TVWS (where there is no attempt to manage the usage of channels by different WSDs) and coordinated use of TVWS (where a central coexistence entity is employed to efficiently use the spectrum and avoid or mitigate harmful interference between WSDs from different systems using the same white spaces). The architecture is specified in ETSI TS 103 145 [90]. For the coordinated case, the interface between the WSD and the so-called spectrum coordinator (SC) is being defined as an European norm in ETSI EN 303 387-1 [91].

Additionally, ETSI has produced the harmonized standard ETSI EN 301 598 [85] covering the essential requirements of the R&TTE Directive concerning the use of WSD in the in the 470–790-MHz band. It's worth noting that this standard basically focuses on the uncoordinated case since requirements to facilitate coexistence between WSDs were not considered essential with respect to compliance with Article 3.2 of Directive 1999/5/EC. However, this point may be subject to further review in the future.

Within IEEE, a number of working groups in the IEEE 802 LAN/MAN standards committee [92] have also triggered multiple activities related to TVWS usage:

- IEEE 802.22 is a standard for wireless regional area networks (WRANs) in TVWS [93]. This standard allows broadband point-to-multipoint wireless access within VHF and UHF TV bands between 54 and 862 MHz utilizing CR techniques for fixed and portable user terminals. The WRAN aims to provide wireless access (e.g. Internet access) in the distance up to tenths of kilometres for underserved and un-served rural communities. This standard is sometimes called as the 'Wi-FAR'. There is an amendment (IEEE 802.22.1-2010, Standard for the Enhanced Interference Protection of the Licenced Devices) that specifies a beaconing network intended to protect low-power, licenced devices operating in the TV bands (e.g. wireless microphones) from harmful interference from licence-exempt devices, such as WRANs. Recently, a new study group was established, named Spectrum Occupancy Sensing (SOS) Study Group, to consider the standardization of technology for optimizing the usage of RF spectrum for wireless broadband services.
- IEEE 802.11af is a standard that provides similar services to the traditional IEEE 802.11 (e.g. IEEE 802.11a/b/g) but utilizes CR and operates in the TVWS bands. It supports 6-, 7- and 8-MHz wide TV channels for global applicability and allows for concatenation of up to four UHF channels, either contiguously or in two non-contiguous blocks. Each (8 MHz) channel supports up to 35.6 Mb/s. IEEE 802.11af standard, sometimes referred to as 'super Wi-Fi', is already published [94].
- IEEE 802.15.4 specifies standards for low-rate WPAN technologies. Targeted applications include sensor, smart grid/utility and machine-to-machine networks. In order to enable WPAN to take advantage on the TV band spectrum, IEEE 802.15.4 formed IEEE 802.15.4m to specify the enabling technologies for low-rate WPANs in TVWS, primarily for optimal and power-efficient command and control applications.

- IEEE 802.19.1 is a standard for TVWS coexistence methods that enables the line of IEEE 802 wireless standards to effectively utilize the TVWS by providing standardized coexistence methods among dissimilar or independently operated TVWS networks (WPAN, WLAN, WRAN, WMAN). The IEEE 802.19.1 standard is intended to help in achieving a fair and efficient spectrum sharing. The coordination for interference from one WSD to another WSD is provided by a coexistence discovery and information server (CDIS), coexistence manager (CM) and coexistence enabler (CE). GLDS can have a set of these entities in addition to the core GLDB function. This standard was published in 2014 [95].

Still within the IEEE, the Dynamic Spectrum Access Networks (DySPAN) standards committee (IEEE DySPAN-SC) develops standards related to TVWS. The scope of the DySPAN-SC includes [96] dynamic spectrum access radio systems and networks with the focus on improved use of spectrum, new techniques and methods of dynamic spectrum access including the management of radio transmission interference and coordination of wireless technologies including network management and information sharing among networks deploying different wireless technologies. There are several working groups. The baseline IEEE 1900.1 standard that defines the terms and definitions in the field of dynamic spectrum access and related technologies and the IEEE 1900.4 standard that specifies the architectural building blocks to enable network-device distributed decision-making for optimized radio resource usage in heterogeneous wireless access networks are already published [97]. Further activities within IEEE DySPAN-SC are conducted by IEEE 1900.5 Working Group on 'Policy Language and Policy Architectures for Managing Cognitive Radio for Dynamic Spectrum Access Applications', IEEE 1900.6 Working Group on 'Spectrum Sensing Interfaces and Data Structures for Dynamic Spectrum Access and other Advanced Radio Communication Systems' and IEEE 1900.7 Working Group on 'White Space Radio Working Group', this latter aimed at specifying a radio interface for white space dynamic spectrum access radio systems supporting fixed and mobile operation.

Within ECMA International, an industry association dedicated to the standardization of ICT and consumer electronics (CE), the standard ECMA-392, 'MAC and PHY for Operation in TV White Space' [98], has also been produced (it was released as early as 2009 and later revised in 2012). ECMA-392 specifies a medium access control (MAC) sub-layer and a physical (PHY) layer for cognitive wireless networks operating in TVWS bands. The standard also specifies a multiplexing sub-layer to enable the coexistence of concurrently active higher layer protocols within a single device and a number of incumbent protection mechanisms which may be used to meet different regulatory authorities requirements, which themselves are outside of the scope of the standard.

Within the IETF, a Protocol to Access White Space (PAWS) databases to address interaction with the databases has been standardized [99]. This specification defines an extensible protocol that can be used to obtain available spectrum from a geospatial database by a device with geo-location capability. The work is based on the assumption that the database will be reachable via the Internet and that radio devices too will have some form of Internet connectivity, directly or indirectly. The protocol supports the following main functions:

- Devices connect and register with a database.
- Devices provide geo-location and attributes to the database.
- Devices receive in return a list of available white space spectrum.
- Devices report to the database the anticipated spectrum usage.

Not all of the specified PAWS functions are necessarily mandated in a given regulatory framework. Some functions can be optional or just not used depending on the regulatory domain and database implementation. The IETF has reused existing protocols and data encoding formats where possible for the specification of PAWS. In particular, PAWS is based on the use of HTTP Secure (HTTPS) for information exchange between the WSD and the GLDS, and JSON-RPC request/response objects are used to encode the exchanged information elements.

6.5.3.3 Applicability to PPDR Communications

The access to TVWS by PPDR users can bring additional spectrum for PPDR communications. The long-reaching and highly penetrative signal capabilities of this spectrum make it especially valuable in hard-to-reach areas (e.g. tunnels, building basements) as well as in low or very low populated areas where an important part of this spectrum might remain underutilized.

Based on the prevailing regulatory approach being adopted for the operation of WSDs (described in Section 6.5.3.1), the following challenges shall be considered for the potential exploitation of TVWS by PPDR applications:

- PPDR systems intended to exploit TVWS will have to deal with temporary unavailability of specific channels or groups of channels, due to coexistence decisions or use of the channels by primary incumbents (e.g. licenced wireless microphones).
- PPDR systems intended to exploit TVWS may need to deal with interference from other secondary users in the same band or find mechanisms to coexist with those secondary users so that the bandwidth is still used efficiently by the PPDR systems in the presence of these secondary systems.
- Coverage range can be significantly limited by the relatively low authorized transmit power values so that externally mounted antennas (EMA) solutions may be necessary.

A functional architecture for the exploitation of TVWS for PPDR was developed within the EU research Project HELP [72]. This functional architecture, reproduced in Figure 6.9, is intended to enable and control the use of TVWS spectrum in a PPDR network. LTE is assumed to be the technology in use. The solution encompasses the following functional components:

- **Geo-location Database** (GLDB). As described in Section 6.5.3.1, this is the entity that contains the information about spectrum availability at any given location and time, as well as other types of relevant information related to the white space spectrum. The database can be operated by the spectrum regulator or a third-party entity (e.g. authorized TV band database manager).
- **Network spectrum manager** (NSM). This functional element is a central control point allocated within the NMS of the PPDR network and used to control the access of a number of LTE eNBs to TVWS in a coordinated way. The NSM directly interacts with the WSD database to acquire the information of TVWS spectrum usage status in a given area. Hence, the NSM serves as a CM (i.e. a sort of SC according to the ETSI framework [91]) so that channel availability information obtained can be used to decide on the most appropriate spectrum allocation of the TVWS resources among eNBs. This NSM can also support priority schemes for the allocation of the available spectrum to the eNBs over specific periods of time. Besides information received from the WSD database, the NSM may be

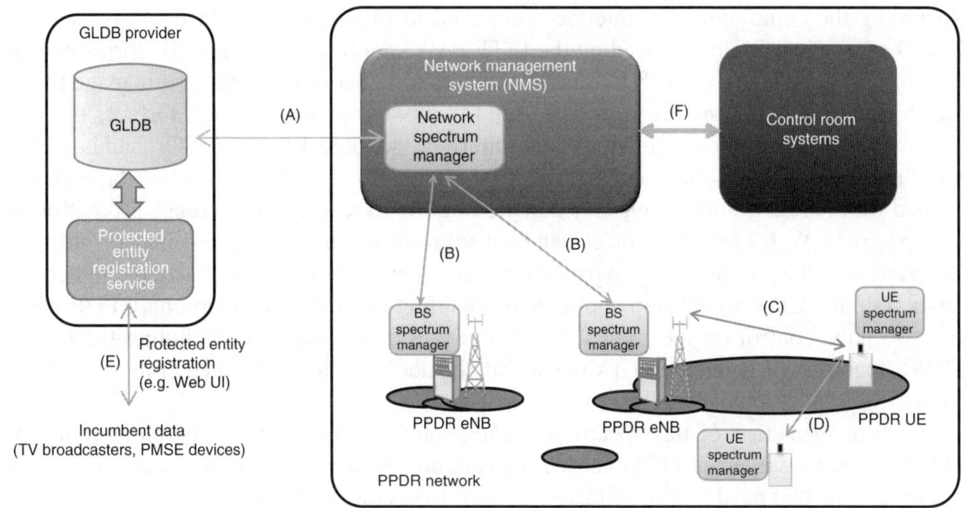

Figure 6.9 Functional architecture for PPDR exploitation of TVWS.

able to collect sensing results from eNBs and terminals and produce a Radio Environment Map (REM) to enhance decision-making. Through the NMS, the PPDR network operator will be able to control and oversee the proper configuration, activation and deactivation of this additional capacity according to PPDR operational needs in an incident area.

- **Base station spectrum manager** (BSSM). This entity is allocated within eNBs to manage the spectrum use of any individual eNB. Two main modes of operation are envisioned for this entity. In one of these operation modes, this entity mainly interacts with the NSM and is in charge of enforcing decisions coming from this central control point. In this case, whenever some triggering situation occurs (e.g. a temporary base station to be switched on, congestion threshold reached in licenced spectrum, etc.), the BSSM contacts the NSM to get the usable TVWS frequencies. In the other operation mode, this entity provides full autonomy to the eNB for TVWS access. This mode can be exploited in case that the NSM is not deployed or unreachable for whatever reasons, providing an inherent level of redundancy. In this autonomous mode, the BSSM can directly contact the GLDB to get the information of TVWS spectrum usage status and decide on spectrum utilization locally. Spectrum sensing functionalities in eNBs can also be used to improve decision-making. In any case, regardless of both possible operation modes, the BS is always assumed to control all RF parameters of its attached terminals (frequency, EIRP, modulation, etc.) in a 'master–slave' relationship.
- **UE spectrum manager** (UESM). This entity is mainly needed to support UE-to-UE communications over TVWS when there is no BSSM to control the allocation of this spectrum. This functionality allows a UE to behave as a master node for the allocation of RF channels for UE-to-UE communications for, for example, ad hoc network deployment in cases that the PPDR network infrastructure is missing. Hence, UEs embedding this functionality are able to access the WSD database on their own by means other than PPDR network access (e.g. through commercial networks or satellite Internet connection). The UESM can also support relying or proxying functions for other master UEs that cannot reach the database by themselves.

Regarding the communication interfaces depicted in Figure 6.9, interface (A) is used for GLDB access and could be based on the IETF PAWS protocol. Interface (B) forms part of the protocols used by the NMS to remotely access to network elements and manage them, thus leveraging/extending the protocols used for network management. Interfaces (C) and (D) are technology dependent (e.g. LTE interface adapted for use in TV bands, IEEE 802.11af interface) as they actually support the radio transmissions. Interface (E) is needed for the registration of entities that are eligible to receive interference protection in TV spectrum. Web-based interfaces can be a solution for this interface. Finally, interface (F) represents the Application Programming Interfaces (APIs) that are reachable in control room systems (CRS) to develop applications enabling tactical and operational PPDR managers to have control on the capacity and coverage of the network, including the use of TVWS spectrum (e.g. the use of TVWS within the network can be enabled or disabled from control room positions).

The implementation of this spectrum sharing solution requires that the configuration of LTE hardware installed in eNBs can be adjusted to operate in TV UHF bands. There are certain aspects that need to be considered for LTE to operate in TVWS:

- The allowable modes of operation for LTE. Most likely, deployments of LTE in TVWS will be either TDD mode (either in TVWS stand-alone operation or through carrier aggregation with the primary carrier provided via non-TVWS frequencies) or FDD mode in the context of carrier aggregation where the primary carrier is provided via non-TVWS frequencies. In the FDD case, both DL and UL operation should be supported in TVWS via DL component carriers or UL component carriers.
- Robustness of control signalling and data transmission. The operation in unlicenced bands will naturally require a system to function (either temporarily or for a long period of time) under some levels of interference that are caused by other secondary users. Control signalling and data transmission of any system need to take this into account. Silencing gaps in the LTE transmissions can be employed for coexistence purposes with other secondary systems [100].

Indeed, the implementation of this spectrum sharing solution could leverage the technological enhancements being targeted under 3GPP Release 13 for the operation of LTE in unlicenced spectrum [101], commonly referred to as LTE-U. These enhancements are expected to introduce a so-called Licenced-Assisted Aggregation (LAA) operation mode in LTE to aggregate a primary carrier (using licenced spectrum, to deliver critical information and guaranteed QoS) and a secondary carrier (using unlicenced spectrum, to opportunistically increase capacity and data rates). The secondary carrier operating in unlicenced spectrum could be configured either as downlink-only carrier or contain both uplink and downlink. While current work in LTE Release 13 is on the use of LTE in the 5-GHz band, the underlying principles and technologies could be directly extended to other unlicenced bands (e.g. 2.4 and 3.5 GHz, TVWS).

The opportunistic nature of the usage of the TVWS does not fit well with the deployment of services that require provisioning of QoS as there is uncertainty about the availability of spectrum resources and there is no protection from harmful interference from other WSDs. However, different measures could be considered by the regulatory authority to increase the degree of reliability and QoS guarantees on the exploitation of TVWS for PPDR:

- Allowing higher transmit powers for PPDR devices. This approach would be aligned with the ECC/REC/(08)04 recommendation for the allocation of a BBDR band at 5150–5250 MHz, where PPDR devices are expected to coexist with other devices (e.g. Wi-Fi devices) though higher transmit power limits are established for the operation of PPDR devices. The more controlled nature of PPDR equipment and its operation may be turned into less stringent incumbent protection levels requirements for those devices (and so higher allowed transmit powers) as compared to commercial devices.
- Supporting priority access for PPDR WSDs. In this case, if access to radio resources is requested by several WSDs concurrently, the GLDB could treat PPDR WSD devices with a higher priority (i.e. higher precedence than ordinary devices but always remaining below the precedence of the current incumbents).
- Reserving a number of TV channels for PPDR use when an emergency situation is declared by a competent authority. In this situation, GLDB could be mandated to guarantee that a fraction of the available capacity in a given region is excluded from the channel availability information provided to conventional WSDs and only advertised to PPDR WSDs.
- Registering PPDR base stations in the TV bands database and establishing protection criteria to prevent interference to these stations when an emergency situation is declared by a competent authority. In this case, PPDR equipment would form part of the incumbent radio services/systems authorized for operation with a regulatory priority (e.g. PPDR devices will be registered similarly as it could be done for PMSE equipment). Another option could be to create a second, high-priority tier of WSD above the general authorization tier. This would require that the regulator defines rules for the coexistence of WSD so that higher tier devices do not get interfered. Access to the higher-priority tier could be through a licence. Indeed, the realization of a three-tiered shared access system is currently being pursued by the FCC in the United States to create a new Citizens Broadband Service (CBS) in the 3550–3650-MHz band managed by an SAS based on the use of geo-location-based opportunistic access technology [48]. In this case, the first tier (denoted as incumbent access) includes authorized federal users and grandfathered fixed satellite service licencees. These incumbents are afforded protection from all other users in the 3.5-GHz Band. The second tier (denoted as priority access) includes critical use facilities, such as hospitals, utilities, government facilities, and public safety entities that are afforded quality-assured access to a portion of the 3.5-GHz band in certain designated locations. The third tier (denoted as general authorized access) includes all other users, including the general public, that have the ability to operate in the 3.5-GHz band subject to protections for incumbent access and protected access users and can use the spectrum when incumbent and priority access users are not using it.

References

[1] Martin Cave, Chris Doyle and William Webb, 'Modern Spectrum Management', Cambridge/New York: Cambridge University Press, 2007.

[2] ITU-InfoDev ICT Regulation Toolkit. Available online at http://www.ictregulationtoolkit.org/en/home (accessed 30 March 2015).

[3] ITU-R Radio Regulations. Available online at http://www.itu.int/pub/R-REG-RR (accessed 30 March 2015).

[4] ITU-R Study Groups. Available online athttp://www.itu.int/dms_pub/itu-r/opb/gen/R-GEN-SGB-2013-PDF-E.pdf#page=19&pagemode=none (accessed 30 March 2015).

[5] ERC Report 25, 'The European Table of Frequency Allocations and Applications in the Frequency Range 8.3 kHz to 300 GHz (ECA Table)', May 2014. Available online athttp://www.efis.dk/ (accessed 30 March 2015).

[6] Electronic Communications Committee (ECC) within the European Conference of Postal and Telecommunications Administrations (CEPT). Available online at http://www.cept.org/ecc (accessed 30 March 2015).

[7] Decision No 676/2002/EC of the European Parliament and of the Council of 7 March 2002 on a regulatory framework for radio spectrum policy in the European Community (Radio Spectrum Decision). Available online at http://eur-lex.europa.eu/legal-content/EN/NOT/?uri=celex:32002D0676 (accessed 30 March 2015).

[8] European Commission, 'Radio Spectrum Policy Program: the roadmap for a wireless Europe'. Available online at http://ec.europa.eu/digital-agenda/en/radio-spectrum-policy-program-roadmap-wireless-europe (accessed 30 March 2015).

[9] European Commission, 'Regulatory framework for electronic communications in the European Union: situation on December 2009', European Union, 2010.

[10] ECC Report 205, 'Licensed Shared Access (LSA)', February 2014.

[11] ECC Report 46, 'Report from CEPT to the European Commission in response to the Mandate on inclusion of information on rights of use for all uses of spectrum between 400 MHz and 6 GHz', March 2013.

[12] Directive 2002/20/EC of the European Parliament and of the Council of 7 March 2002 on the authorisation of electronic communications networks and services (Authorisation Directive) as amended by Directive 2009/140/EC. Available online at http://eur-lex.europa.eu/legal-content/EN/TXT/?uri=CELEX:32002L0020 (accessed 30 March 2015).

[13] ECC Report 132, 'Light licensing, license-exempt and commons', June 2009.

[14] Broadband Series, 'Exploring the value and economic valuation of spectrum', ITU Telecommunication Development Sector, April 2012.

[15] Radio Spectrum Policy Group (RSPG), 'Report on Collective Use of Spectrum (CUS) and other spectrum sharing approaches', November 2011.

[16] Johannes M. Bauer, 'A comparative analysis of spectrum management regimes', 2006.

[17] Resolution 645 (WRC-2000), 'Global harmonization of spectrum for public protection and disaster relief', The World Radiocommunication Conference, Istanbul, 2000.

[18] Report ITU-R M.2033, 'Radiocommunication objectives and requirements for public protection and disaster relief', 2003.

[19] Annex 6 to Working Party 5A Chairman's Report, 'Working document towards the preliminary draft CPM text for WRC-15 agenda item 1.3', Document 5A/TEMP/163(Rev.1), November 2013.

[20] CEPT ECC Report 199, 'User requirements and spectrum needs for future European broadband PPDR systems (Wide Area Networks)', May 2013.

[21] National Public Safety Telecommunications Council (NPSTC), 'Public Safety Communications Assessment 2012–2022: Technology, Operations, & Spectrum Roadmap', Final Report, 5 June 2012.

[22] Canada Defence R&D Canada – Centre for Security Science (DRDC CSS), '700 MHz spectrum requirements for Canadian public safety interoperable mobile broadband data communications', February 2011.

[23] ETSI TR 102 628, 'Additional spectrum requirements for future Public Safety and Security (PSS) wireless communication systems in the UHF frequency', August 2010.

[24] J. Scott Marcus, John Burns, Val Jervis, Reinhard Wählen, Kenneth R. Carter, Imme Philbeck, Peter Vary, 'PPDR Spectrum Harmonisation in Germany, Europe and Globally', December 2010.

[25] John Ure, 'Public Protection and Disaster Relief (PPDR) Services and Broadband in Asia and the Pacific: A Study of Value and Opportunity Cost in the Assignment of Radio Spectrum', June 2013.

[26] ETSI TR 102 485 V1.1.1, 'Technical characteristics for Broadband Disaster Relief applications (BB-DR) for emergency services in disaster situations; System Reference Document', July 2006.

[27] IDATE Research, 'Public safety spectrum: how to meet the broadband needs of public safety users?', March 2014.

[28] LS Telcom, 'Spectrum – LS telcom Customer News Magazine: Special Edition PMR', 2014.

[29] ECC Decision (08)05, 'Harmonisation of frequency bands for the implementation of digital Public Protection and Disaster Relief (PPDR) radio applications in bands within the 380–470 MHz range', June 2008.

[30] ECC Recommendation (08)04, 'Identification of frequency bands for the implementation of Broad Band Disaster Relief (BBDR) radio applications in the 5 GHz frequency range', October 2008.

[31] European Communications Office (ECO), Documentation Database. Available online at http://www.erodocdb.dk/default.aspx (accessed 30 March 2015).

[32] Radio Spectrum Policy Group (RSPG), European Commission, 'RSPG report on strategic sectoral spectrum needs', RSPG13-540 (rev2), November 2013.

[33] The Council of the European Union, 'Council Recommendation 10141/09 on improving radio communication between operational units in border areas', Document 10141/09 ENFOPOL 143 TELECOM 116 COMIX 421, June 2009.
[34] Public Safety statement from Law Enforcement Working Party – Radio Communications Expert Group (LEWP – RCEG), October 2012.
[35] Huawei, 'Ukkoverkot and Huawei to deploy world's first commercial LTE 450 MHz network in Finland', June 2014.
[36] Airbus Defence and Space, 'Technical feasibility of 733–743/748–758 MHz products', FM49 – Radio Spectrum for Broadband PPDR, Helsinki, 11–12 November 2014.
[37] Motorola Solutions, 'Some technical considerations for a 2 × 10 MHz broadband allocation in the centre gap of the 700 MHz CEPT band plan', FM49 – Radio Spectrum for Broadband PPDR, Helsinki, 11–12 November 2014.
[38] Pascal Lamy, 'Results of the work of the high level group on the future use of the UHF band (470–790 MHz)', Report to the European Commission, August 2014.
[39] Donny Jackson, 'PSCR official: public-safety LTE devices may include multiple bands, Android', Urgent Communications, 30 December 2013.
[40] Headlines and outcomes of 37th ECC Meeting, Aarhus, Denmark, 24–27 June 2014. Available online at http://www.cept.org/ecc/37th-ecc-meeting,-aarhus,-denmark,-24-27-june-2014 (accessed 30 March 2015).
[41] NPSTC Public Safety Communications Report, '4.9 GHz National Plan Recommendations', October 2013.
[42] Thomas Welter, 'Assessing the potential of the 700 MHz band for PPDR', Critical Communications Europe, Amsterdam, 11 March 2014.
[43] TeleSíntese Portal de Telecomunicações, Internet e TICs, 'Segurança pública também vai ganhar espectro de 700 MHz', November 2013. Available online at http://telesintese.com.br/index.php/plantao/24644-seguranca-publica-tambem-vai-ganhar-espectro-de-700-mhz (accessed 30 March 2015).
[44] IEEE Std 1900.1-2008, IEEE Standard Definitions and Concepts for Dynamic Spectrum Access: Terminology Relating to Emerging Wireless Networks, System Functionality, and Spectrum Management, 2008.
[45] J. Scott Marcus, 'The need for PPDR Broadband Spectrum in the bands below 1 GHz', Report for the TETRA + Critical Communication Association, October 2013.
[46] European Commission, 'Standardisation mandate to CEN, CENELEC and ETSI for Reconfigurable Radio Systems (RRS)', European Commission mandate M/512 EN, November 2012.
[47] European Commission Communication COM(2012) 478, 'Promoting the shared use of radio spectrum resources in the internal market', September 2012.
[48] Federal Communications Commission, 'FCC 12-148, NPRM & Order on Enabling Innovative Small Cell Use In 3.5 GHz Band', December 2012. Available online at http://www.fcc.gov/document/enabling-innovative-small-cell-use-35-ghz-band-nprm-order (accessed 30 March 2015).
[49] Jon M. Peha, 'Sharing Spectrum through Spectrum Policy Reform and Cognitive Radio', in Proceedings of the IEEE, April 2009.
[50] R. Ferrus, O. Sallent, G. Baldini and L. Goratti, 'Public Safety Communications: Enhancement Through Cognitive Radio and Spectrum Sharing Principles', Vehicular Technology Magazine, IEEE, vol. **7**, no. 2, pp. 54, 61, June 2012.
[51] CEPT ECC Report 169, 'Description of practises relative to trading of spectrum rights of use', May 2011.
[52] Ofcom, 'Simplifying spectrum trading: reforming the spectrum trading process and introducing spectrum leasing', 15 April 2010. Available online at http://stakeholders.ofcom.org.uk/binaries/consultations/simplify/statement/statement.pdf (accessed 30 March 2015).
[53] Federal Communications Commission, 'Promoting efficient use of spectrum through elimination of barriers to the development of secondary markets', Second Report and Order on Reconsideration and Second Further Notice of Proposed Rule Making, 2004.
[54] William Lehr and Nancy Jesuale, 'Spectrum pooling for next generation public safety radio systems', 3rd IEEE Symposium on New Frontiers in Dynamic Spectrum Access Networks (DYSPAN), October 2008.
[55] M.M. Buddhikot, 'Understanding dynamic spectrum access: models, taxonomy and challenges', 2nd IEEE International Symposium on New Frontiers in Dynamic Spectrum Access Networks, 2007 (DySPAN 2007), April 2007.
[56] S. Chan, 'Shared spectrum access for the DoD', 2nd IEEE International Symposium on Dynamic Spectrum Access Networks, 2007 (DySPAN 2007)17–20 April 2007.
[57] J. Bradford, T. Cook, D. Ramsbottom and S. Jones, 'Optimising usage of spectrum below 15 GHz used for defence in the UK', 2008 IET Seminar on Cognitive Radio and Software Defined Radios: Technologies and Techniques, 18–18 September 2008.

[58] C.-S. Sum, G.P. Villardi, M.A. Rahman, T. Baykas, Ha Nguyen Tran, Zhou Lan, Chen Sun, Y. Alemseged, Junyi Wang, Chunyi Song, Chang-Woo Pyo, S. Filin and H. Harada, 'Cognitive Communication in TV White Spaces: An Overview of Regulations, Standards, and Technology', Communications Magazine, IEEE, vol. 51, no. 7, pp. 138, 145, July 2013.
[59] Nancy Jesuale, 'Lights and sirens broadband – how spectrum pooling, cognitive radio, and dynamic prioritization modeling can empower emergency communications, restore sanity and save billions', IEEE International Symposium on Dynamic Spectrum Access Networks (DySPAN), May 2011.
[60] M.J. Marcus, 'Sharing Government Spectrum with Private Users: Opportunities and Challenges', Wireless Communications, IEEE, vol. 16, no. 3, pp. 4–5, June 2009.
[61] European Commission's Radio Spectrum Policy Group, 'Report on CUS and other spectrum sharing approaches: "Collective Use of Spectrum"', 2011.
[62] RSPG Opinion on Licensed Shared Access, RSPG13-538, November 2013.
[63] Ingenious Consulting Networks, 'Authorized Shared Access (ASA): an evolutionary spectrum authorization scheme for sustainable economic growth and consumer benefit', Report commissioned by Qualcomm and Nokia, 2011.
[64] Ericsson White Paper, 'Spectrum sharing', Uen 284 23-3205, October 2013.
[65] EC Mandate on 'Harmonised technical conditions for the 2300-2400 MHz ('2.3 GHz') frequency band in the EU for the provision of wireless broadband electronic communications services', April 2014.
[66] Draft CEPT Report 55, Report A from CEPT to the European Commission in response to the Mandate on 'Harmonised technical conditions for the 2300–2400 MHz ("2.3 GHz") frequency band in the EU for the provision of wireless broadband electronic communications services', June 2014.
[67] ECC Decision 14(02) on 'Harmonised technical and regulatory conditions for the use of the band 2300–2400 MHz for Mobile/Fixed Communications Networks (MFCN)', June 2014.
[68] ECC Recommendation (14)04 on 'Cross-border coordination for mobile/fixed communications networks (MFCN) and between MFCN and other systems in the frequency band 2300–2400 MHz', May 2014.
[69] ETSI TR 103 113 V1.1.1, 'Electromagnetic compatibility and Radio spectrum Matters (ERM); System Reference document (SRdoc); Mobile broadband services in the 2300 MHz – 2 400 MHz frequency band under Licensed Shared Access regime', July 2013.
[70] ETSI TS 103 154 V1.1.1., 'Reconfigurable Radio Systems (RRS); System requirements for operation of Mobile Broadband Systems in the 2300 MHz – 2400 MHz band under Licensed Shared Access (LSA)', October 2014.
[71] ETSI TS 103 225 v0.0.4 (Work in progress), 'System Architecture and High Level Procedures for operation of Licensed Shared Access (LSA) in the 2300 MHz – 2400 MHz Band', November 2014.
[72] Gianmarco Baldini, Ramon Ferrús, Oriol Sallent, Paul Hirst, Serge Delmas, Rafał Pisz, 'The evolution of Public Safety Communications in Europe: the results from the FP7 HELP project', ETSI Reconfigurable Radio Systems Workshop, Sophia Antipolis, France, 12 December 2012.
[73] ETSI TR 103 217 v0.0.3, 'Reconfigurable Radio Systems (RRS); Feasibility study on inter-domains synergies; Synergies between civil security, military and commercial domains', April 2014.
[74] ECC Report 159, 'Technical and operational requirements for the possible operation of cognitive radio systems in the 'white spaces' of the frequency band 470–790 MHz', January 2011.
[75] M. Nekovee, 'Quantifying the availability of TV white spaces for cognitive radio operation in the UK', IEEE International Conference on Communications (ICC) 2009, 14–18 June 2009.
[76] A. Mancuso, S. Probasco and B. Patil, 'Protocol to Access White-Space (PAWS) databases: use cases and requirements', RFC 6953, May 2013.
[77] Federal Communications Commission (FCC), 'Second memorandum opinion & order – unlicensed operation in the TV broadcast bands', Document FCC 10-174, September 2010.
[78] Federal Communications Commission (FCC), 'Third memorandum opinion & order – unlicensed operation in the TV broadcast bands', Document FCC 12-36, April 2012.
[79] Allen Yang (FCC, USA), 'Overview of FCC's New Rules for TV White Space Devices and database updates', ITU-R SG 1/WP 1B Workshop: Spectrum Management Issues on the Use of White Spaces by Cognitive Radio Systems, Geneva, 20 January 2014.
[80] OFCOM, 'Regulatory requirements for white space device in the UHF TV band', 4 July 2012.
[81] OFCOM website. Available online at http://stakeholders.ofcom.org.uk/spectrum/tv-white-spaces/ (accessed 30 March 2015).
[82] ECC FM Project Team 53 on Reconfigurable Radio Systems (RRS) and Licensed Shared Access (LSA). Available online at http://www.cept.org/ecc/groups/ecc/wg-fm/fm-53 (accessed 30 March 2015).

[83] ECC Report 185, 'Complementary Report to ECC Report 159 – further definition of technical and operational requirements for the operation of white space devices in the band 470–790 MHz', January 2013.
[84] ECC Report 186, 'Technical and operational requirements for the operation of white space devices under geo-location approach', January 2013.
[85] ETSI EN 301 598, 'Wireless Access Systems operating in the 470 MHz to 790 MHz TV broadcast band; Harmonized EN covering the essential requirements of article 3.2 of the R&TTE Directive', April 2014.
[86] Emmanuel Faussurier, ANFR Chairman CEPT/WGFM Project Team FM53, 'Introduction of new spectrum sharing concepts: LSA and WSD', ITU-R SG 1/WP 1B Workshop: Spectrum management issues on the use of white spaces by cognitive radio systems, Geneva, 20 January 2014.
[87] Draft ECC Report 224, 'Long Term Vision for the UHF broadcasting band', September 2014.
[88] ETSI TS 103 143 V0.0.7 (2014-11), 'Reconfigurable Radio Systems (RRS); System Architecture for WSD GLDBs', November 2014.
[89] Draft ETSI EN 303 144 V0.0.5, 'Enabling the operation of Cognitive Radio System (CRS) dependent for their use of radio spectrum on information obtained from Geo-location Databases (GLDBs); Parameters and procedures for information exchange between different GLDBs', December 2014.
[90] ETSI TS 103 145, 'System Architecture and High Level Procedures for Coordinated and Uncoordinated Use of TV White Spaces', January 2015.
[91] Draft EN 303 387-1 V0.0.4, 'Reconfigurable Radio Systems (RRS); Signalling Protocols and information exchange for Coordinated use of TV White Spaces; Part 1: Interface between Cognitive Radio System (CRS) and Spectrum Coordinator (SC)', November 2014.
[92] IEEE 802 LAN/MAN Standards Committee. Available online at http://www.ieee802.org/ (accessed 30 March 2015).
[93] IEEE 802.22-2011 'Standard for Information technology – local and metropolitan area networks – specific requirements – Part 22: Cognitive Wireless RAN Medium Access Control (MAC) and Physical Layer (PHY) specifications: policies and procedures for operation in the TV Bands', 2011.
[94] 802.11af-2013, 'IEEE Standard for Information technology – telecommunications and information exchange between systems – local and metropolitan area networks – specific requirements – Part 11: Wireless LAN Medium Access Control (MAC) and Physical Layer (PHY) Specifications Amendment 5: Television White Spaces (TVWS) Operation', 2013.
[95] IEEE 802.19.1-2014, 'IEEE Standard for Information technology – telecommunications and information exchange between systems – local and metropolitan area networks – specific requirements – Part 19: TV White Space Coexistence Methods', 2014.
[96] IEEE DySPAN Standards Committee (DySPAN-SC). Available online at http://grouper.ieee.org/groups/dyspan/index.html (accessed November 2014).
[97] IEEE 1900.4, 'Architectural building blocks enabling network-device distributed decision making for optimized radio resource usage in heterogeneous wireless access networks', February 2009.
[98] ECMA-392, 'MAC and PHY for Operation in TV White Space', June 2012.
[99] V. Chen, S. Das, L. Zhu, J. Malyar and P. McCann, 'Protocol to Access White-Space (PAWS) databases', IETF Internet-Draft draft-ietf-paws-protocol-20 (Approved as Proposed Standard), November 2014.
[100] ETSI TR 103 067, 'Feasibility study on Radio Frequency (RF) performances for Cognitive Radio Systems operating in UHF TV band WS', May 2013.
[101] 3GPP Work Item Description RP-141664, 'Study on Licensed-Assisted Access using LTE', 3GPP TSG RAN Meeting #65, Edinburgh, Scotland, 9–12 September 2014.

Index

Mobile Broadband Communications for Public Safety: The Road Ahead Through LTE Technology, First Edition.
Ramon Ferrús and Oriol Sallent.